ROUTLEDGE LIBRARY EDITIONS:
PSYCHOLOGY OF EDUCATION

Volume 22

INTERPERSONAL RELATIONS AND EDUCATION

INTERPERSONAL RELATIONS AND EDUCATION

DAVID H. HARGREAVES

Routledge
Taylor & Francis Group

LONDON AND NEW YORK

First published in 1972 by Routledge & Kegan Paul Ltd

This edition first published in 2018
by Routledge
2 Park Square, Milton Park, Abingdon, Oxon OX14 4RN

and by Routledge
711 Third Avenue, New York, NY 10017

Routledge is an imprint of the Taylor & Francis Group, an informa business

© 1972 David H. Hargreaves

British Library Cataloguing in Publication Data
A catalogue record for this book is available from the British Library

ISBN: 978-1-138-24157-2 (Set)
ISBN: 978-1-315-10703-5 (Set) (ebk)
ISBN: 978-1-138-29399-1 (Volume 22) (hbk)
ISBN: 978-0-415-30001-8 (Volume 22) (pbk)
ISBN: 978-1-315-22899-0 (Volume 22) (ebk)

Publisher's Note
The publisher has gone to great lengths to ensure the quality of this reprint but points out that some imperfections in the original copies may be apparent.

Disclaimer
The publisher has made every effort to trace copyright holders and would welcome correspondence from those they have been unable to trace.

Interpersonal relations and education

David H. Hargreaves
Department of Education, University of Manchester

Routledge & Kegan Paul
London and Boston

First published 1972
by Routledge & Kegan Paul Ltd
Broadway House, 68–74 Carter Lane,
London EC4V 5EL
9 Park Street,
Boston, Mass. 02108, U.S.A.
Printed in Great Britain by
Western Printing Services Ltd,
Bristol

ISBN 0 7100 7245 7

Contents

		page
	Acknowledgments	vii
	Preface	ix
1	Introduction	1
2	The self	6
3	Perceiving people	31
4	Roles	69
5	Interaction	93
6	Teacher–pupil interaction	130
7	Discipline	228
8	Friends	267
9	Groups	289
10	Youth, youth culture and the school	337
11	Changing attitudes	373
12	Staff relationships	402
	Addendum to Chapter 6	421
	Bibliographical Index	427
	Subject Index	444

Acknowledgments

The author and the publishers gratefully thank the following for permission to include extracts from their work: Evans Brothers Limited for extracts from H. F. Ellis: *The Papers of A. J. Wentworth*; Victor Gollancz Limited for extracts from John Partridge: *Middle School*; Prentice-Hall Inc. for extracts from Solomon E. Asch: *Social Psychology* © 1952 the Publishers; Harcourt Brace Jovanovich Inc. and Jonathan Cape Limited for extracts from © 1968 Kingsley Amis: *I Want It Now*; Little, Brown and Company and A. D. Peters & Company for extracts from Evelyn Waugh: *Decline and Fall*; Scott Meredith Literary Agency, Evan Hunter & Constable & Company Limited for extracts from Evan Hunter; *The Blackboard Jungle*; Penguin Books Limited for extracts from Leila Berg: *Risinghill*; Peter Terson for extracts from his article in the *Guardian*.

Preface

In writing this book I have relied heavily on a very large number of people for their help, stimulation and criticism. In particular I am indebted to several friends and colleagues for their exciting conversation over several years, especially Harold Entwistle, Michael Greenhalgh, Eric Hoyle, Philip Jackson and John Powell. I am also grateful to Mohan Mankikar for constructing all the diagrams and for his help with proof reading.

My greatest debt is to all those 'students' who have followed my courses in social psychology and education in the Department of Education in Manchester University with such enthusiasm and critical diligence, and to all those student teachers, teachers and head-teachers who have listened patiently to my lectures at conferences and then so actively discussed the relevant educational or social psychological issues. It is they who have sustained my own enthusiasm and interest and who have confirmed my belief that there is room in education for the social psychologist. It is to these fellow learners that I dedicate this book.

1 Introduction

This book is about education and about social psychology. It is written because I am a social psychologist with a profound interest in education. Each of these two areas can shed considerable light on the other. Quite intentionally I have not called this book a 'social psychology of education' because I have tried to avoid one of the main faults of books on the human sciences applied to education. The books entitled 'the psychology of education' or 'the sociology of education' or 'the philosophy of education' are often unattractive to their readers because they seem to be loaded with technical language or jargon. Such books aim, often within a relatively short space, to give the reader some insight into how a psychologist, sociologist or philosopher approaches and analyses educational issues and problems. In so doing, the expert has to use some of the basic concepts in his discipline. These terms and concepts have a much wider meaning and significance for the expert. The technical language in which the concepts are couched are devised, not to be obscure or esoteric or pseudo-scientific, but in order to use words with a more specific meaning than is given in everyday language and in order to bring together under the umbrella of one concept a very wide range of phenomena. When such terms are used in a treatise applied to education, the attention is focused on educational phenomena. The reader, who is often quite unfamiliar with the parent discipline, cannot always see the need for the concept since he is unaware that the concept can be usefully applied to a wide range of phenomena that are irrelevant to or quite distinct from educational phenomena. Thus to him the book seems to be unnecessarily and heavily loaded with jargon. In a real sense he is right, for the book has obviously failed to make the wider value of the concept clear. His charge that such books merely state the obvious in a complex way is not without foundation. A second, but associated danger of writing books such as these is that eventually

1

the reader goes away with relatively little understanding of the discipline and its major concepts, yet also without his understanding of educational phenomena having been increased. He may, for examination purposes if he is a student, remember some of the most important names and concepts for a short period, but the general approach is forgotten with the factual knowledge. The reader has learned nothing of the psychological, sociological or philosophical perspective.

In this book I have tried to avoid both these dangers, though the reader alone can judge with what success. I have taken the view that the reader will have to learn quite a lot of social psychology before he can understand and then evaluate what a social psychologist can say about education. For this reason the book is rather longer and more daunting than I had originally hoped. The book deals with many traditional problems examined from the perspective of social psychology, a discipline that has not been popularly or extensively applied to education. Such an examination does, I believe, shed light on the old problems as well as throwing up new problems, or at least problems which are not adequately acknowledged, formulated and discussed. This is perhaps the most important task of the application of the human sciences to education—at least from the point of view of the educationist. If it fails in this task, it fails indeed, and the human scientists should keep their musings about education strictly among themselves. In view of this, large sections of the book are devoted to an exposition of some basic social psychological assumptions, perspectives and concepts, to empirical research and to theoretical formulations. Inevitably the book is selective in that some major areas of social psychology have been excluded as have even larger areas of the field that we call education.

Broadly speaking, social psychology is the study of interpersonal relations and small group behaviour. The boundaries of the discipline, which lies between the two giants of the human sciences, sociology and psychology, are fortunately very blurred. The two giants have often left that area lying uncomfortably between them relatively unexplored with respect to education, though in recent years there has been a growing recognition of its importance, indicated by the increasing willingness of psychologists and sociologists to spread their nets in this direction. The importance of the area rests on the fact that one of the most central features of education is its social quality. Education occurs within the relationships of teachers and pupils, pupils and pupils, teachers and teachers. It takes place within groups, both formal like the classroom and informal like the friendship group. Psychology deals with intrapersonal dynamics and structures; sociology with institutions and organizations. Social psychology deals with the exciting area of human behaviour, the structure and dynamics of relationships between people, or interpersonal relations. But to treat the interpersonal aspects of educational processes and events, we need a broader understanding of the interpersonal in general. It is for this reason that so much of the book, especially in the early chapters, is concerned with social psychology rather than with specific educational problems.

Broadly speaking, the approach taken in this book is that which is usually referred to as 'symbolic interactionist' and thus the two dominant concepts that pervade most of the chapters are *interaction* and *social influence*. Since interaction and social influence are self-evidently important aspects of relationships between people in education contexts, it is easy to see what led me to write this book. It is difficult to say whether I am attracted to these concepts because they seem relevant to the analysis of interpersonal relations in education or whether I am interested in education because I am fascinated by these two concepts. Of more relevance to the reader is the fact that a social psychological approach to education could be made on the basis of different concepts. Peter Kelvin's *The Bases of Social Behaviour*, an excellent introduction to social psychology, uses *order* and *value* as its central concepts. If one were to use these for a social psychological analysis of education, the selection of educational phenomena and the interpretation and analysis of them would probably be rather different.

I must also confess to the reader that with some notable exceptions I find many textbooks, both on the social sciences and on the applications of social science to education, exceedingly tedious. Like all writers I have tried to avoid boring or confusing the reader. In places I have tried to introduce ideas or illustrations that are intrinsically of interest rather than simply relevant to the point under discussion. I have also tried, quite intentionally, to be provocative. I have made little effort to be 'objective' or 'coolly detached'; it is difficult to be either with respect to the passionate subjects of human relations and education. Nor have I heeded the academic warnings against being prescriptive. I have always found the dichotomy between objective analysis or description and prescription rather false. As Lionel Trilling's incisive comments on the Kinsey Report show, it is when one tries hardest to be non-evaluative and non-prescriptive that one's concealed assumptions and prescriptions become most dangerous. Also, in order to avoid boring the reader and making the book too unwieldy, I have not always given detailed references to the relevant social psychological theory or research, nor have I always added to my summaries or statements about social psychological work those qualifications that are strictly in order. Whilst I have not sought to mislead, I have assumed that the interested reader will follow up the social psychological aspects by referring to the many competent textbooks in the field.

It is also important for me to advise the reader that this book is not about *personal relationships* in the sense of encounters between persons. The main focus of the book is on *interpersonal relations*, as the title suggests. This distinction is a difficult but necessary one, largely because these aspects of human behaviour represent somewhat different theoretical and empirical approaches. Both approaches are necessary in education, but they are not easily combined in one book. In learning about interpersonal relations the reader will soon enough realize how much needs to be done in the area of personal relationships in education.

It would have been much easier to write a book such as this if the audience for the book were better known. Basically it is written for teachers, though it is well known that only a very small minority of teachers ever reads a book like this. I would also like to think that it would reach that mysterious person we call the 'intelligent layman', but since I know so little about him I cannot tailor the book to his needs. It is not written for academics, either educational or social psychological; we write too many highly specialized books for one another as it is. At the same time I have sought to make a real contribution to both education and social psychology and am content to be subjected to the critical scrutiny of my academic colleagues in this respect. However, my main audience, I am forced to conclude, will be formal students of education. This is a very mixed audience, containing at one extreme the first year college of education student and at the other the middle-aged teacher with many years of classroom experience taking a Master's degree in educational studies. If one speaks largely to the former, one is open to accusations of condescension or labouring the obvious from the latter. If one speaks largely to the latter, one is open to charges of incomprehensibility from the former. Since I wanted to speak to both, I have tried to aim somewhere between the two, hoping that the more sophisticated reader will generously bear with the lengthy expositions and frequent repetitions, and hoping that the less sophisticated reader will not be disheartened by those sections which appear to be rather difficult or obscure. I trust that both will forgive me for being unwilling to write two books, one for each, on the main themes.

A social psychologist who is no longer a classroom teacher in a school cannot pretend to produce many answers from his study of those educational issues which are of interest to him. Where I am heavily prescriptive or critical, the views are offered in the spirit of an outsider whose position has some advantages of less immediate involvement as well as the obvious disadvantages of being out of touch and unrealistic, and perhaps even unfair. In many places I have tried to make it clear that I am giving a highly personal view, based on either a personal selection or reading of the evidence or on no evidence at all, by using the personal pronoun *I* as a preface to my remarks.

In the last analysis it is the members of schools who will make the changes, by analysing their problems and finding solutions to them. This book seeks to shed light on some of the problems and some of the possible solutions. It is not my analyses or my solutions that are important, but rather the reaction of the reader to them. If it helps to clarify the reader's thinking, if it boosts a failing idealism, if it disturbs some taken for granted assumptions, then the book will have reached one of its goals. In the end it remains a book, and for teachers and social scientists alike, it is worth remembering La Rochefoucauld's maxim: *Il est plus nécessaire d'étudier les hommes que les livres.*

2 The self

It would be easier to begin this book if the teaching and learning of social psychology could be approached in a similar way to the teaching and learning of mathematics or a language. In the case of these latter subjects there are fairly obvious simple and primary elements that must be mastered before moving on to more advanced areas. It would be unusual to begin a book on mathematics with calculus and end it with simple algebraic equations, just as it would be strange to teach the pluperfect of irregular verbs in French before teaching the present tense of regular verbs. The teacher of these subjects can assume in a broad way a linearity in the subject. All disciplines are really spherical rather than linear, for all the elements of a discipline are inter-related in complex ways, but some subjects are more readily adaptable to linear treatments than others. Social psychology does not lend itself easily to such an approach—the standard text-books take very different starting points and paths through the subject.

Since the chapter structure of a book implies some degree of linearity, even though a spherical shape has no obvious beginning or end, some justification is required for entering the sphere at a particular point. We shall enter the discipline of social psychology by looking at a concept which is closer to the middle of the sphere than to its outer edge, or, to change to a more popular metaphor, by plunging the reader into the deep end. One of the more central conflicts within social psychology is the tension between the 'individual' and the 'social'. It is a conflict within other social sciences too, but there it is not so central. Sociology claims that its central concerns are with social phenomena, social facts and social explanations. The 'individual' aspect is of secondary importance. Thus Albert Cohen, whose book *Delinquent Boys* (1955) is one of the most important and influential sociological approaches to delinquency, writes: 'We are not primarily interested in explaining why this boy adopts a delinquent solution to his problems and why another boy does not . . .

6

This book does, however, have implications for the explanation of the individual case.' The task which Cohen dismisses is often regarded as the task of psychologists, for psychology is concerned with the 'individual' rather than the 'social'. It takes as its central focus individual phenomena, (cognition, perception, motivation etc.) and their internal structure and dynamics. The research, theories and explanations of psychologists are rooted in individual behaviour, in its prediction and control. Psychologists recognize the importance of the 'social' but they are interested in it not in its own right but in so far as it affects the individual's behaviour.

The tension between the 'social' and the 'individual' is at its most acute in social psychology, which takes as its central focus interpersonal relationships. Social psychology is social for it seeks to analyse the relationship between two or more persons; it is individual in that it takes account of what an individual contributes to a social interaction or a relationship and how he is affected by it. The dichotomy between the 'individual' and the 'social' becomes largely artificial, though it remains a useful and convenient distinction.

We shall begin this book by trying to come face to face with (but not with resolving) this central tension and it is nowhere more obvious than in the concept of the 'self'. At first sight this seems to be an 'individual' rather than a 'social' concept, but on further analysis such a simple view is immediately undermined and we are forced to recognize the tension which is inherent in all social psychology. Because the concept of the self is toward the centre of the sphere of the discipline of social psychology, it is linked with all the other concepts in a very complex way. It is one of several threads that run through all the chapters, implicitly or explicitly. It cannot be dealt with and then dismissed in one chapter as if it were an isolated bit. This is the danger of the implied linearity of books, a danger for the author in his desire to carve up the field into convenient segments, but an even greater danger for the reader who may come to perceive these convenient segments as the inherent structure of the discipline. For this reason the reader should perhaps re-read this chapter when he has finished the book. He can then consider the concept of the self from a more sophisticated position, recognize the over-simplification necessary in an early chapter, and, armed with greater understanding and insight, acknowledge that the imposed structure has been convenient, arbitrary and tentative. In the end it is simply one of many possible approaches.

The self and the body

A baby still within the womb leads a totally dependent existence. Its needs are automatically satisfied by the mother's bodily processes.

7

At birth, the point of physical separation between child and mother, the activation of a number of basic mechanisms, for example breathing, guarantees independent physical functioning to the child, whilst in other respects, for example feeding, his dependence on mother persists. One of the most dramatic changes arising from birth is the fact that the baby's sense-organs now begin to receive stimulation from the new environment—sight, sound, touch, smell and taste. Perhaps, as William James suggested, the new-born baby is confronted by 'one big blooming buzzing confusion'. Of the many problems which the baby must learn to solve, three are of special interest to social psychology since they are problems involving other people which persist into adult life. First he must make sense out of all the apparent chaos which surrounds him, structure into a meaningful pattern all the data which impinge on his sense-organs and which come to make up his experience of the world. Second, he must achieve a degree of autonomy and independence whereby he can exercise some control over the environment, both interpersonal and non-personal. Third, he must learn to develop satisfying interdependent personal relationships with other people.

These problems are aspects of learning to cope with the external and social world. Clearly a prerequisite of this is the development by the baby of an ability to distinguish himself from that environment, which is no longer co-extensive with him as was the case in the womb. He must learn that certain objects 'belong' to him whilst others do not: his hands and toes are part and parcel of him in a way that mother's breast is not. Mother, her nipple, his toys and his cot exist outside himself, independently, beyond the control of his whims. Many of the baby's earliest frustrations originate in his inability to control such objects in the satisfaction of his needs, but it is because the environment is frustrating that the baby needs to come to terms with it at all. Also he must learn that objects which are not physically part of him have a permanent existence of their own and that they do not cease to exist when they disappear over the edge of the cot beyond the range of his sense-organs.

The primary element of becoming a person consists in coming to terms with the environment in terms of learning to differentiate between self and not-self. Initially, the problem is a physical one, learning to distinguish one's own body from everything else. The body continues to be of great importance to a person's self conception, or to what is often called the *self-image*. One's conception of who and what one is cannot be unrelated to one's bodily characteristics since we all learn that certain characteristics are socially valued. A woman's self-image is enhanced if she is beautiful as is a man's if he is physically large and strong. The booming cosmetic industry and the Charles Atlas courses testify to our social valuations. Adolescent girls have to learn to be embarrassed by their acne. The relation between the self-image and the body is often part of our taken-for-granted experience, because we adjust ourselves to our bodies. But those who are suddenly smitten with some physical defect or disfigurement tend to experience severe problems associated with their image of themselves (e.g. Goffman, 1963a).

The point becomes dramatically clear in the experience of John Griffin (1962) who, in his desire to see the life of the American Negro at first hand, intentionally took a drug to change the colour of his skin. When the drug has taken effect he looks at himself in the mirror. Instead of seeing himself reflected in the mirror, the white man disguised as a negro, he sees a total stranger, who seems to bear no relationship to himself. In a state of acute distress, he experiences himself as two people, the one who observes and the stranger in the mirror with a different personality, history and culture. John Griffin had ceased to exist, for he would no longer be recognized by family and friends; in his place was a negro, 'a man born old at midnight into a new life'.

The self: social and individual

Some writers have emphasized the fact that in a very fundamental sense the self is a product of a person's interaction with others. It is the *social* environment which is seen as fundamental. The central idea, in short, is that a person's self develops in relation to the reactions of other people to that person and that he tends to react to himself as he perceives other people reacting to him. That is to say, the self-system is not merely a function of a person's manipulation of the environment, but a function of the way in which a person is treated by others. The self is a social product. This line of thought is an important element in many of our great theorists, such as Piaget and Freud, but its most notable exponents come from another tradition.

The American sociologist Charles Horton Cooley (1902) coined the important phrase the *looking-glass self*. Just as a man considers his reflection in a mirror and uses the reflection to acquire information about his physical nature, so he uses other people's attitudes to him as a measure of what he is really like. The mirror provides 'feedback': he can corroborate by means of the reflection whether or not his appearance is acceptable to him. Similarly, the reactions of others provide a man with feedback on how he appears to them and he can then discover whether or not this is in accord with the way in which he perceives himself.

Cooley never fully developed this line of thought, but the task was taken up by another American, George Herbert Mead (1934). The major part of Mead's work was not published until after his death, but it has subsequently exerted a profound influence on many areas of thought in social psychology and sociology. Mead's great contribution was the recognition that the self is a social structure which rises through communication. In his terms, the primary element of communication is the *gesture*. For example, an animal in a state of anger will bare its teeth. In Darwinian terms such a gesture has adaptive value for survival, since the teeth are ready for action or for counteraction. This gesture is really the first part of a series of acts and can come to serve as a *sign* for the whole act.

The situation in which each of two animals responds to the actions of the other may be referred to as a 'conversation of gestures'. A good example is the instinctive mating behaviour of the three spined stickleback as described by the ethologist Tinbergen (1951). The female with her swollen abdomen swims casually through the water. When the male sees her, he begins to perform his zig-zag dance. At this point, the female approaches the male, who goes to the nest he has previously built, closely pursued by the female. He demonstrates the nest to his prospective mate, who enters. At this moment, the male begins to quiver (with excitement?) which induces spawning in the female. Finally, the male fertilizes the eggs. This complex behaviour pattern is really a sequence of actions and reactions. The action of one is the stimulus to the other to respond in a certain way, and this reaction becomes a stimulus to the first fish to react in his turn. Each stimulus 'releases' a reaction which then becomes itself a stimulus 'releasing' another reaction, and so on. Courtship among the sticklebacks is a 'conversation of gestures'.

Yet a 'conversation of gestures' is not the same thing as communication. According to Mead, the 'conversation of gestures' becomes communication when the gestures become *significant*, that is, when they arouse in the organism making the gesture the same response that the gesture arouses, or is intended to arouse, in the other organism. Communication is considerably facilitated by the use of language. Communication does not demand language, for as John Macmurray (1961) reminds us, communication is not the offspring of language, but its parent. The process, however, is best illustrated by the use of speech. When I speak to another person, I am seeking to call out in him a response to my linguistic stimulus. At the same time, I call out the same response in myself. The value of language hinges upon the fact that I get the meaning of what I say. So if I make a request of another person I arouse in myself the very response I am trying to evoke from him. This is a very important step, because if I arouse in myself the response I am trying to evoke from someone else, then without the other having to respond at all I can anticipate the other's reaction to my behaviour and, by extension, I can anticipate my own reaction to the other's anticipated reaction. The individual can anticipate the other's reaction by the process which Mead calls *taking the role of the other*, by considering oneself from the standpoint of the other. Thus on Mead's view the self is essentially reflexive; it is a social structure. This is because the self can be both subject, which Mead calls the 'I', and object, which Mead calls the 'me'. There becomes possible a kind of internal conversation by which the individual can, without action, rehearse possible courses and consequences of action. This means that the individual's behaviour ceases to be determined by the *actual* response of the other person, as was the case with our amorous sticklebacks. By anticipating the other's reaction to my action and also my reaction to his reaction, I am able to examine and evaluate several possible courses of action and from these choose one particular course of action. In a social interaction a person's behaviour is influenced not by the acts of the other as such but by the meaning (intentions, motives, etc.) which the person assigns to the other's acts. Individual action is constructed in relation to the other, not simply released by the other's acts. As Blumer (1962, 1966) has pointed out, Mead's argument means that the individual's behaviour is not simply determined by outside forces (as in some sociological views) or simply evoked by internal impulses (as in some psychological views).

In short, the self is not inborn, nor could it appear in the individual isolated from his fellows. The self arises from the social experience of interacting with others. The self has an important reflexive quality: it is both subject and object. In interaction, a man learns to respond to himself as others respond to him. He becomes object to himself when he takes to himself the reactions of others to him. He acquires a self by putting himself in the shoes of others and by using their perspective of him to consider himself. It is to this ability of man to take to himself the attitudes of others to him that Mead refers in his famous phrase 'taking the role of the other'.

On this view the distinctive attributes of man arise not so much from the superiority of his nervous system as such (the biological view) but from the kind of social relationships man holds with others of his species. A person is a being with a self, and that self arises through the social relationships of that being. Macmurray (1961) sums it up when he writes:

> the unit of personal existence is not the individual, but two
> persons in personal relation; and . . . we are persons not by
> individual right, but in virtue of relation to one another. The
> personal is constituted by personal relatedness. The unit of the
> personal is not the 'I' but the 'You and I'.

The notion of the person totally independent of others is a fiction.

Mead's major contribution, then, rests on his suggestion that a self develops only when a person begins to 'take the role of the other', when a person takes to himself the attitudes that others take to him. One of the important ways in which the child learns to do this is in play, as Mead and Piaget have stressed. Whether the play is alone or with others, whether in fantasy or with toys, it serves the same function. Let us consider the case of a child playing at being another person, either a parent or some childhood hero. Frequently the child will play two roles in interaction, as in the game of 'house' where the child is both mother and father. In such play or fantasy situations the child is exploring the reactions of other people to one another and to him, as well as his reactions to them. It is in this way that the child learns to see himself from the perspective of others. The organization of the responses of others is an essential part of the process of self development.

Self and role

In the early socialization of the child, then, the learning of roles and the development of the self are inseparably fused (cf. Maccoby, 1959, 1961). Both are achieved in the essentially social context of interpersonal relationships. But from the complex process of socialization let us tease out the strand of role learning for analysis. Clearly the young child's interpersonal environment usually consists of the home. It is with his parents that the child most frequently interacts and on whom he is so dependent. Mother and father are so important to the child that they may, in Sullivan's (1940) phrase, be regarded as his *significant others*. Kuhn (1964) has defined significant others as

(*a*) the others to whom the individual is most fully, broadly and basically committed, emotionally and psychologically;

(*b*) the others who have provided him with his general vocabulary, including his most basic and crucial concepts and categories;

(*c*) the others who have provided and continue to provide him with his categories of self and other and with the meaningful roles to which such assignments refer;

(*d*) the others in communication with whom his self-conception is basically sustained and/or changed.

Thus it is the parents who dominate the child's world, it is their role relationships with him that must be learned and also explored in fantasy and play, it is their concepts, categories and attitudes that the child must take to himself. As the child does take the attitudes of his parents to himself, he begins to acquire a conception of the role of the child. His parents expect certain forms of behaviour from him and when he conforms to these expectations he is rewarded by them either materially or with love and approval. If he deviates from these expectations, he is punished. The child is learning how a child should behave. He is learning the role of a *child* in relation to his *parents*, and then more specifically the role of a *son* in relation to his *mother* and his *father*. Concomitant with his understanding of the son role is his understanding of the roles of mothers and fathers. He is beginning to discover that relationships between people are *structured* and that the role structure is independent of particular individuals. In time he will appreciate the relationships that his friends have with their parents. He will be in a position to understand, and to respond appropriately to, other adults who also have children. They are mothers and fathers like his own parents, but their parent roles are held in relation to their own children, not to him. The learning of these roles and role relationships is slow and complex, and certainly slower than simply learning to name the roles correctly.

When the child learns the content of his own and other people's roles, he is also learning to facilitate his relationships with them. As he learns what his parents expect of him he also learns what he can expect of them. When he learns to take their attitudes to himself, he becomes capable of anticipating and predicting their reactions to him. This has at least two important consequences. Firstly, by anticipating the reactions of the others, the child learns how to reduce potential conflict by gearing his behaviour to the role relationship. In role learning we shape and oil the cogs so that the social machine will run smoothly. Secondly, by taking the attitudes of his significant others to himself, their expectations can be effective in their physical absence. For when the child takes the attitudes of his significant others to himself, he internalizes controls which were originally external. It represents a movement from 'Mummy will smack me (not love me) if I . . . ' to 'I am a bad boy if I . . . ' to 'It is bad to . . .' We all learn to make the jump from the specific 'Mummy will smack me (not love me) if I . . . ' to the normative and conventional 'One does not . . . ' which has cut free from its specific interpersonal origins. We become what others want and expect us to be by making *their* views and *their* rules and *their* ways *our* views, rules and ways. More than this, we can soon come to view our (i.e. theirs and mine) views, rules and ways as the *only* and the *right* ones.

Whilst role learning is one of the basic features of socialization within the family, it is by no means the only important feature. We have suggested that much of the basic role learning is acquired through the child's interactions with his primary significant others, his parents, because his dependence on them makes him sensitive to their expectations, their evaluations and their rewards and punishments. It is this same dependence and sensitivity in the child which makes the parents the primary source of the child's knowledge, beliefs and values. In short, the 'culture' of a society is initially bestowed upon the child through his parents, who channel and filter the culture to the child in their interactions with him. And among the attitudes which the child takes to himself from his parents are his parents' knowledge, beliefs and values. The child's conception of 'reality'—what society, social institutions, his fellow men are like, what is 'important' or 'proper' or 'good' or 'right'—all this is socially mediated to the child, and in the first instance by taking over his parents' attitudes.

As the child grows older he moves out into a wider social environment and makes relationships with a wider range of people. The nature of the interaction, too, tends to change. One such situation is the organized game. The young child does not really play such games with other children, as Piaget (1932) has shown. Rather, he plays a private fantasy game alone but in the company of others. Knowledge of the rules is in itself not sufficient to ensure true game playing. An organized game represents a complex interactional system, for the roles of all the players have definite relationships with one another. To play the game requires a reciprocity and co-operation of which the ability to take the role of the other is an essential component. Indeed, the child can participate in the game only when he allows his own actions to be related to the actions of all the other players. For the fullest development of the self, according to Mead, a person must take to himself the role of the 'generalized other' or the attitudes of the group or community of which he is a member. Just as in an organized game the player learns to synthesize in a general way the attitudes of all the other players, that is the collective goals and attitudes of the team-as-a-whole, so the growing child synthesizes the attitudes of members of groups of which he is a part and to whose influence he is subject, whether the group be small, like the family or friendship group, of intermediate size, like the neighbourhood or religious group, or large like the nation or culture. In Mead's own words:

> The individual experiences himself as such, not directly, but only indirectly, from the particular standpoints of other individual members of the same social group or from the generalized standpoint of the group as a whole to which he belongs. For he enters his own experience as a self or individual, not directly or immediately, not by becoming a subject to himself, but only in so far as he first becomes an object to himself just as other individuals are objects to him or are in his experience; and he becomes an object to himself only by taking the attitudes of other individuals toward himself within a social framework or context of experience and behaviour in which both he and they are involved . . . If the given human individual is to develop a self in the fullest sense it is not sufficient for him merely to take the attitudes of other human individuals toward himself and

toward one another within the human social process and to bring that social process as a whole into his individual experience merely in these terms. He must also, in the same way that he takes the attitudes of other individuals toward himself and toward one another, take their attitudes toward the various phases or aspects of the common social activity or set of social undertakings in which, as members of an organized society or social group, they are all engaged.

In this way Mead brilliantly demonstrates the connection between learning to take the role of the other and the process of socialization, both in the sense of becoming what is distinctively human and in the sense of internalizing the norms and values of one's membership groups, both at the cultural and sub-cultural level.

Mead is regarded as the father of the important school of thought within the social sciences that is called Symbolic Interactionism because he first asserted that both human nature and the social order are products of communication. The implications of this assertion are far reaching, and will be felt throughout this book. The self is one of the most central concepts in social psychology because of Mead's fundamental contribution. For the moment we must content ourselves with a hint of the implications of a symbolic interactionist position. In Shibutani's (1961) words:

> From this standpoint, behaviour is not regarded merely as a response to environmental stimuli, an expression of inner organic needs, nor a manifestation of cultural patterns. The importance of sensory cues, organic drives, and culture is certainly recognized, but the direction taken by a person's conduct is seen as something that is constructed in the reciprocal give and take of interdependent men who are adjusting to one another. Furthermore, a man's personality—those distinctive behaviour patterns that characterize a given individual—is regarded as developing and being reaffirmed from day to day in his interaction with his associates. Finally, the culture of a group is not viewed as something external that is imposed upon people, but as consisting of models of appropriate conduct that emerge in communication and are continually reinforced as people jointly come to terms with life conditions. If the motivation of behaviour, the formation of personality, and the evolution of group structure all occur in social interaction, it follows logically that attention should be focused upon the interchanges that go on among human beings as they come into contact with one another.

With increasing age the child comes into contact with many persons outside the immediate family, with relatives, neighbours and local children. When he goes to school he is supervised by teachers and other adults, and he becomes a member of a peer group. With the expansion of the child's social world, the range of potential models available to him increases. (In this respect television may have a significant impact on the social conceptions of small children, since through this medium they may be introduced to a variety of people and roles with which personal contact would normally be impossible.) More important, the great diversity of interactional situations into which the child is brought means that new and complex roles and role relationships must be learned. Of these, the roles of pupil and friend are particularly important.

Parsons (1959) has implied that the role of pupil may be learned only with difficulty, for there are permanent and fundamental differences between the role of son or daughter and the role of pupil. This is related to the fact that though mother and teacher are (usually) both women, the role relationships with the child are of a different order. For example, it is part of the teacher's job that she should evaluate the children differentially on the basis of scholastic performance. The mother, on the other hand, is concerned with offering emotional support to the child's needs and the child is less likely to be evaluated in terms of his capacities for achievement. Any experienced nursery or infant teacher has a fund of entertaining and moving stories which illustrate the problems experienced by the child in learning to differentiate, and adjust to, the new role relationships required by the school and classroom situation.

As the child's social world expands, as the range of roles he is called upon to perform increases, so also arise changes in the child's 'significant others'. Though it may well be true that the child's basic identity, personality and values are indelibly established by his parents, as childhood recedes parents lose their unique pre-eminence among the significant others. Every new group the child enters—the school, the gang, the club, the job, the marriage—bring new significant others to exert their influence, just as they bring new roles to be learned. Often there will be relatively little conflict between the old and the new in the transitions between groups and roles—if only because we actively avoid the transitions where the disjunction between the old and the new is so radical that our conceptions of our self and 'reality' are seriously threatened. But every transition involves some degree of change, and in accommodating to the changes in roles and significant others our selves become progressively elaborated and differentiated.

These changes in one's self-image are not noticed by oneself because they take place so slowly and gradually. People who have not seen us for many years may comment on the changes which have escaped our notice. Christopher Isherwood (1962) remarks on this

16

phenomenon when as an older man he begins to recount his youthful adventures.

And now before I slip back into the convention of calling this young man 'I', let me consider him as a separate being, a stranger almost, setting out on this adventure in a taxi to the docks. For, of course, he *is* almost a stranger to me. I have revised his opinions, changed his accent and his mannerisms, unlearned or exaggerated his prejudices and his habits. We still share the same skeleton, but its outer covering has altered so much that I doubt if he would recognize me on the street. We have in common the label of our name, and a continuity of consciousness; there has been no break in the sequence of daily statements that I am I. But *what* I am has refashioned itself throughout the days and years, until now almost all that remains constant is the mere awareness of being conscious.

Sometimes the changes in one's self-image can suddenly be brought to mind by an event which compels one to recall what one was like in former days. Thus in Tolstoy's *Resurrection* when Prince Nekhlyudov meets Katusha after an interval of three years he is forced to recognize the changes in his attitudes, values and behaviour, his role and self-image. Before he was 'an honest unselfish lad'; now he is 'a dissolute accomplished egoist'. He knows that the change is related to his entry into the army and the changing expectations of significant others who now approve different forms of conduct.

Then he had regarded his spiritual being as his real self; now his healthy virile animal self was the real *I*.
All this terrible change had come about simply because he had ceased to put this faith in his own conscience and had taken to trusting in others. . . . When he trusted his own conscience he was always laying himself open to criticism, whereas now, trusting others, he received the approval of those around him.
Thus when Nekhlyudov used to think, read and speak about God, about truth, about wealth and poverty, everyone round him had considered it out of place and in a way ridiculous, and his mother and aunt had called him, with kindly irony, *notre cher philosophe*. But when he read novels, told *risqué* stories, drove to the French theatre to see absurd vaudevilles and gaily repeated the jokes—everybody admired and encouraged him. When he considered it his duty to moderate his needs, and wore an old overcoat and abstained from wine, they had all thought it odd, as if he were being eccentric in order to show off; but when he spent large sums on hunting, or on the appointments of a special or luxurious study for himself, everybody praised his

17

taste and gave him expensive presents. While he had been chaste and had intended to remain so till he married, his family had been afraid for his health, and even his mother was not distressed but rather pleased when she found out that he had become a real man and had taken a certain French lady away from one of his comrades.

There is a sense in which we can be said to have as many social selves as we have roles and relationships, for in different relationships different aspects of the self are brought to the fore. As William James (1890) put it:

> Properly speaking, a man has as many social selves as there are individuals who recognize him and carry an image of him in their mind . . . But as the individuals who carry the images fall naturally into classes, we may practically say that he has as many different social selves as there are distinct *groups* of persons about whose opinion he cares.

A man at work can be a very different person from a man at home, for the roles of father and husband are very different roles from occupational roles. Roles are performed in relation to different others and with reference to different significant others. Thus teachers' reports about a pupil on Parents' Evening are sometimes received with great surprise. The parental claim that 'He's never like that at home' may well be true.

When a person changes his role and social situation and is subjected to new expectations and pressures which differ from the original ones, two conflicting selves may develop in response to the two roles and situations. Several writers have noted how working-class pupils in grammar schools lived in effect in two very different social worlds, the school and the home in a working-class neighbourhood.

> the Bethnal Green girls soon became bilingual (for they would have been as much criticized for speaking Cockney at school as they would for 'putting on airs' and speaking 'posh' at home). . . (Young & Wilmott, 1957)

> there were children who became bilingual, speaking B.B.C. English at school but roughening up when they got home . . . Some kept the new accent at home as well as at school, and though this was approved by parents paying for elocution lessons, it thrust a touch of discord into other working class homes. (Jackson & Marsden, 1962)

> such a boy is between two worlds of school and home; and they meet at few points. Once at the Grammar school, he quickly learns to make use of a pair of different accents, perhaps even

two different characters and differing standards of value.
(Hoggart, 1957)

In the adult the self-image has become progressively structured and refined in terms of the roles he performs. The human self-identity does seem to consist of the social categories to which he is assigned combined with the role relationships that he maintains with others. The adult, unlike the small child, occupies a large number of role categories simultaneously and consecutively and for the most part he is able to integrate these with relatively little conflict into a unified whole. The way in which a person's identity is bound up with his roles becomes clear if you ask someone to describe who they are in ten or more words or phrases. A typical response would be:

1. Man	6. Play golf
2. Teacher	7. Good driver
3. Middle-aged	8. Tall
4. Husband	9. Intelligent
5. Father	10. Friendly

People tend to describe themselves in such role terms (sex, age, occupational and leisure roles etc.) before they give information about their feelings or attitudes to themselves, and for obvious reasons children are less likely to lay such emphasis on roles (Kuhn & Mc-Partland, 1954). Because adults specify themselves in these conditions in terms of roles with greater frequency than do children cannot be taken as clear evidence that it is the roles which most contribute to an adult's sense of identity. When we are asked to describe ourselves we are implicitly asked to differentiate ourselves from others in the same society. To specify ourselves in terms of the configuration of roles we occupy is the easiest and most obvious way of making such a differentiation. Other differences are very difficult to verbalize and we turn to the more difficult differentiating factors only after we have exhausted the easier course. However, when we do try to express the less tangible aspects of the self, we do rely on the evaluations of others. 'Good driver' is a role category plus an evaluation of role effectiveness that relies on socially established criteria, and such personal factors or personality traits as 'intelligent' and 'friendly' are self-applied as a result of the reactions of others to us. We claim certain personality characteristics because others tell us that we possess them or respond to us as if we did. Our identity, in whatever terms we may construe this, is socially mediated. Identity is social in its origin, social in its maintenance, and social in its development and change.

We are left with the problem of a person's 'real self'. Which of

James's many social selves or which of our many role relationships form the 'real self'? This is a problem in which few contemporary social scientists outside psychiatry have shown much interest, but there is no doubt that it can be a very real experiential problem to some people. The problem might arise when a person is forced to execute two or more role relationships which are experienced as inconsistent or incompatible. Most of us are successful in avoiding role commitments that are radically inconsistent, but, as in the case of the working class pupil in a grammar school, this is not always possible. Even here the conflict can be reduced by abandoning the less attractive role—leaving the grammar school or cutting one's neighbourhood ties. When the roles are separated in time and space, they can be divorced in experience by giving each a distinctive rationale and justification which are not allowed to conflict subjectively. Thus it was possible for the Nazi guard at Auschwitz to lead a happy and conventional home life. Alternatively, concern about one's 'real self' may arise from the necessity of erecting a façade in order to fulfil certain role obligations. The sense of façade can spring from a variety of sources such as an incompatible personality or from deeply held social or moral values that conflict with the role demands. A solution can be sought in withdrawal from the role or by an adjustment of personality or values. Where solutions such as these are impossible the façade may have to be maintained, even at the cost of considerable personal anxiety.

Whilst the experience of losing one's sense of identity is common in various forms of neurosis and psychosis, it can also arise in the 'normal' person, as the novelist Gore Vidal (1947) portrays. The hero of the novel, Robert Holton, has played all the necessary roles, living up to what others had expected of him. Every person sees him in a different way, not merely because all relationships bring out different facets of the person but because he had intentionally yielded to the widely differing roles assigned to him by other people. But now, when he wants to be himself, he cannot find his identity among the series of masks and the myriad faces that make up his life.

Although most of us escape severe forms of such conflicts, we all have a problem of our 'real self' to some degree in that we feel that we are more truly ourselves in some situations and relationships than others. For a minority the 'real self' emerges in a non-social activity e.g. painting a picture or meditating, but for the majority social relationships are the medium of its expression. Although the precise conditions are difficult to define, it can perhaps be said in general terms that we can be our 'real self' when (a) we know what is expected of us, (b) we are able to live up to these expectations to the gratification of ourselves and the other persons involved, and (c) we feel that we are accepted for what we are without conditions and reservations.

Such conditions are not easily attained and it is perhaps in default of such relationships that we turn to power and status of various kinds as alternatives.

The social sex role

We shall consider in more detail one aspect of role learning, that of the social sex role, or the means by which we come to adopt the behaviour which our society regards as appropriate to being either male or female. Small children usually show a definite interest in the factors which distinguish males and females. 'He' and 'she' is one of the earliest differentiations the child learns to make among the human beings around him. The initial focus of this interest is often physical: a boy has a penis, but a girl does not; mothers have breasts, but fathers do not. It is natural that the child should be interested in the physical differences between the sexes since these are so obvious, but this interest is probably nourished by the tendency of many parents to shroud sexual differences with mystery or embarrassment. The more subtle distinctions between the sexes are our concepts of masculinity and femininity, or the qualities we ascribe to men and women. The process whereby children learn to perceive themselves as masculine or feminine, and acquire the appropriate patterns of behaviour, is complex.

Many adults in the Western world feel that the Western conceptions of the social sex roles are entirely 'natural', as if they sprang from innate biological differences and that no social learning is involved. Anthropological studies have shown that this view is untenable. Many of our values, beliefs and attitudes are taken over from our culture because they are accepted by all or most people within that culture. Indeed, the principal way we can assert the validity of our values, beliefs and attitudes is on the basis that they are shared by everyone else. For this reason, the concept of *consensual validation* (Sullivan, 1940) is a very basic one in social psychology. It indicates that we can validate our conceptions and interpretations of the world and our daily experiences only by achieving consensus with other people about our conceptions and interpretations. It is this consensus which validates our sense of what is real. Thus our values, attitudes and beliefs become 'proper' or 'right' or 'true' when others share them. That so much of our experience of the world is treated by us as 'natural', is taken for granted and is not perceived as problematic should not blind us to the fact that the definitions which guide and organize our actions have been built up through our interpersonal environment. Symbolic Interactionism, as Herbert Blumer has tirelessly reminded us, suggests that action is *constructed* through the meaningful *interpretation* of situations which confront human beings

21

—notions which stem directly from the work of G. H. Mead. As Blumer (1956) puts it:

> We can and, I think, must look upon human group life as chiefly a vast interpretative process in which people, singly and collectively, guide themselves by defining the objects, events, and situations which they encounter. Regularized activity inside this process results from the application of stabilized definitions. Thus, an institution carries on its complicated activity through an articulated complex of such stabilized meanings. In the face of new situations or new experiences individuals, groups, institutions and societies find it necessary to form new definitions. These new definitions may enter into the repertoire of stable meanings. This seems to be the characteristic way in which new activities, new relations, and new social structures are formed. The process of interpretation may be viewed as a vast digestive process through which the confrontations of experience are transformed into activity.

Frequently, our judgments are *ethnocentric*—we judge others by our own particular standards, by the stabilized meanings and definitions of our own culture or subculture. When we find a human social system that differs from our own, we evaluate it from our own taken for granted frame of reference, with the consequence that we view their way of doing things as strange and very often also as wrong. This human tendency has been vividly described, for example, by novelists giving accounts of the work of Christian missionaries in unfamiliar cultures (March, 1952; Michener, 1966). Of course, in every known culture there are features which differentiate men from women, but there is an enormous diversity of known patterns. In fact, there is no single trait which we in our society ascribe to males or females which is not ascribed to the opposite sex in some other society. Societies do exist which almost entirely reverse our own social sex codes. All this evidence compels us to regard our concepts of masculinity and femininity as products of cultural learning.

Parents are the major sources of influence in the child's acceptance of the 'correct' social sex role. They have been taught to accept certain characteristics as belonging to one sex or the other and they will try to train their children accordingly. From an early age boys and girls are dressed differently. Even the new-born infant is sometimes swathed in a colour (pink or blue) which is sex appropriate. As the child grows older, different kinds of clothing (or the same clothing with different styles of decoration) and different hair styles are used to distinguish the sexes. Social sex role training really begins when the infant has passed the rattle-and-teddy stage, for it is at this stage that sex-appropriate toys are given. Trains, cars, and tools are suitable for boys, whereas dolls, prams, and tea-sets are pastimes for girls. In short, boys are presented with toys involving mechanical, manipulative and constructional skills, but girls' toys stimulate domestic feelings and skills. Toys thus reflect, if vaguely, the roles the children will be expected to assume when they grow up.

Social sex role training also includes the encouragement in the child of personality traits and behaviour patterns which are regarded as peculiar to one sex. Boys are told that only girls cry when they are upset, frustrated or slightly hurt. This is the psychological 'toughness' which has its physical counterpart in fighting ability. The tendencies of children to acquire the social sex role of the opposite sex are discouraged or, if the child is older, ridiculed. The labels 'cissie' and 'tomboy' are marks of disapproval to the deviants we wish to bring back into line.

One important aspect of social sex role learning is the process by which the child comes to identify with the parent of the same sex, that is, use the same sex parent as a model for behaviour. Typically parents reward, by subtle signs of approval, the child who imitates the behaviour of the same sex parent. But this is only part of the story. As Freud recognized, the problem of learning to identify with the same sex parent is more difficult for boys than for girls. The reason is obvious. Mothers are the principal child-rearing agents in our society. It is mother who looks after the child for the majority of the first five years of life. Father is absent at work for the major proportion of the young child's waking life. Children of both sexes are likely to develop deeper emotional and affectional ties with mother than with father. What then leads the boy to turn away from mother and towards father as a model?

Freud believed he had found the key to this problem in the so-called Oedipus situation. As the young boy's emotional attachment to his mother deepens, he begins to see father as a rival who constantly intrudes into his blissful relationship with mother. When father comes home in the evening, it is he who becomes the focus of attention. And it is father, not son, that mother sleeps with all night. The boy becomes more possessive of mother and more hostile to father. How much more pleasant life would be if father were out of the way! The boy, however, is distressed by his own hostile feelings toward father, whom he also loves. He solves this problem by believing that it is father who is being hostile towards him, rather than the other way round. This, of course, does not really help matters. Psychoanalysts believed that the boy resolves the situation by coming to identify with the father. To model himself on the father's behaviour should neutralize the aggression the boy has projected onto the father. Secondly, if he becomes like father he should acquire greater power over mother, a power similar to that which father appears to exert over her. 'If you can't beat them, join them.'

There is little doubt that the Oedipus situation does appear in many parent-son relationships in *some* form, even though its classical form may be an extreme case. Unfortunately it is very difficult to verify or falsify much of Freud's theorization, so the Oedipus complex has remained a suggestive but not widely accepted idea. More recent workers have been concerned to specify the nature of the relationship between father and son which fosters social sex role development. It can now be said fairly confidently that where father is absent very frequently or where the father-son relationship lacks warmth, then the boy will tend to experience difficulties in acquiring a normal masculine sex role (Mussen and Distler, 1959; Payne and Mussen, 1956; McCord, McCord and Thurber, 1962).

Several other factors facilitate the boy's adoption of the appropriate social sex role. Our culture is still largely dominated by males who in a number of ways possess great status and privilege. Such enhanced attractiveness of the male role may help the boy to display the appropriate masculine behaviour. (Girls may be restrained from rebellion by a culturally induced passivity.) Further, boys tend to incur greater social punishment from deviating from their social sex role than do girls. A boy who is girlish is less socially acceptable than a girl who is boyish. The greater social pressures exerted on boys to conform closely to the cultural stereotype may help to compensate for the relative lack of availability of fathers to their sons. Girls, who can model their behaviour on their almost constantly present mothers and primary school teachers, have no need of such strong supportive pressures from outside the family.

Among adolescents in contemporary society there is evidence of a movement towards a breaking down of earlier sharp differentiation of the social sex roles, at least in respect of clothing. Many boys have grown their hair long, whilst some girls have adopted short hair styles. Lace and frills and flower designs, which in the earlier part of the twentieth century have been associated with femininity, have become fashionable on men's shirts. Trouser suits are increasingly fashionable for women, even women teachers who are not normally in the forefront of modish attire. Girls have taken to wearing jeans, which were formally an adolescent male uniform. In the last few years we have seen the development of 'unisex' clothing which is appropriate to members of either sex.

Many of these new styles, particularly among boys, have not been given a warm reception by the teaching profession. Some teachers have taken these changes as signs of a general decline in standards and have resisted the trends to what they regard as effeminacy. Grammar school headmasters have been known to expel boys with long hair. In many respects such a reaction on the part of teachers is a curiously narrow view. They do not fully appreciate that the innovations they so much resent are a continuation of a gradual process of change towards less sharply defined social sex roles, a process which began more than half a century ago. One of the effects of giving women greater rights and a higher level of education has been the creation of a much larger work force. Women no longer expect to be drudges and frequently expect to follow a full career except for a short period devoted to the raising of a family. As a result many of the domestic tasks which were formally exclusively female duties must now be shared by husband and wife. Men in contemporary society are likely to do things—push prams, go shopping, change nappies, visit the launderette—which would have earned nothing but contempt from their fathers or grandfathers. Few would regard this change as undesirable, for they are a necessary concomitant of the emancipation of women. Young people are simply pushing this trend a little further. To promote an inter-generational conflict on this basis seems absurd in the extreme.

My own view is that hopefully the trend will reach to much deeper levels than clothing and the sharing of domestic chores. I should be happy if we gave up our conceptions that mechanical and mathematical skills are essentially masculine. I know of no biological reason why women should not make successful engineers and physicists; indeed, they are likely to make a highly significant contribution to these areas. But at present most people show surprise, and sometimes implicit disapproval, towards a girl with interests in such matters. Similarly it seems to me that much harm can spring from the conception that men are not very emotional and/or that men should not express their emotions if they want to be truly masculine. Many men are emotionally retarded because of their reluctance to accept the strength of their emotions and because of their socially induced inability to cope with deep emotion in themselves. Women almost invariably show greater ability to cope emotionally in times of stress, and this is perhaps because they have learned to do so since they were expected to do so as part of their feminine social sex role. Perhaps if we would abandon these prejudices fewer adolescent males would need to commit acts of violence in order to prove their masculinity.

The ideal self

The self-image includes all that a person considers himself to be at a particular time. The sources of the self-image we have so far examined are two-fold. Firstly, a man learns to perceive himself as others perceive him: the attitudes of others to him are incorporated into each man's attitudes to himself. Those with whom a person interacts most frequently and on whom he models his behaviour—usually but not always the same people—become his 'significant others'. It is on their reactions to his behaviour that he most depends for feedback. Secondly, how a man sees himself is related to the roles he performs. All the roles which a man plays contribute to his self-identity, though some roles, such as the occupational role and the sex role, play a more central part than others.

A man's self-image is thus a product of past experience, especially the history of his relationships with others. The self is a pattern imposed on experience, a meaningful organization of an infinite number of past experiences and relationships. The self is that which unifies the variety and complexity of the past into a consistent present and which acts as a point of departure into the future.

Yet a man does not move into the future merely in terms of the present structured by the past. He has a perception not only of what he *is*, and what he *has been*, but also of what he plans to do, of what his goals and projects are, and of what he *would like to be*. The fact that the future is to some degree moulded by what a person thinks he would like to be or should be—a man's aspirations and ideals—is one of the ways in which human behaviour is greatly more complex than that of subhuman animals.

The ideal self, like the other aspects of the self, is probably derived from parents in the first instance. For the older child, friends, other adults, film and 'pop' stars, political or religious figures, may be taken as models of the sort of person he would like to be. Eventually hero-worship will probably give way to a fusion of several such figures, a fusion which involves an adaptation more related to one's own potentialities.

The distinction between the sort of person one would like to be and the sort of person one feels one ought to be is very fine. Clearly this second aspect, involving a feeling of *ought*, is related to the notion of conscience. Freud recognized this by making the ego-ideal and conscience two elements within the super-ego. It might be said—with a certain inaccuracy—that the conscience is largely negative, in that it represents an internalization of what parents (and others) consider to be bad, behaviour to be shunned: the ideal self is largely positive, in that it represents an internalization of what parents (and others) consider to be good, goals which must be strived after.

It is quite remarkable how little we know about either the means whereby human beings develop self-ideals or the impact that the self-ideal has on behaviour. Most people, however, do seem to retain a self-ideal. If the ideal disappears or is very much higher than the actual self, problems of adjustment are likely to be experienced.

Self-esteem

The attitudes a person takes to himself involve an evaluative dimension. A man views himself, to a greater or lesser extent, as good or bad, favourable or unfavourable, acceptable or unacceptable. Self-esteem concerns a person's feeling of *worth*. High self-esteem arises when the attitudes taken to self are positive, low self-esteem when they are negative.

In taking the attitudes of others to himself, a person automatically acquires a certain amount of self-esteem. If parents show love and appreciation of their child, he will perceive himself as someone possessing good qualities, so his self-esteem will be high. If the teacher constantly tells the pupil that he is stupid and lazy, and treats him as such, the pupil will tend to perceive himself as stupid and lazy. Since laziness and stupidity are negatively valued in our culture, such a pupil will have low self-esteem. Some teachers have found that by treating the pupil as if he did possess talent, even though there may have been little evidence of any hitherto, they can bring the pupil to perceive himself in positive terms and thus be motivated to use his abilities in the pursuit of scholastic goals. Probably more of human behaviour than we imagine follows the pattern of the 'self-fulfilling prophecy': people tend to become what we tell them they are.

Most people at some stage during their lives, especially between late childhood and early adulthood, elevate their ideal self to a height which makes fulfilment of the ideal impossible. The normal person adjusts his ideal to the demands of reality. Because of our natural limitations and environmental restrictions we can rarely become what we would like to be. Unless we alter our ideal to a level which is potentially or partially attainable, we take the risk of losing our self-esteem. If our actual achievements fall a long way short of the ideal then we are likely to indict ourselves as failures and develop feelings of worthlessness and even self-hatred. Religious belief may create or support an elevated ideal self which effectively reduces the believer's self-esteem (Cf. Williams, 1965). It has been argued by Karen Horney (1951) that the central core of neurosis lies in the tendency of the neurotic to judge himself in terms of an unattainable ideal.

Because our self-esteem does depend so much on the reactions of others to us, we tend to reveal to others those aspects of ourselves which will meet with an approving reaction. At the same time, we hide those aspects of ourselves which are likely to provoke disapproval from others. When we are not sure of the reaction, we often make tentative and exploratory remarks to probe out the reactions of others. To reveal oneself with a carefree spirit is to risk the censure and rejection of others—and it is difficult to retain a high level of self-esteem when we are rejected by others, especially our 'significant others'. A man may find his opinions totally rejected by a casual acquaintance in a railway carriage and forget the incident once he has left the train; but if his wife shows the same 'lack of understanding', his self-esteem may suffer an irreparable blow. Perhaps this is why casual and teasing remarks made by teachers to their pupils can have a greater impact than the teacher ever imagines. To the pupil, the teacher may be one of the most significant of his 'significant others', whereas the reverse is much less likely to be true.

Unless people possess a basic core of self-esteem they may experience great difficulty in coping with the social environment. A man requires great self-esteem if he is to stand by what he believes. Those with low self-esteem are likely to be cautious, shy of initiative and excessively conformist. Without the possession of an internal locus of self-evaluation, conformity to the beliefs of others becomes a major source of positive evaluation.

Stability of the self

As a man grows older his self concept becomes more stable and more resistant to change. It is difficult to teach an old dog new tricks because to incorporate the new may involve a radical change in the central aspects of the self. There are a number of reasons why the self does become more stable with age. Perhaps the most important is the relative stability of the social environment in which most people operate. A man tends to perform the same major roles and interact with the same people, or the same sort of people. In addition, most people probably actively strive after consistency. Such consistency is attained by several techniques. The most common is the tendency, to associate with those people who confirm a man's feelings about, and attitudes to, himself. Such friends and associates treat a man as he is accustomed to be treated and so confirm his self to him. Moreover, mixing with those who share his attitudes, beliefs and values confirms a man in believing that he is 'right' to be what he is. The stable social environment reflects a stable image. To live in an unstable social environment would be as disturbing as walking round the Hall of Mirrors in a fair. Most people have at some time in their lives had the experience of being caught within a social situation in which they have felt a complete outsider. To be surrounded by people with different values and beliefs produces feelings of uneasiness and tension, a fear of participating and a desire to escape. In extreme cases the experience may have a nightmare quality. Restoration to the usual social environment is accompanied by a feeling of relief, of normality and sanity. It is only the great leader and the incurably insane who are prepared to persist in public proclamations of beliefs which are rejected by the rest of the world. It is hard to believe that all of the world is totally wrong all of the time.

Another means of promoting stability in the self is to misperceive in the direction of stability. Most people overestimate the extent to which others agree with them about the sort of person they are. If this normal trend becomes excessive—the woman who believes that her husband loves her and cannot live without her, though he is asking for a divorce—we regard it as mental derangement. The pathological should not blind us, however, to the fact that such distortion occurs in most people to some extent. Sometimes, of course, our misperception of how others feel about us is a direct result of their behaviour to us. 'Tact' is where we act towards people in a way which does not undermine their attitudes to themselves. Tact is also frequently an excuse for behaving in a way towards others that avoids the revelation of one's true feelings. The image of the 'blunt Yorkshireman' is both favourable and unfavourable precisely because frankness ensures accurate feedback.

Once a person has developed a reasonably stable self-image he may resist the adoption of new roles, values and attitudes not only because it would involve radical changes to the self-image but also because it might involve a form of personal treachery. A man is most open to change when he is very dissatisfied with his present self-conception and admires another way of life that is open to him. But most people are trapped within their roles and social environments and a sense of identity can be anchored and stabilized with the belief that this way of life is 'right' and perhaps even the 'best' way of life. Other ways of life, and thus other forms of self-image, can be dismissed as inferior or wrong. We use our own self-image as a basis for evaluating ourselves and others, for justifying ourselves and those like us and denigrating those who are different. Tolstoy puts the matter very clearly.

> Whatever a man's position may be, he is bound to take a view of human life in general that will make his own activity seem important and good. People usually imagine that a thief, a murderer, a spy, a prostitute, knowing their occupation to be evil, must be ashamed of it. But the very opposite is true. Men who have been placed by fate and their own sins or mistakes in a certain position, however irregular that position may be, adopt a view of life as a whole which makes their position appear to them good and respectable. In order to back up their view of life they instinctively mix only with those who accept their ideas of life and of their place in it. This surprises us when it is a case of thieves bragging of their skill, prostitutes flaunting their depravity or murderers boasting of their cruelty. But it

surprises only because their numbers are limited and—this is the point—we live in a different atmosphere. But can we not observe the same phenomenon when the rich boast of their wealth, i.e. of robbery; when commanders of armies pride themselves on their victories, i.e. on murder; and when those in high places vaunt their power—their brute force? We do not see that their ideas of life and of good and evil are corrupt and inspired by a necessity to justify their position, only because the circle of people with such corrupt ideas is a larger one and we belong to it ourselves.

The tragedy is that in bolstering our self-image with notions of rectitude we not only reinforce the stability of the self-image but also make it static and the associated dogmatism and intransigence inhibit development and change. In addition, it becomes difficult to perceive matters from any other perspective except one's own. The legitimacy of other views based on different assumptions and premises is denied. Intolerance reigns.

Recommended reading

The self
GEORGE HERBERT MEAD, *Mind, Self and Society*, University of Chicago Press, 1934.
ANSELM STRAUSS (ed.), *George Herbert Mead: On Social Psychology*, Phoenix Books, University of Chicago Press, 1956.

General introduction to social psychology
M. ARGYLE, *The Psychology of Interpersonal Behaviour*, Pelican, 1967.
P. KELVIN, *The Bases of Social Behaviour*, Holt, Rinehart & Winston, 1970.
TAMOTSU SHIBUTANI. *Society and Personality*, Prentice-Hall, 1961.
W. J. H. SPROTT, *Human Groups*, Pelican, 1958.

Role-learning in childhood
O. G. BRIM, 'The parent-child relation as a social system'. *Child Development*, vol. 28, 1957, pp. 343–64.
E. E. MACCOBY, 'Role-taking in childhood and its consequences for social learning'. *Child Development*, vol. 30, 1959, pp. 239–52.

3 Perceiving people

This book is primarily concerned with human interaction and the ways in which human beings behave in face-to-face relationships. An analysis of human interaction clearly involves a consideration of the ways in which persons perceive one another. It is the perception of just who and what the other person is which steers one man's behaviour to another. At a very elementary level the knowledge that I am going to interact with an old lady rather than a small boy is in itself sufficient to evoke certain sorts of behaviour and inhibit others. All interactions are affected by the basic physical characteristics of the other—age, sex, colour etc. Interaction is also guided by our evaluations of others. We accord a different treatment to our friends than to our enemies, to our superiors than to our subordinates, to our fellow countrymen than to foreigners. Our behaviour to others depends not only on their objective characteristics but on our perception of these characteristics, and it is these perceptions which in part create our opinions of them and our feelings towards them. Interaction is not structured simply by the behaviour of the two participants but by the ways in which the participants perceive each other. If we wish to understand an interaction we cannot stand on the outside as a neutral observer of the participants' perceptions of each other and these perceptions are not directly observable. So before we can turn to the more direct study of interaction we must examine some of the basic mechanisms of person perception.

Suppose we ask someone (whom we shall call Person) to go into a room to interview another person (whom we shall call Other). Person has not been told anything at all about Other. Person goes into the room and *sees* Other. Person's first action is to make a visual perception of Other. If we ask Person what he has seen, he will tell us that he has seen a young woman. If we ask him how he knows this he will tell us that it is obvious since Other is wearing a dress and

31

has a well developed figure. Person did not use all the information at his disposal but *selected* certain aspects of the multiplicity of available data. He has selected information relating to Other's age and sex, but neglected information that might permit additional conclusions about Other. Further, he has not noted all the information that might be used as evidence for Other's femaleness. He can 'tell at a glance' that Other is female. The means whereby he has selected some information about Other is rapid and subconscious. He is not aware of his selection of some data as a basis for reaching a conclusion until we ask him how he reached that conclusion. The primary perceptual phenomenon is thus the selection of data. Some information passes through the filter and some information does not (Fig. 3.1). Data relating

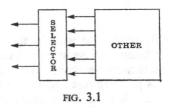

FIG. 3.1

to the sex and age of Other are selected at once not only because they are the most obvious characteristics of Other that can be obtained through a visual inspection but also because they represent basic aspects of Other that will help Person to know how he should behave towards Other and how he might expect Other to behave towards him.

Implicit in our illustration of the selection process is the process of *organizing* the data. In concluding that Other is a woman Person has organized his perception. In organizing the data, Person is trying to make sense of what he sees. Perception, to use the words of Bartlett (1932), is an effort after meaning. To organize the information is essentially to make *interpretations* and *inferences* about the information. It is because Other is wearing a dress that Person infers that it is a woman. The two processes of selection and organization are only theoretically distinct, for they are related in complex ways. It is the need to organize which leads Person to select in the first place. Once the data have been organized this may influence the selector by filtering in information which is congruent with the interpretation made and promote the selection of further data to confirm or resolve a tentative or ambiguous interpretation.

Most people are prepared to make interpretations and inferences on the basis of a very small amount of information. The knowledge that Other is wearing a dress is not sufficient in itself to make an inference about Other's sex absolutely certain, although it will be highly probable. We do not normally regard more extensive research for confirmatory evidence as necessary because we have learned that the wearing of a dress correlates highly with femaleness and that baldness and pipe-smoking correlate highly with maleness. Indeed normally there is such a superfluity of information available about Other's basic attributes that we need to pay attention to only a small number of cues which are sufficient to ensure correct inference making.

Suppose, however, that when Person went into the room to meet Other, Other was smoking a pipe. Person would notice, i.e. select, this information immediately because it would not be congruous with the other evidence e.g. the wearing of a dress which could be used as the basis for inferring that Other was a woman. The process of organization and inference making would become more explicit in Person's thinking because one element has interfered with the normal process of making inferences at great speed. He would be forced to consider several possible inferences. Is this a woman who is exhibitionistic in her emancipation from conventional mores? Or is it really a man who for some reason is dressed to look like a woman? The inferences would now be tentative and could be resolved only through the selection of additional information. An active search for confirmatory evidence is now necessary (Fig. 3.2).

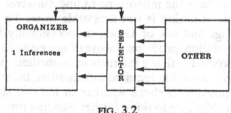

FIG. 3.2

It is because most people are prepared to make inferences on the basis of the most slender evidence that so many of our initial inferences about other people are misleading and sometimes completely false. The mistaken identity situation is a common result of false inference making. Who has not at some time mistaken a complete stranger on the street for a friend? An elderly teacher once boxed the ears of a student teacher on his first day of teaching practice for running along a corridor. He had inferred from the student teacher's age that he must be a sixth form pupil.

The actor in a social situation, then, uses in his perception what Schutz (1932) has called 'interpretive schemes' to make sense of the data. Some of these interpretive schemes are shared by all the members of a culture; others may be restricted to members of subcultures or groups; others may be unique to the individual, arising through his unique biography. Any experience is referred to these interpretive schemes as it occurs. The selectivity of perception hinges on the fact that not all the available data is *relevant* to the actor. Further some of the information that is relevant to the actor may not be available to him. Schutz (1970) speaks of 'systems of relevancies' which are related to the actor's interests, goals, roles and so on. For instance, in the example we have used so far, Person has to find out a number of things about Other before he can interact with Other in the light of his goal. It is this problem which becomes most relevant and he has to suspend his plan of interviewing Other. Having recognized this new problem, he then has to make interpretations about Other to resolve the ambiguities. To this end appropriate interpretive schemes are activated. When he has made the necessary interpretations and inferences, he can then return to his original goal, namely that of interviewing Other, and a new system of relevancies and appropriate schemes of interpretation are brought to bear.

The first step, then, is to gather as quickly as possible as much information about Other so that Person will have some basis on which to structure his relationship to Other. The primary information will be of physical order. Person will absorb instantly the basic cues of age and sex. This is such an automatic process that we rarely think about it at all unless we have the misfortune to find ourselves in a situation in which such information is not accessible to us. To be unable to determine the age and sex of Other, as occasionally happens in a telephone conversation, produces a sense of unease and an uncertainty about how to proceed with the interaction. Sometimes further inspection of Other will reveal the required information. In interaction with a woman it is possible to check whether or not one should address her as 'Miss' or 'Mrs', by looking for her wedding ring. If such basic information is not available, if for example the woman in our illustration is wearing gloves, we warily make tentative steps to try to tease out the information we want without directly asking for it. Most people have had the experience of being warmly greeted by a person whom they cannot for the moment recognize. It is a fascinating illustration of the subtleties of human interaction when one person tries to carry on a conversation with another person without betraying his lack of recognition.

Once Person has acquired the basic information about Other's age and sex, which is in itself enough to provide quite an elaborate structure for their interaction, Person will seek out more subtle information about Other from the cues that Other gives off. The facial expression of Other is a major source of further information. From the face Person can discover whether Other is friendly or hostile, happy or sad. Other aspects of Other and his behaviour are selectively

34

used by Person. Posture, gestures and other physical movements offer a rich source of information. Clothing reveals much more than Other's sex, for it can give indications about taste, wealth and social background, personality, sexual attractiveness, occupation. Uniforms convey information of one's profession and often of one's status within that profession.

When a person speaks, the richest source of potential information becomes available. The voice can convey much, purely in the way the words are spoken. Tone and volume can reveal important clues about the emotional state of the speaker, and accent may yield information about a person's origins, both regional and social, as well as his educational level. The principal source of information comes, of course, from the content of what is said. In the words he speaks Other discloses much of what Person wants to know about Other. In a new interaction it is customary for Other to provide Person with certain basic facts about himself. When Other introduces himself to Person he is giving a basic structure to the early phases of the interaction and unless Other is willing to furnish Person with relevant verbal information the interaction is unlikely to proceed smoothly.

The process of inference and perceptual organization has been brilliantly demonstrated by Solomon Asch (1946), who showed that from a list of isolated traits we experience no difficulty in describing the sort of person who is purported to possess these traits. In this classic experiment Asch read to his subjects a list of traits which were said to belong to an unknown person. One such list contained the following terms: energetic—assured—talkative—cold—ironical—inquisitive—persuasive. These subjects were then requested to write a short characterization of the person to whom these traits belonged. It is interesting that the subjects readily organized and integrated these discreet characteristics into a unified whole. One subject described the unknown person as follows:

> He seems to be the kind of person who would make a great impression upon others at a first meeting. However as time went by, his acquaintances would easily see through the mask. Underneath would be revealed his arrogance and selfishness.

The subject not only converts the bare terms into a rounded portrait, but also makes inferences about characteristics which are not actually mentioned in the original list. He does not seem to be satisfied with the pieces of evidence at his disposal but goes beyond the evidence in elaborating a more complex whole.

In a second experiment by Asch the subjects were given the following list of traits: intelligent—skilful—industrious—warm—determined—practical—cautious. Another group of subjects was given the same list except that 'cold' had replaced 'warm'. Changing one of the traits made an enormous difference to the portraits painted by the subjects. One subject given the 'warm' list wrote: A person who believes certain things to be right, wants others to see his point, would be sincere in argument and would like to see his point won. In contrast, one subject with the 'cold' list wrote: 'A rather snobbish person who feels that his success and intelligence set him apart from the run-of-the-mill individual. Calculating and unsympathetic.' A change in just one of the stimulus traits can produce a deep and pervasive change in the whole impression. Indeed Asch shows that the subjects with the 'warm' list are much more likely than the subjects with the 'cold' list to see the person as exhibiting such characteristics as wise, humorous, popular, imaginative, humane, generous, happy, good-natured and sociable. It seems that knowledge of the warmth or coldness of a person leads one to make inferences about temperament (e.g. happy—unhappy) and about relations to others (e.g. popular–unpopular). All these inferences go beyond the known information.

It seems that some traits, such as the warm-cold dimension, are more 'central' than others in their ability to provoke strong associations and to transform other known traits. Some characteristics have more weight in forming the total impression than others. It is known for example that knowledge of whether a person is polite or blunt has little effect on the other traits or on the inferences we make. In this sense the polite-blunt dimension is 'peripheral'. However, even central dimensions such as warm–cold do not affect *all* other characteristics. In the above experiment the knowledge of whether a person is warm or cold did affect the inferences about many characteristics such as popularity and generosity but failed to influence the inference about other properties such as honesty, strength, seriousness, persistence or reliability.

It cannot be disputed that the experimental procedure of Asch is somewhat artificial. In everyday life we do not have to form an impression of someone from a list of traits read out to us. Perhaps reading a confidential reference about a person is the closest approach to Asch's experimental procedure, but in this case one is normally supplied with additional information, such as the person's age and sex, and the traits designated by the referee are likely to be related to specific activities and situations. Veness and Brierley (1963), have repeated Asch's experiment in a form less divorced from reality. In their version the subjects listened to a tape-recording of a woman's voice in which she described her work as a veterinary surgeon. The stimulus traits were implied in the content of the talk, though none was mentioned directly. A short extract from the tape-recording illustrates the procedure.

My day begins with the arrival of the postman at half-past six and, very often, with the ringing of the telephone, as farmers are early risers and crises so often happen during the night (INDUSTRIOUS). Surgery over by ten o'clock, I set out in my car for the neighbouring and distant farms. The problems of weather, transport, distances, bad roads and difficult hills are very great in a country practice, but my assistant and I have managed not to let such difficulties deter us from our purpose of reaching a sick animal at all costs (DETERMINED).

The warm–cold variable did not form part of the content, but the speaker adapted her voice to give an impression of either warmness or coldness. Asch's original findings on the warm–cold dimension were substantially confirmed in this experiment.

Many interesting questions are raised by the experiments of Asch. Firstly, if Person knows that Other possesses a given trait, what inferences about Other's other traits is he likely to make on the basis of this knowledge? Secondly, which traits are most central, that is, which traits tend to lead to relatively large numbers of inferences about other traits? As yet no definitive answers to these questions are available, although some dramatic progress has been made (Bruner, Shapiro and Tagiuri, 1958; Wishner, 1960).

From what has been said it is clear that people have an 'implicit personality theory' or a tendency to believe that certain traits go together (Bruner and Tagiuri, 1954). It is the implicit personality theory which is brought into action when we make inferences and when we organize our perceptions and impressions of other people. It is possible that different persons have different theories about what traits go together and develop implicit personality theories both from the common elements of our culture and from the various forms of social learning peculiar to the individual. If it is true that people vary in their theories of what traits hang together—and there is very little research on this topic—then this would have important effects on any single Person in his inferences about Other.

The second element in the organization of our perceptions of others is the process of *attribution*. On the basis of the inferences he makes about Other, Person tends to attribute certain qualities to Other. Paul Secord (1958) using photographs showed that from certain facial characteristics of Other, Person is prepared to make judgments about Other's emotional state and personality. For example, if Other is a woman with a full mouth and considerable lipstick, Person infers that she is sexy and attributes the trait of sexiness to her. If Other is a man with a dark skin, Person makes inferences which lead him to attribute to Other such qualities as hostility, boorishness, conceitedness, dishonesty, shyness and unfriendliness. Thus from cues given off by Other, Person attributes to Other what Fritz Heider has termed 'dispositional properties' or stable characteristics such as personality traits, attitudes and abilities.

The process of attribution can take place at various levels of complexity. On the simplest level Person attributes to Other one central trait such as 'reserved' or 'aggressive' which somehow seems to sum up Other. On the next level Other is described in terms of a number of traits which are congruous, which somehow seem to 'hang together'. For instance, to describe a person as having the following characteristics:

aggressive
cold
ambitious
self-confident
shrewd
mean

is likely to bring to mind a definite sort of person whose qualities are meaningfully consistent. At the most complex level Other is described in terms which are, to the outsider but not to Person, incongruous. In this case the properties of the person do not seem to 'hang together' in any obvious way. To describe a person as

kind
cautious
quick-tempered
sociable
blunt
wise

is to offer a picture of a person which is not easily conjured up in the imagination.

It seems to me rather doubtful whether people possessing a large number of completely incongruous traits actually exist. If they did, they would be monsters of inconsistency and unpredictability. The level at which we describe people is probably a function of how well we know them: the deeper our knowledge of a person, the higher will be the level of complexity with which we describe him. With someone we have met for but a few moments, for example at a party or on a bus journey, we are often left with the impression of just a single dominant trait. In the case of people we know rather better, our appreciation of the complexity of their personalities is limited by the formality of the relationship and the specificity of the situation in which we interact with them. But the people whom we know intimately can be perceived as possessing traits which are not entirely congruous because we see them in a variety of social situations and conditions in which different aspects of their personalities are brought into play. People are probably much less congruous than we imagine, in spite of the tendencies towards consistency operating on the self, for there is a strong tendency within every Person to over simplify every Other, to try to make sense of Other in an economical and meaningful way by stressing those elements which give him consistency and by ignoring those aspects which are at variance with our picture of him.

In this respect E. M. Forster's (1927) distinction between 'flat' and 'round' people is of interest. Characters in novels he divides into these two types.

Flat characters were called 'humours' in the seventeenth century, and are sometimes called types, and sometimes caricatures. In their purest form, they are constructed round a single idea or quality: when there is more than one factor in them, we get the beginning of the curve towards the round. The really flat character can be expressed in one sentence such as 'I will never desert Mr. Micawber.' There is Mrs. Micawber—she says she won't desert Mr. Micawber: she doesn't, and there she is.

From this description it is clear that 'flat' characters are at the lowest two levels outlined above. And flat characters, like persons who are perceived in terms of a single dominant trait or a cluster of congruous traits, have, as Forster points out, the advantage of being easily recognized and easily remembered. In novels, and perhaps in life too, such characters never need re-introducing and never need to be watched for development. They move through life unaffected by the chances and changes of this fleeting world. From the writer's (or perceiver's) point of view, all but the two or three dominant facets of them are disregarded. The recipe for their construction involves a small number of very obvious ingredients. In Forster's view almost all the characters in the novels of Charles Dickens and H. G. Wells are essentially flat.

'Round' characters, in contrast, cannot be summed up in a neat phrase (or a few traits). They wax and wane as they pass through life and they are constantly modified by their experience of it. In Forster's words: 'The test of a round character is whether it is capable of surprising in a convincing way. If it never surprises, it is flat. If it does not convince, it is a flat pretending to be round. It has the incalculability of life about it'. Forster regards Jane Austen and Russian writers such as Tolstoy and Dostoyevsky as specializing in round characters. In the works of such writers even the minor parts are essentially round.

It might be interesting to ask which type of novelist, the one describing largely round characters or the one describing largely flat characters, more accurately portrays real people in his work. It may be that the question is misleading in such a form, in that the difference between the flat writer and the round writer is at root one of a difference of perspective. The flat novelist may reflect the human tendency to simplify other people into a cluster of congruous traits, whilst the round writer reflects in his characters those people who exemplify complex clusters of not entirely congruous traits. Perhaps people in real life fall along a continuum from flat to round. My own experience is that whilst most people seem rather flat at first sight, they come on closer acquaintance to acquire greater rotundity. So I conclude that most of us are in fact rather round. If this is the case, then the work of psychologists of personality may be more complex than they sometimes allow, for many such psychologists seem to conceive of the human personality in essentially flat terms. (It is, incidentally, worth considering just how much of psychology as a whole, in its aim of arriving at generalizations about man, is inherently flat in its approach to man.)

The dispositional properties which Person attributes to Other do not consist only of personality traits. Of rather greater importance is the fact that from his inferences Person may attribute to Other certain *motives* or *intentions*. If Person sees a man rushing towards him with a dagger in his raised hand, Person is likely to attribute to the man not only the trait of anger or aggression but also certain motives such as murderous intentions. It is from these dispositional properties which Person imputes to Other that Person feels that he can 'explain' Other's behaviour. To work out a motive or intention behind Other's behaviour is to explain that behaviour. To be unable to attribute a motive or intention to Other makes Other's conduct somewhat puzzling to Person. Thus he may say, 'Other is crazy: I don't understand him at all.' We search for possible intentions and motives in order to make other people's behaviour comprehensible and meaningful. Most of the time we are able to impute definite intentions and motives to other people and most of the time our attributions correspond to Other's actual motives and intentions. But such is not always the case. One of the most common sources of the 'misunderstandings' which arise in human relationships is the attribution to Other of motives and intentions which are incorrect. The art of the confidence trickster is based on his ability to lead his prey into making false inferences and attributions of intent.

A good example of the interactional consequences of the process of making wrong inferences and attributions is provided by Laing, Phillipson & Lee (1966). A recently married couple were sitting at a bar in a hotel on the second night of their honeymoon. When the wife struck up a conversation with a couple sitting next to them, the husband refused to join the conversation and remained aloof and generally antagonistic. The wife bitterly resented her husband's reaction. Tempers rose and eventually a bitter quarrel took place. Some years later the misunderstanding was brought to light. The wife had initiated the conversation with the other couple because she was excited by her newly acquired status as a wife and was keen to relate to other people as a wife with her husband. The husband, however, viewed the honeymoon very differently. To him the honeymoon was a time when he and his wife could withdraw from normal social relationships into a private world where he and his wife could be alone with each other. To the wife, the husband's aloof behaviour was insulting because he was apparently refusing to allow her to fulfil her anticipated joy of mixing with other married couples. The husband felt insulted by the wife because her behaviour implied that his company was not adequate for her. Each had made a false inference about the other and in consequence each felt justified in accusing the other of being inconsiderate. In situations of this sort most of us are regrettably so absorbed in inventing the next insult or in nursing our wounded pride that we have not time to consider the grounds on which we based our initial inferences and attributions.

It is through the attribution of dispositional properties such as personality traits or intentions that Person is able to perceive Other's behaviour as consistent. When Person ascribes an enduring and persistent personality characteristic to Other he is drawing together and unifying a wide variety of past incidents in Other's behaviour. And once Person has created a consistent picture of Other by attributing certain dispositional properties to him, then he will tend to acquire expectations of how Other will behave in new situations in the future. Other's behaviour becomes, within limits, predictable. It is in this way that the attribution process can structure and facilitate future interaction between Person and Other. If Other has lied to and stolen from Person on several occasions, then Person will infer that Other is dishonest and he will tend to impute to Other the general characteristic of dishonesty, and expect this characteristic to manifest itself again, even though the time, place and form of its expression may be different. The attribution of dishonesty may exert a strong influence on future interaction between Person and Other in that Person may treat Other as untrustworthy.

Once Person has developed a fairly consistent picture of Other he will tend to resist new information which threatens this consistency. First impressions may be important precisely because on the basis of the initial information we create an impression of another person which is not easily changed. When Person is faced with information about Other which does not seem consistent with his impression, Person can resort to some common methods for resolving the psychological strain involved. Person can assume, for instance, that some factor unknown to Person must be influencing Other, a factor which has upset Other's normal pattern of behaviour. Person can say to himself, 'Other would never behave like that of his own free will. Someone else must be making him behave in this way.' Thus when a prisoner-of-war condemns his own side and eulogizes his captors in an enemy broadcast, his countrymen can dispel their psychological discomfort by assuming that his conversion is not genuine but the result of pressures exerted by the enemy on the prisoner. Alternatively, Person can maintain his consistent picture of Other by concluding that Other's behaviour is only apparently inconsistent and that Person himself must have misperceived or misunderstood Other's actions. It is also possible for Person simply to ignore those aspects of Other's behaviour for which he is unable to account, either because it appears to be inconsistent or because it does not appear to be governed by rational principles.

A further aspect of the attribution process is the tendency to see persons as origins of actions, to which Fritz Heider (1944) has drawn our attention. In this respect we tend to overestimate the contribution of a person to his actions and underestimate the other factors at work. Thus we tend to see the causes of a person's success or failure in his personal characteristics ('because he worked so hard' or 'because he is just bone idle') and to ignore other contributory factors in the environment. Nietzsche made the same point when he remarked that 'Success is the greatest liar'. One of the most useful contributions by sociologists of education has been to increase our knowledge of the role of the home in providing values and conditions conducive to academic attainment. Yet many teachers still hold their pupils wholly and personally responsible for their attitudes and achievements.

The motives which Person attributes to Other in order to 'explain' his behaviour will, of course, be related to other information which Person has about Other. A clever experimental verification of this notion has been undertaken by Thibaut & Riecken (1955). In this experiment three subjects arrived at the laboratory. The experimenter introduced himself and the three subjects to one another. One of the subjects, neatly dressed in a coat and a tie, said that he had just received the Ph.D. and was a member of the teaching staff. The second subject, in shirt sleeves and without a tie, said that he had just completed the first year of his undergraduate course. The experimenter then announced that for the experimental task one of the subjects would act as a communicator and the other two as an audience. The communicator would try to persuade the audience to donate blood for a Red Cross drive. The subjects then drew lots to decide who should play the part of the communicator. Actually the first two subjects already described were confederates of the experimenter, and they were playing the pre-arranged high status (Ph.D. teacher) and low status (undergraduate) roles. The third subject was entirely unaware of the deception and had no reason to be suspicious when he drew the lot which assigned him to the role of the communicator. At the end of the game, when the two confederates had simultaneously complied with the naive subject's requests that they should donate blood, he was asked certain questions about the motives of the audience in accepting his persuasive communication. Most of the students who participated as naive subjects believed that the high status confederate, to whom they ascribed greater power to resist influence, was 'persuaded' because he wanted to donate blood anyway. The reasons for his compliance were 'internal'. But the low status person, who was seen as having lower potential resistance to influence, was believed to have accepted the persuasion because he was forced to do so by the pressures exerted on him, that is, the cause is seen as 'external'. In this way Thibaut and Riecken brilliantly demonstrate that the motives which Person attributes to Other to account for his conduct depend on the *status* of Other as perceived by Person. Incidentally, at the end of the experiment the naive subjects were found to have increased their liking for the high status but not for the low status subjects. The reader might like to speculate on reasons for this. Unfortunately there is little systematic research on the ways in which the attributions of motives by Person to Other are influenced by the information which Person possesses about Other, though great advances have been made by several investigators, especially Pepitone (1958).

A related experiment, which has rather disturbing implications for teachers, is that of Lloyd Strickland (1958). The subjects in this study acted as supervisors to two people working on monotonous problems. The supervisor had the power to fine the workers, by making deductions from their pay, if they failed to meet his output demands, and he was also able to monitor or check on the rate at which they were working. However, he was able to monitor the work of one worker more often than the other—on ten occasions for worker A as against two occasions for worker B during the work period. At the end of the work period both workers had completed exactly the same amount of work. In the second part of the experiment, the supervisor was informed that he could now monitor A and B as often as he wished during the second work period.

The fascinating result is that during this second work period the subject-supervisor actually checked on the progress of A more often than B, even though they had done the same amount of work in the past. Strickland demonstrates that because the supervisor has monitored the work of A more than B in the past, he comes to believe that A is doing the work *because* he is under surveillance (an 'external'

43

reason) whereas B is regarded as having worked because he wished to do so (an 'internal' reason). Once the supervisor has attributed a motive which 'explains' why A and B have worked for him, then he comes to treat A and B in accordance with the attributed motives. A cannot be trusted and so must frequently be checked on, whilst B is dependable so the need for surveillance of his work becomes less essential. In this way the supervisor is really victimized by his own previous supervisory behaviour. By subjecting a worker to constant surveillance, the supervisor may deny himself the opportunity to test whether or not this supervision is necessary. Once the supervisor infers that the workers work because of the control he exercises over them, and in consequence attributes untrustworthiness to them, then he may persist in his behaviour and attitudes because of what Strickland calls a 'self-perpetuating information loss'. He cannot test the loyalty of the workers until he revises his attitudes sufficiently to provide the workers with an opportunity for disloyalty.

The attribution of intentions and motives is a crucial aspect of person perception which has not as yet been extensively investigated (Jones & Davis, 1965). Experience suggests that Person constantly strives to attribute intentions and motives to Other in his attempt to give meaning to Other's behaviour. Thus the meaning which Person gives to Other's behaviour varies as a function of the attributed intentions and motives. Thus the millionaire's donations to charity acquire different meanings, 'generosity' or 'conscience money', according to the motives attributed to him. Sometimes the results are somewhat tortuous. In Gore Vidal's novel *Washington D.C.* (1967) a United States senator describes his own experience.

> He had learned very early that to do any good in the Senate one must first present it as an act of self-interest since to do good for its own sake aroused suspicion. Mrs. Roosevelt had been genuinely hated for what seemed to many to be a genuine lack of self-interest. Ultimately she had been effective only because certain Senators, disliking her, decided that she was at heart a superb Machiavelli, using the public money to attract the unwashed to the Roosevelt banner. Self-interest acknowledged, even she could occasionally work a miracle in the Congress.

The discussion of interpersonal perception in terms of a process of selection and organization has tended to focus on the dynamics of perception in situations where Person and Other are strangers to each other. It is when two strangers meet that the dynamics of interpersonal perception are most obvious and most easily analysed, both to the psychologist and to Person and Other themselves. In an ongoing interaction where Person and Other have met on a number of previous occasions perceptual factors continue to play a vital role, but a more subtle one. Their present perceptions are influenced by their past perceptions and actions. It is difficult for the experimental psychologist to study the perceptual factors at work because he is unable to control or observe these factors, and this is true *a fortiori* where the extent and nature of past interaction is unknown. It is for this reason that many investigations of person perception rest on rather contrived and artificial situations in which there is no interaction at all, and often no stimulus person. It seems paradoxical that many of the classic experiments on person perception, like Strickland's experiment described above, do not involve the perception of one actual person or another. Though many of these experiments are extremely ingenious, it is difficult to know how much of the results can be transferred to ordinary interactions and perceptions. This is the perennial problem for the social psychologist. He must try to contrive a situation in which he can reduce the complexity of social relationships into a form where he can control and observe, where he can analyse and test restricted hypotheses. But once the experiment is completed, once the social psychologist feels he has teased out the one aspect of human relationships with which he is specially concerned, then the question arises of the extent to which his findings or theories can be reapplied to man outside the controlled atmosphere of the laboratory experiment. Too often, social psychologists have followed one experiment by another, and then another, and another *ad nauseam* without ever raising the issues of how far our understanding of man in his natural environment is being advanced. As the experiments snowball after a pioneering study, they become in the eyes of the layman 'of academic interest only'—and I am not entirely convinced by the psychologist's defence that this is either the best or the only way in which we can progress our knowledge. The 'naive' question of the layman or student is often so deadly that it hurts, but the reaction of the psychologist is often to shield his vulnerability with yet another layer of protective armour.

On the basis of Person's selection of certain data about Other and the means by which these data are organized and interpreted through inference and attribution, Person creates what might be called his conception of Other. Whenever Person thinks of Other or interacts with him, this conception of Other is activated. This conception of Other is constructed slowly as Person's perception of Other becomes more extensive and complex. When Person believes that he knows Other very well, this conception becomes relatively stable and influences the selection and organization of any further information about Other. It is used as a basis for predicting Other's future behaviour. In its most primitive form Person's conception of Other is based on his 'first impressions' of Other. First impressions are really a temporary working hypothesis about Other. As interaction proceeds or as additional information about Other becomes available from other less direct sources, this working hypothesis can be accepted and refined or rejected and replaced (Fig. 3.3).

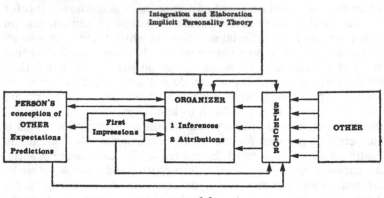

FIG. 3.3

A crucial problem for the psychology of interpersonal perception is not simply the processes which underlie the nature of first impressions of others, but the impact these impressions have on extended interaction. Most laymen believe that first impressions count. The research on this rather basic question is limited, but there does seem to be some evidence to support the widely held belief that first impressions may have repercussions on the ensuing interaction. Again, one of the earliest experiments in this field is that of Asch (1946). He read the following list of traits to a group of subjects: intelligent—industrious—impulsive—critical—stubborn—envious. A second group of subjects was given the same list of traits, but they were read out in reverse order. For Group A subjects the trait list opens with highly valuable qualities (intelligent, industrious) but proceeds to rather dubious qualities (stubborn, envious). For Group B subjects the movement is from negatively valued to positively valued qualities. When the groups were asked to write brief characterizations of the persons possessing these traits, the two groups produced very different portraits, despite the identical content of the two lists of traits. For example, a member of Group A wrote:

> The person is intelligent and fortunately he puts his intelligence to work. That he is stubborn and impulsive may be due to the fact that he knows what he is saying and what he means and will not therefore give in easily to someone else's idea which he disagrees with.

In contrast, a member of Group B, who received the traits in the negative-to-positive order wrote:

This person's good qualities such as industry and intelligence are bound to be restricted by jealousy and stubbornness. The person is emotional. He is unsuccessful because he is weak and allows his bad points to cover up his good ones.

It seems that the first terms which the subject hears set up a direction which then exerts a continuous influence on the succeeding terms. The traits at the end of the list are not simply summated with the other known traits, but are incorporated into the direction which is already established by the first traits. The Group A subject assimilates the negative traits of stubbornness and impulsiveness into his primary positive impression and converts them into an inference that the person is determined and not easily persuaded. The Group B subject regards the person's good qualities of intelligence and industry as muted by an essentially emotional and unstable make-up. An alternative interpretation (Anderson, 1965) would be that the later adjectives are simply given less weight than the earlier ones in the creation of the total impression. Such an experiment is, of course, no more than an analogue of perceptual processes in everyday life. Hearing a list of traits read out by an experimenter in a laboratory does not adequately correspond to the ways in which Person receives various bits of information about Other in a real interactional context. But it does not seem unreasonable that in normal interaction our initial inferences should colour any information that subsequently becomes available. Indeed, it may be on the basis of such 'common sense knowledge' that Asch was led to conduct the experiment. If this is a universal phenomenon it may explain the not uncommon experience whereby on deeper acquaintance with a person we sometimes feel, 'How funny, I never thought X was like this at all!' Massive doses of corrective information, which can come with further acquaintance, may be needed before we jettison erroneous inferences based on the directional influence of our first impressions.

An experiment on the effects of first impressions which is a little closer to everyday experience is that of Abraham Luchins (1957). His subjects were presented with two paragraphs of information. In the first the central character is represented as an extraverted person.

Jim left the house to get some stationery. He walked into the sun-filled street with two of his friends, basking in the sun as he walked. Jim entered the stationery store which was full of people. Jim talked with an acquaintance while he waited for the clerk to catch his eye. On his way out, he stopped to chat with a school friend who was just coming into the store.

47

Leaving the store he walked toward school. On his way out he met the girl to whom he had been introduced the night before. They talked for a short while, and then Jim left for school.

In the second paragraph, Jim is represented as an introverted person.

After school Jim left the classroom alone. Leaving the school he started on his long walk home. The street was brilliantly filled with sunshine. Jim walked down the street on the shady side. Coming down the street toward him, he saw the pretty girl whom he had met on the previous evening. Jim crossed the street and entered a candy store. The store was crowded with students, and he noticed a few familiar faces. Jim waited quietly until the counterman caught his eye and then gave his order. Taking his drink, he sat down at a side table. When he had finished the drink he went home.

There were two main groups of subjects. Group I was given the E or extravert passage first, followed by the I or introvert passage. This we may call the EI group. Group 2 was given the same two passages but in reverse order, so that we may call this the IE group. When the subjects had read the passages, they were asked to do various short tests which included the writing of a description of their impressions of Jim. The analysis of these tests showed quite clearly that the information received first exerted a stronger influence on the subjects' impressions of Jim than did the second. The EI group perceived Jim as more extraverted than introverted, the opposite being true for the IE group. The second dose of information, although somewhat at variance with the first, was not completely ignored by the subjects. The IE group saw Jim as less introverted than a control group which was given simply the I passage alone, and the EI group saw Jim as less extraverted than did the control group which was given E only. The later information is regarded by the subjects as somehow less fundamental to Jim's character than is the primary information and is used to modify the first impression created by the paragraph which was presented first.

Because the subjects in Luchins' experiment do not observe or interact with a real person, it is difficult to know how far we may transpose the findings to everyday situations outside the laboratory, though the experiment does suggest that the form and order of written information, which we frequently receive through newspapers, confidential references etc., may strongly influence the impression we gain of someone. When the layman speaks of 'first impressions' he is usually referring to the impressions gained through a brief interaction with a real person. In this respect an experiment by Harold Kelley (1950) is particularly relevant. At the beginning of a psychology class some students at the Massachusetts Institute of Technology were informed that their usual instructor was out of town and that a substitute would take his place. The students were then given, in the form of a note, some pre-information about the replacement instructor.

Mr. —— is a graduate student in the Department of Economics and Social Science here at M.I.T. He has had three semesters of teaching experience in psychology at another college. This is his first semester teaching Ec.70. He is 26 years old, a veteran, and married. People who know him consider him to be a rather cold person, industrious, critical, practical and determined.

The last set of traits in this pre-information recall the lists used by Asch in his experiments. This is indeed the source of the pre-information. For half the students in the class the pre-information contained the words 'very warm' instead of 'rather cold'. At the end of class the students were asked to rate the instructor on a number of traits and it was found that the original Asch findings on the warm-cold variable were substantially confirmed. Once again the 'warm' instructor was rated as more considerate, more informal, more sociable, more popular, better natured, more humorous and more humane than the 'cold' instructor. In this experiment the power of the warm-cold dimension to influence impressions in an interactional situation is fully supported.

But did the impression the students gained of the instructor influence their behaviour towards him? This is an important question, for if the impression we have of another person does *not* influence our behaviour towards him in any way, then the study of impression formation may be much less significant than the other experiments have indicated. Since the psychology class took the form of a leader-centred discussion, Kelley was able to record the number of times any student initiated verbal interaction with the instructor. Fifty-six per cent of the students with the 'warm' pre-information participated as against only 32 per cent of the 'cold' students. It is a striking fact that two words in the pre-information should influence the students' perception of and interaction with the instructor to such a remarkable extent. It shows that under certain circumstances our expectations about Other may be much more influential in determining our behaviour to him than Other's actual characteristics.

The impact of the first impression may not be permanent, especially when subsequent information is at variance with the first impression. Luchins has shown that the first impression effect in his own experiments is easily destroyed. By warning the subjects against first impressions, he was able to overcome the effect. Also when he introduced a time-lag between the two paragraphs about Jim, Luchins found that the subjects were more influenced by the second account. In the Kelley experiment a comparison of the data from two different persons who played the part of the instructor revealed that, although the warm-cold variable produced differences in the same direction for both, there were also differences in the ratings which reflected differences in the actual personalities of the instructors. It is as if the subject were making a compromise between the pre-information and the actual behaviour of the instructor. In real life it is probably true that first impressions count in the short term, but in the case of an interaction which extends over a period of time it seems likely that Person's assessments of Other are subject to constant revisions as he gains more information about Other. Probably the revisions get progressively smaller as Person consolidates his picture of Other, unless Other behaves in an unexpected way or is seen by Person in a new situation. For most of us, first impressions of another person are no more than a working hypothesis to which we temporarily subscribe, but which we are prepared to alter in the light of new information. In this sense we are all detectives of the human personality.

Impressions of others serve an important function in preparing the ground for and thus facilitating human interaction, but they also serve the function of organizing a vast amount of information that is available about others. Person cannot possibly remember all the details about Other's appearance and behaviour: to do so would be extremely wasteful. So he condenses what he has learned about Other into an 'impression' of him. In this way, impression formation has an important *economic* function. One of the most basic labour-saving devices of the human mind is the ability to organize all the incoming data into *categories*.

> To categorize is to render discriminably different things equivalent, to group the objects and events and people around us into classes, and to respond to them in terms of their class membership rather than their uniqueness. (Bruner, Goodnow & Austin, 1956)

Unless we invented categories, we could have no concepts at all. No two apples are exactly alike, but unless we invent the category of 'apple' then we must react to every apple as a unique object. Once we invent the category of 'apple' we can identify examples of the category, and thus reduce the complexity of the environment and the need for learning. Knowing that an object belongs in a category also gives advance direction of our behaviour towards an object. To know that an object is an apple is to know its physical appearance and that it is edible.

It is the same with people. When Person meets Other for the first time Person seeks to discover the information on the basis of which he can categorize Other. Is Other in the male or female category? in the young or old category? in the friend or enemy category? Once we have made these basic categorizations, we are to some extent prepared for a particular sort of interaction with Other.

Roles, which we shall examine in the next chapter, are really categories of human beings. Person's interaction with Other is considerably facilitated if he can discover the roles of Other. In fact in normal interaction we respond to others as unique persons much less frequently than we imagine, but more in terms of the mutual expectations controlling the role relationship. When Person is playing the role of customer in a store and Other the role of the assistant behind the counter, both Person and Other are unlikely to possess, or want to possess, information about the personality or the motives of each other. (Some intentions and motives are, of course, relevant within the role relationship. The customer may believe that the sales-girl is anxious to sell him a tie to increase her commission and thus suspect her comments that the tie suits him. Whether or not the sales-girl has such a motive, she will, if she infers that the customer is attributing such a motive to her, have to behave in such a way that the customer's fears are allayed.) They are able to carry on a satisfactory interaction if both customer and assistant are aware of and conform to the rules which govern their role relationship. It is the roles which structure and guide the interaction. So seeking out information about other people's roles is typical of the early stages of an interaction between strangers. The correct perception of the role relationship is often facilitated for the participants. In our example, the store assistant may wear a label on her lapel which states her own name or the name of the store so that her role is easily perceived by the customer. Uniforms of various kinds are a common method of revealing roles. Where there are no indications of role, the participants in an interaction will actively seek information about each other's roles by subtle probes or even direct questions.

Earlier we used the 'mistaken identity' situation to illustrate a point about person perception. This same situation was used by Abravanel (1962) to illustrate the importance of role factors in perception. In this experiment, subjects listened to one end of a telephone conversation. Some of the subjects were told that the speaker was a college student and others were told that he was the chairman of an academic department. The person at the other end of the line, who could not be heard, was supposed to be a college instructor. It was clear that the speaker was unhappy about the instructor's teaching. The subjects who thought that the speaker was a student thought him aggressive, ambitious and egotistical; those who thought the speaker was the chairman of the department termed him hesitant, compassionate

51

and indecisive. This experiment is a clever demonstration of an obvious point with which we are all familiar in experience. Our categorization of Other in terms of roles exercises a profound influence on the way in which we assign meaning to Other's behaviour and attributable personality traits or motives to him. The content of Other's communication is interpreted in relation to Other's roles.

The role *relationship* between Person and Other also has an important influence on the way in which Person perceives Other. This is neatly demonstrated in an experiment by Jones & de Charms (1958). The subjects who were naval air cadets, listened to a tape recording of an interview with a sergeant who had acknowledged signing propaganda petitions against the United States while a prisoner of war in Korea. The subjects were instructed to assume different roles in listening to the interview and in forming an impression of the sergeant as a person. Subjects in the first group imagined themselves to be a member of the psychological board set up to examine why the sergeant had behaved in this way. The role for the subjects in the second group was that of a member of a preliminary court of legal inquiry to determine whether or not and on what basis the sergeant should be charged. In the third group, the subjects played the role of a soldier like the sergeant and were asked to find out whether they would like the sergeant or not. How the subjects perceived the sergeant and the traits which they ascribed to him was shown to be related to the role the subjects played in relation to the sergeant. In other words, Person's interpretation of Other's behaviour is a function of Person's roles as well as Other's roles.

The mistaken identity situation has become a standard comic ploy of dramatists. Indeed no Whitehall farce is complete without at least one such occurrence. When Person believes that he is holding a certain role relationship with Other but Other believes that he is holding a quite different role relationship with Person, the interaction between them begins to take strange turns. Neither can understand the strange behaviour of the other, but to the audience, who must of course be in on the secret, their inappropriate behaviour, puzzlement and blunderings are extremely funny. In Oliver Goldsmith's *She Stoops to Conquer*, Charles Marlow is under the impression that he is staying at an inn, whereas in reality he is at the house of his prospective father-in-law, Mr Hardcastle. Marlow accords Hardcastle the treatment appropriate to the landlord of an inn, and Hardcastle's own behaviour is that appropriate to a future son-in-law. Each is astonished by the behaviour of the other. But the consternation of the actors is high entertainment to the audience.

It is because of the universal human need to place other people in categories that the phenomenon of *stereotyping* arises. A social stereotype refers to a set of characteristics which are held to be common to members of a category. For example, to say that 'Germans are arrogant and aggressive' or 'Negroes are lazy and dirty' is to suggest that by virtue of his membership in an ethnic or national group a person is likely to possess certain characteristics. To say 'Professors are absent-minded' is to do the same with an occupational group. The definition of a social stereotype seems fairly straight forward, but this is not the case. Presumably not every person in the world supports any particular stereotype, so the consensus is not total. Even within any subgroup it is unlikely that every member would offer his support. A social stereotype arises when there is *some* agreement among a given population that the members of a given category possess certain characteristics. But so far as I know no-one has ever suggested how much agreement is required for a stereotype to arise. Moreover even those people who seem to accept the stereotype would probably not accept the notion that every single member of the category possesses the alleged characteristics. Someone, for instance, who would give his assent to the proposition 'Politicians are charlatans' would not typically believe that there are no exceptions at all. Most people would, I imagine, see their stereotypes as propositions with a general level of validity but would also allow for exceptions. Indeed, the exception may be felt to prove the general rule. So in practice stereotypes arise when some people believe that some members of a given category possess certain characteristics. These stereotypes are frequently positive or negative or a combination of both elements. The view that 'Negroes are lazy and dirty' is negative for it implies criticism or denigration. But to say 'Nurses are considerate' is a positive stereotype since it implies approval and commendation. Sometimes stereotypes are neutral in their evaluations.

Stereotypes are one form of categorization and generalization among humans. Like any form of categorization they may serve an important and useful function in that they reduce the need for learning and help a person to anticipate how he might react to a member of the category. The ability to categorize is one of the fundamental aspects of human intelligence and it is important that he categorizes people as well as things. The categories of 'Negroes' and 'politicians' are as essential as the categories of 'apples', 'fruits' and 'edible things'. The stereotype goes beyond categorization for it ascribes certain characteristics to members of the category and the drawback is that in so doing stereotypes become neither totally true nor totally false. To suggest that 'Spinsters are sexually frustrated' is as dangerous as saying that 'Fruits are edible'. Because one cannot put the word 'all' before most social stereotypes, they have a limited value, unless it can be shown that it is true for the vast majority of the members of the category. But since we have little evidence on the truth or falsity of most social stereotypes, they are potentially dangerous generalizations.

Some stereotypes do contain some truth. For example, the stereotype that 'Jews are ambitious and industrious' could to some extent be supported by the fact that Jews, as against Protestants and Roman Catholics, are more likely to have higher achievement motivation, to be over-achievers, to enter further education, and to have higher rates of social mobility. But at the same time it is doubtful whether the differences between Jews and other religious groups are of a sufficient magnitude to give the stereotype any great value in facilitating human interaction between particular individuals.

Stereotypes are at best oversimplifications of the facts, even when they contain some truth, and they ignore individual variations and differences. To apply the stereotype to the individual member of the category, to assume that a particular individual will exhibit the alleged characteristics because he is a member of the category, is thus to ascribe to the individual properties which he may in reality not possess at all. To anticipate, on the basis of the stereotype, that an individual Negro is dirty and lazy is to approach interaction with him on the basis of predetermined, but untested, ascriptions—and this is likely to blind one to the *actual* qualities possessed by the Negro. A prejudiced person, because he believes the stereotype, ignores the evidence when it does not fit the stereotype, and when it does— presumably some Negroes *are* dirty and lazy—uses the evidence as further support for his belief in the stereotype. The eureka quality of 'There you are!' or 'Didn't I tell you so?' is a remarkable testimony to the selectivity of perception. Believing the stereotype may also lead one to expect the members to behave in accordance with the stereotype, and sometimes the members may actually fulfil the expectations, not because they are what the stereotype alleges them to be but because they are expected to be so. It is not difficult to become what the people around us expect us to be, to take to ourselves the attitudes that others take to us.

A false stereotype, particularly a negative one, is also dangerous because it is so easily disseminated and used to prop up already prejudiced attitudes. Prejudiced people usually refer to the group towards whom they are prejudiced in terms of stereotypes. And their prejudice may prevent interaction with the individual members of the category which is part of the process of dissolving the prejudice. Prejudice has an enormous capacity to survive because it insulates itself against the forces which can destroy it.

Brown (1965) has brought attention to another objectionable feature of stereotypes namely their ethnocentric nature. As we discussed in Chapter 2 with reference to the social sex role, we tend to judge others by our own particular standards. When we come across a culture that is different from our own, we perceive it as strange and often as somehow wrong. We evaluate the difference negatively. For this reason we can take the view that 'The Italians are religious' but that 'The Chinese are superstitious'. Certain sorts of belief are called 'religious' when they are common to our own culture, but 'superstitious' when such beliefs take a form which is not institutionalized in our own culture. When two cultures with strong ideological differences comment on one another, each makes a negative evaluation of the other and reveals the ethnocentric basis of the stereotype. Thus both the Americans *and* the Russians can refer to each other as 'repressive' or 'cruel' or 'undemocratic'. To think in such ethnocentric stereotypes is to inhibit a consideration of the objective facts, understanding of the other's point of view and the possibility of cooperative enterprises.

Another phenomenon of person perception, in some ways similar to stereotyping, is that of the 'halo effect'. This refers to the tendency to be generally favourable or unfavourable in one's reaction to others. Once Person has decided that Other has certain favourable characteristics, this impression misleads him into giving Other a very high rating on many other characteristics. On the other hand if Person's initial impression of Other is unfavourable, Person is likely to under-rate Other on other dimensions. The 'halo effect' is really another example of the tendency of persons to overestimate the homogeneity of other people, to perceive consistency at the expense of reality. It is also an illustration of the tendency of persons to see certain characteristics as belonging together, to see people in essentially 'flat' terms as a bundle of congruous traits. The 'halo effect' is particularly strong in interviews and rating scales, though its repeated demonstration does not seem to have affected the addiction of most organizations to the interview as the basis of selection. I can only conclude that the interview as a selection device serves a highly important function for the interviewer himself, in that it convinces *him* that he has chosen the best person, even though it may do nothing of the kind. Most interviewees are aware of the 'halo effect'. I have seen many undergraduates waiting for interviews with Unilever and I.C.I. in clean white shirts, formal ties and dark suits instead of their usual hoary sweaters and ragged jeans. Presumably they hope that their spruce appearance will lead the interviewer to attribute intelligence, efficiency, reliability etc. to them. I don't suppose the interviewer is so easily taken in: the 'halo effect' can operate in much more subtle ways. But of course if the interviewee is an attractive member of the opposite sex, the 'halo effect' may not be so subtle after all (Fig. 3.4).

One important link in our model of person perception, which is clearly implied in the earlier exposition, must now be made explicit. Person's conception of Other is one of the most important factors affecting Person's behaviour towards Other. With reference to an interaction between Person and Other the model must be completed with the mirror image of the model—a similar model involving Other's perception of Person. Only then can we be said to have a model of *inter*personal perception. We shall consider this more fully in a later chapter devoted to a more specific analysis of social interaction processes. This chapter has been relatively little concerned with a direct discussion of the impact of perception on behaviour or with illustrations from educational situations. So it is appropriate to conclude this chapter with the examination of a concept which relates perception to behaviour and which has significant applications to

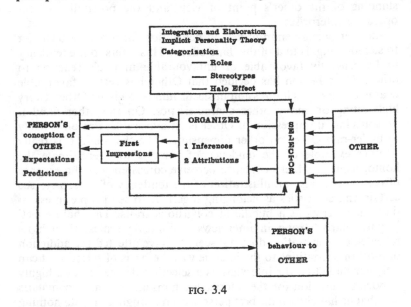

FIG. 3.4

teacher–pupil relations. It is the concept of the self-fulfilling prophecy.

The concept of the self-fulfilling prophecy is not new, though the most famous exposition and analysis is that of Robert Merton (1949). It is related to a wide variety of other ideas and concepts within the social sciences. Certainly interpersonal perception is highly relevant, though most analyses do not emphasize this aspect. In a memorable sentence W. I. Thomas (1928) once said, 'If men define situations as real, they are real in their consequences.' What counts in human behaviour is not so much the objective features of a situation but the way in which we perceive or define situations. For instance, it may not be true that you dislike me, but if I believe that you dislike me it will have important repercussions on my behaviour towards you. In Carl Rogers's words, 'It is the perception, not the reality, which is crucial in determining behaviour.' The relationship between perception and reality is very complex. It is not simply a question of perception being accurate, i.e. a true reflection of reality, or inaccurate, i.e. having no foundation in reality. As Lemert (1962) has shown in a brilliant paper, the beliefs of the paranoid person that he is being persecuted and excluded are by no means groundless.

We have observed that one important aspect of person perception is the development of a conception of Other which is used as a basis for developing expectations or predictions about Other's future behaviour. If we make predictions with respect to objects the predictions do not influence subsequent events. If I predict that it will rain tomorrow, my forecast does not and cannot affect the weather. In the social world, however, it may be that under certain circumstances my expectations and predictions about Other's behaviour will exert some influence on Other's subsequent behaviour. In the self-fulfilling prophecy the prediction so influences events that the prophecy comes true as a direct result of that prophecy. For example, a bank is conducting business as usual. The bank's directors and manager are confident about current business and future prospects. A rumour spreads that the bank is in difficulties and in danger of being declared insolvent. Large numbers of clients hurry to the bank to withdraw their deposits before the rumoured collapse. Because so many believe that the bank is failing and try to withdraw money at one time, the bank does indeed collapse. Or, a man is about to get married. The teasing remarks of his male friends arouse anxieties that he might prove to be impotent on his wedding night. His fears are really quite groundless but he becomes so worried about his sexual competence that he does indeed prove to be impotent on his honeymoon.

There have been a number of studies of the self-fulfilling prophecy in educational contexts, but the most famous study is that of Robert Rosenthal & Lenore Jacobson (1968), social psychologists at Harvard University. Rosenthal is concerned with the positive self-fulfilling prophecy, that is, when children improve as a result of a prediction that they will do so, though the *negative* form, whereby pupils deteriorate in work or behaviour as a result of the teacher's expectation, is obviously of at least equal educational significance. The results of Rosenthal's experiment are very complex in detail, but the main findings can be briefly summarized. The pupils of an American elementary school were given the 'Harvard Test of Inflected Acquisition.' The teachers were told that this test would single out those pupils who could be expected to make dramatic academic progress. Actually the test was simply an intelligence test. Rosenthal selected at random twenty per cent of the pupils, who, the teachers were informed, were the spurters. After the teachers were given the names of the alleged spurters no further action was taken until the pupils were all re-tested for intelligence at a later stage. If the teachers believed that the alleged spurters would indeed bloom academically then they might so influence events that the prophecy would come true. We would then have a self-fulfilling prophecy. The results supported such an interpretation. After one year the children nominated as spurters did make marked gains in intelligence. Spurters in the first grade gained on average fifteen IQ points more than did other pupils and in the second grade the spurters gained on average nine IQ points over the rest of the pupils. In these two grades one fifth of the pupils who were not nominated gained over twenty IQ points whereas almost half the spurters did so. The spurters also made marked gains in reading as assessed by the teacher.

To demonstrate the existence of the self-fulfilling prophecy we must exclude at least two alternative explanations. First, we must exclude the possibility of coincidence or correct prediction. A child might gain in IQ not as a result of the teacher's prediction but for a wide variety of reasons quite unconnected with the teacher's prediction. The fact that the teacher predicts such a gain accurately does not of course mean that the prediction caused the gain. This possibility seems excluded by the fact that the spurters had been nominated at random yet the gains were statistically significant. Second, we must have evidence that the children did in fact bloom. It is possible that the teacher, believing that the pupil would make dramatic academic progress, might be led to interpret the child's behaviour in line with the prediction and evaluate the child and assign grades accordingly. The possibility seems excluded by the fact that the intelligence test score was independent of the teacher and could not be falsified by the teacher's perception. The superiority of the spurters in reading could be explained in terms of teacher misperception, but there is little evidence in support of such bias. As Rosenthal reports, the teachers' gradings seemed to show a negative halo effect. In other words, the special children were graded by the teachers more severely than were the ordinary pupils.

We cannot take Rosenthal's finding entirely for granted in the light of the methodological and statistical criticisms that have been levelled at his work and in the light of the failure of at least one replication (Claiborn, 1969). For the rest of the discussion, however, we must assume its existence, for if such self-fulfilling prophecies do arise in educational contexts, then the real problem is the mechanics and dynamics by which the process of the self-fulfilling prophecy operates. Two main alternatives are available. It could be that the teacher, accepting the prediction that the spurters would bloom, altered her behaviour to these pupils by according them special and differential treatment, for example by devoting more of her time and attention to them. We have good reasons for supposing that once the teacher perceived the spurters in a new way she would bring her behaviour into line with her perception. The pupil might then respond to this special treatment by an appropriate improvement in performance. Alternatively, the explanation could be that the teacher communicates her beliefs and expectations about the spurters' ability to the pupils—and the communication of the prophecy could be achieved in very subtle ways—so that the pupils change their own beliefs about themselves and alter their behaviour in line with these new beliefs and so improve. These two alternatives are not mutually exclusive and may well occur in combination.

Rosenthal himself cannot be sure how the self-fulfilling prophecy was effected, nor can he explain why some of the nominated children made very high gains (up to sixty-nine IQ points) whilst other spurters did not (one actually lost six IQ points). Any explanation of the self-fulfilling prophecy must account for individual differences which are masked in the reporting of the average gains of the spurters. Rosenthal was unable to test the differential treatment hypothesis directly, though he cleverly uses indirect evidence to show that more time was not devoted to the spurters. Certainly the teachers themselves in their reports about the time spent with individual pupils gave no evidence for such an explanation. Rosenthal himself is inclined to the view that it is quality of the teacher–pupil interaction that varies between the two groups, for example by the teacher being more friendly and encouraging and by watching them more closely. Probably both are involved. Recent work by Silberman (1969) shows that teachers give more positive evaluations and are more receptive to pupils whom they like, whereas children who are rejected by the teacher receive a larger number of negative evaluations. Further, where the teacher expresses concern for the child there is a definite tendency to greater contact with the child, whereas when the teacher feels indifference towards the child contact is significantly reduced.

Recent support for the differential treatment hypothesis is available in the work of Brophy and Good (1970) in an intensive interaction analysis of the behaviour of teachers and first grade elementary school pupils of whom the teacher has high expectations ('highs') or low expectations ('lows'). The highs tend to receive much greater praise, whereas the lows receive greater criticism from the teacher. The highs also initiate greater contact with the teacher. One of the most significant findings was that

the teachers consistently favored the highs over the lows in demanding and reinforcing quality performance. Despite the fact that the highs gave more correct answers and fewer incorrect answers than did the lows, they were more frequently praised when correct and less frequently criticized when incorrect or unable to respond. Furthermore, the teachers were more persistent in eliciting responses from the highs than they were from the lows. When the highs responded incorrectly or were unable to respond, the teachers were more likely to provide a second response opportunity by repeating or rephrasing the question or giving a clue than they were in similar situations with the lows. Conversely they were more likely to supply the answer or call on another child when reacting to the lows than the highs.

Whilst there were notable individual differences between teachers in conforming to this pattern, it is still very striking that such differential treatment should emerge in relation to pupils of six years of age. As the authors comment: 'If the differential teacher treatment leads to differential reciprocal behaviour by the children, the classroom behaviour of highs and lows should become progressively more differentiated as the school year progresses.' The expectation may soon become established into a more general reputation, and as we shall discuss shortly, be formally institutionalized within the school.

As to the communication to the pupil of the teacher's changed beliefs and expectations Rosenthal has surprisingly little to say. Yet from our knowledge of social psychology certain conditions under which we might expect the prophecy to be communicated and accepted can be suggested. Following Symbolic Interactionism, we might hypothesize that it is those pupils who regarded the teacher as a 'significant other' who would be the ones most inclined to take to themselves the teacher's attitude to them. It is known that many pupils do see the teacher as a significant other. Wright (1962) for example, has shown that fourth-year secondary modern children, who are not usually thought to be the most teacher oriented of pupils, do see the teacher as closer (than parents) to their ideal self in such dimensions as clever, skilful, persevering or knowledgeable. Brookover, Thomas & Patterson (1964) show that pupils perceive teachers' estimates of their general ability as closer to their own than the estimates of parents or peers. We might predict, then, that the more a pupil perceives the teacher as a significant other, especially with reference to his ability, the more likely he would be to accept those expectations communicated by the teacher. Thus given that the prophecy is communicated we have a means of explaining the differential response of pupils to that communication.

A second hypothesis to explain the acceptance of a communicated expectation would involve the pupil's conception of his own ability and the teacher's conception of the pupil's ability. We might hypothesize that where the teacher has a clear and stable conception of the child's ability he would be less susceptible to accepting new ideas about the child's ability. The situation in Rosenthal's experiment is very unusual in that psychologists intrude and give an authoritative statement about the pupil's ability to develop intellectually at a dramatic rate. Information of this sort and with such authority is not usually given to teachers. But it does not seem unreasonable to suppose that where the teacher did not have a fixed conception of the child's ability, the more readily the teacher would accept such information. We have no evidence on the degree to which the teachers accepted Rosenthal's own communication of the prophecy about the spurters. The prophecy may well not have been accepted with reference to all the spurters *in toto*. We might further hypothesize that where the child's conception of his own ability is relatively unstable, he will be more open to accepting the teacher's belief about him once it is communicated. Rosenthal recognizes these possibilities, though he does not develop them, for he introduces them as an explanation of the fact that the self-fulfilling prophecy was much more effective among the younger children in the first two grades (six to eight years of age). He argues that the experimental conditions may have been more effective in the case of the younger pupils simply because their youth makes them easier to change than older ones. Later, he suggests that the younger pupils would also have less well established reputations in the school. This would mean that the teacher would be more susceptible to accepting the prediction from Rosenthal in the case of these children than in the case of pupils about which the teacher had firmly established views.

On this basis we can hypothesize the conditions in the Rosenthal experiment under which the self-fulfilling prophecy is most likely and least likely to be effected. It is most likely to occur if

(*a*) the teacher has an unstable conception of the pupil's ability (or a stable conception that is congruent with Rosenthal's prediction);

(*b*) the pupil has an unstable conception of his own ability (or a stable conception that is congruent with the teacher's conception of his ability);

(*c*) the pupil perceives the teacher as a significant other.

The self-fulfilling prophecy is least likely to occur if

(*a*) the teacher has a stable conception of the pupil's ability and this conception is incongruent with Rosenthal's prediction;

(*b*) the pupil has a stable conception of his ability and this conception is congruent with that of the teacher and incongruent with Rosenthal's prediction;

(*c*) the pupil perceives the teacher as a significant other.

Whilst these conditions are relevant to the Rosenthal experiment and similar situations in which the teacher is given a new and authoritative indication of the pupil's ability (e.g. an intelligence test score or an eleven-plus result), they are less relevant to the more common situation whereby the teacher has acquired a stable conception of the pupil's ability on the basis of his experience with the pupil in the classroom. In this case the likelihood of the pupil's acceptance of the teacher's perception, belief and prediction with respect to his ability are systematically outlined in Table 3.1. It is the pupils in boxes 1 and 11 who are most likely to live up to the teacher's expectation, for it is here that teacher and pupil conceptions are congruent and the pupil perceives the teacher as a significant other. It is the pupils in boxes 6 and 8 who are least likely to live up to the teacher's expectation, for here the teacher and pupil conceptions are incongruent and the pupil does not perceive the teacher as a significant other. When the pupil does not have a stable conception of his own ability, the prophecy is more likely to be fulfilled when the pupil regards the teacher as a significant other (boxes 3 and 9) than when he does not (boxes 4 and 10). We could use boxes 1 to 6 for explaining the marked individual differences in Rosenthal's results if we were in possession of the appropriate data. It would also be interesting to know, for example, whether the pupil in box 5, having a conception of his ability which is incongruent with that of the teacher but regarding the teacher as a significant other, could be persuaded to change his self-conception and make the prophecy come true. Once we ask questions such as this, the situation grows more complex, for the reactions of the pupil in box 5 are likely to be related to his orientation to other significant others, such as parents and friends, and the nature of their conceptions of his ability.

The concept of the self-fulfilling prophecy is open to further refinements. Merton himself noted the 'suicidal prophecy' or the anti-self-fulfilling prophecy—a prophecy which accurately predicts its own opposite. Rosenthal also recognizes this possibility, but does not develop it in educational terms. Yet it may be a very important variant form. Let me illustrate from my own experience. At school I was not very able in Latin, a subject which like most of my fellow pupils I disliked. The teacher believed that many of us would fail in the 'O' level examination, so as examination day drew near, he divided us into two groups. A third of the class was felt to have a chance and were seated at the front for special tuition. The rest, who seemed certain to fail, were relegated to the back to get on with their own work. I was a member of this latter group. But I was extremely annoyed at the teacher's treatment of me, feeling that I had been unjustly rejected and insulted. I was determined to pass the examination, to prove to myself that I could do it. In the event I passed. I suspect that if I had not been put in the failing group I would indeed have failed. In a very real sense the teacher's prediction that I would fail caused me to pass. It is worth noting that the anti-self-fulfilling

TABLE 3.1 *Three variables affecting the self-fulfilling prophecy*

The teacher's conception of the pupil's ability	Bright						Dull					
The pupil's conception of his own ability	Bright		?		Dull		Bright		?		Dull	
Does the pupil perceive the teacher as a significant other?	Yes	No	Yes	No	Yes	No	Yes	No	Yes	No	Yes	No
Box no.	1	2	3	4	5	6	7	8	9	10	11	12

63

prophecy does not fit in with our earlier theorizing about the conditions affecting the self-fulfilling prophecy. Indeed I fall into the category which suggests that the self-fulfilling prophecy should fail. It seems that the conditions likely to promote the anti-self-fulfilling prophecy are the opposite of those conducive to the normal self-fulfilling prophecy.

The forms of the self-fulfilling prophecy that we have considered up to this point are *interpersonal* for they involve the communication of an expectation between at least two people. However, there is another form which I have called the *autistic* self-fulfilling prophecy because it takes place within the one person. It may be that the pupil perceives that the teacher has an expectation of him or a belief about him which the pupil then accepts and fulfils. But this perception of the teacher's expectation or belief may be a misperception in that the teacher does not in fact have such an expectation or belief. Whilst it is probably true that the majority of pupils perceive the teacher's beliefs and expectations with a high degree of accuracy, there are likely to be cases where the pupil misinterprets the teacher's cues and makes false inferences and attributions. As we have noted, it is the perception rather than reality which counts, and if what from the outside is regarded as a misperception directly affects the pupil's behaviour then we have a self-fulfilling prophecy which is autistic because it rests on a misperception.

Our analysis of the processes of person perception and the self-fulfilling prophecy have some clear educational implications. When Person perceives Other a process of categorization is involved and on the basis of this categorization Person develops expectations and predictions about Other's future conduct. When teachers perceive pupils they inevitably categorize them. We shall examine the basis of this categorization in a later chapter. Here I wish to stress that whenever a teacher categorizes a pupil as 'good' or 'succeeding' or 'bad' or 'failing', he is making a prediction that congruent behaviour is to be anticipated in the future. Teachers then have to beware of two dangers. The first consists in the process of categorizing itself and the second consists in the communication of the predictive element of the categorization with its potentiality for stimulating a self-fulfilling prophecy.

One of the principal dangers of categorizing is that once the teacher has categorized a pupil he is likely to resist having to re-categorize the pupil. This is normal and natural in that most categorizations are reasonably accurate and economical and permit the development of a stable conception of Other which will facilitate interaction. We need large amounts of corrective information before we are willing to allow some of the incongruent information which does not fit our conception of Other through the filter and set about the necessary revisions in our conception of Other. Teachers are no exception. Once a pupil has been categorized as 'good' or 'bad' then it may require marked deviation on the part of the pupil from the teacher's conception of him before the teacher is willing to revise this conception. When teachers appoint a Head Girl they do so on the grounds that she has been firmly categorized as 'good' by certain criteria. She will have to misbehave very seriously before the teachers will change their view and revise their categorization. Early evidence that she is not living up to expectations will be misperceived or discounted. As we say, allowances are made. In effect, she 'gets away with things'. A similar process affects 'bad' pupils. Once categorized, the bad pupil can be caught in a heads-I-win-tails-you-lose situation. The consistently rude boy who is now being polite is regarded with suspicion by the teacher; his intentions and motives come under scrutiny. His politeness may be interpreted as 'taking the mick' out of the teacher or as evidence that a crime is being concealed. His motives are regarded as ulterior motives. The boy who turns in a first class piece of work when normally his work is appallingly poor is suspected of copying. When such pupils behave 'out of character' and thus 'out of category' the teacher's reaction may well be inhibiting the process of change which superficially the teacher is anxious to promote. What could be more devastating for a lazy pupil to find that, when he does respond to the teacher's appeal for change, his efforts are greeted with suspicion and disbelief? Would it not be reasonable for the pupil to conclude that the effort is not worth making since this is a game he can never win? Often the teacher's behaviour towards a pupil may in effect be putting him in what Bateson et al. (1956) have called a *double bind*. This arises when a person is given a message with two elements, one of which denies the other, so that a response becomes difficult or impossible. A typical example of the double bind is the injunction of a possessive mother to her son to go out and have a good time with his friends but indicating non-verbally that she will be upset if he leaves her alone. Teachers can be very skilled, though unintentionally so, at such communications. For example, a teacher may overtly encourage a child who is making little progress to work hard but also at the same time indicate that further effort is likely to be of no avail. Or the teacher may verbally give a pupil permission to embark on an activity whilst indicating that if the child does so he will incur the teacher's disapproval.

The teacher, then, should try to avoid making categorizations of pupils which are premature or too sharp and final. He must constantly be open to the possibility of re-categorization. Also he must be suspicious of those forces which encourage him to categorize or seem to confirm categorizations to the point where the categorization becomes virtually permanent. Staff-room gossip about pupils, record cards, intelligence quotients, all these can easily be used as evidence in support of a categorization which, however 'accurate', may become static. For once the categorization becomes static it influences the teacher's behaviour in a profound way and is easily communicated to the pupil with the obvious danger of initiating a self-fulfilling prophecy. All information which is at the teacher's disposal and which from one point of view serves to assist the teacher in helping the pupil to learn is information which can potentially have the reverse effect. There was once a Psychological Myth which suggested that knowledge of a child's IQ would be a valuable aid to the teacher. The dangers of making unwarrantable inferences from the IQ were soon forgotten. An IQ tells us that this child can do at least as well as this on a particular test at a particular time under specific conditions. In practice the IQ was not treated as such. In more recent years we have become prey to what I call the Sociological Myth. Sociologists have amassed a considerable amount of evidence which demonstrates how many factors in the home and community affect the child's capacity to profit from education. The research is usually taught in some depth to student teachers, in the hope that such knowledge will broaden their understanding of the environmental influences affecting the child's behaviour, attitudes and attainments in school. It may serve as a counter-weight to the teacher's tendency to ascribe the pupil's difficulties in learning to 'laziness' or 'awkwardness' which may pay insufficient attention to the child's background. But the way in which teachers use this knowledge is not by any means always in accordance with the intentions of the sociology lecturers. In the Sociological Myth the evidence is distorted by teachers into two propositions: that if the child comes from a 'bad' home, this is a sufficient explanation of the child's difficulties in school; and that because these environmental factors are beyond the school's control, there is little the teacher can do to help the child. Once the teacher has accepted the Sociological Myth he is led to undervalue the extent to which he can rectify deficiencies originating in the child's home environment and to ignore the contribution which life in school may itself make to the child's difficulties.

The teacher must also beware of those organizational devices of schools which are associated with categorization. Streaming is obviously one such device. Once the pupil is assigned to a stream the teacher tends to perceive him on occasions in terms of the collective category of the stream rather than in terms of his individual characteristics. The streaming system encourages the making of generalizations about streams which are then applied to individual pupils in that stream. Expectations about streams soon become expectations of individuals. As I have tried to show elsewhere (Hargreaves, 1967) children in the top streams are expected to work and behave well whereas children in lower streams are expected to misbehave and to make little academic progress. Sometimes this is associated with differential treatment of different streams, especially when the 'better' teachers are assigned almost exclusively to the higher streams and these pupils are given privileges and status which is denied to the pupils in lower streams. In such a situation it is hardly surprising if the pupils take to themselves the attitudes, expectations and predictions exhibited by the teachers and institutionalized in the organizational structure. A self-fulfilling prophecy seems almost inevitable. Nor are these matters easily rectified. Ability grouping within an unstreamed class poses exactly the same sort of dangers. Indeed they may operate here in a much more subtle way.

The more extreme a categorization is, the more dangerous it becomes. Once the teacher has concluded that the pupil is exceptionally 'bad' or 'good', the more likely the teacher is to convey his conviction and its associated expectations to the pupil with the attendant risks of fulfilment. Further, extreme categories like extreme attitudes are the most resistant to change. Rosenthal notes this effect in his own research.

After one year it was the children of the *medium* track [i.e. stream] who showed the greatest expectancy advantage, though the children of the other tracks were close behind. After two years, however, the children of the medium track very clearly showed the greatest benefits from having had favourable expectations held of their intellectual performance . . . It was the children of the medium track who showed the greatest expectancy advantage in terms of reading ability . . . Children of the medium track were the most advantaged by having been expected to bloom, this time in terms of their perceived intellectual curiosity and lessened need for social approval.

and further:

It seems surprising that it should be the more average child of a lower-class school who stands to benefit more from his teacher's improved expectation.

It is not at all surprising when we remember that it is the average child in the middle stream who escapes extreme categorization in the teacher's estimation and in the organizational structure. On the basis of our earlier theorization we would expect the teacher's conception of the pupil's ability to be more open to suggestion from Rosenthal in the case of the middle track. Also the middle stream pupil does not place himself in an extreme category, so it is this pupil rather than the high or low stream pupil who will be most open to the acceptance of a new expectation communicated by the teacher.

Rosenthal also has evidence of the teacher's unwillingness to revise categories when the change in the pupil is incongruent with the categorization. Again, we should expect the pupils in the extreme categories to be most affected. Rosenthal reports that the more pupils in the highest stream who had been nominated as spurters gained in IQ score, the more favourably they were rated by the teachers. Their nomination as spurters and their high stream status were congruent with the teachers' expectations of them, so when they improved they were rated favourably. By contrast the low stream pupils who had not been nominated as spurters met with a different teacher reaction. The more they gained in IQ the more negatively they were viewed by their teachers, for such change was quite incongruent with the teachers' expectations of them. It seems that when teachers categorize their pupils either in their own minds or as part of an organizational differentiation they must make a special attempt not to consolidate or communicate such categorizations. All teachers are committed to the improvement of their children. It seems that improvements can occur, even dramatically and contrary to the evidence, if the teacher can go on believing that the potentiality for improvement is always there within the child waiting to be released. And an important part of promoting the release of these potentialities consists in the teacher's communication of his faith in the pupil to the pupil.

Recommended reading

R. ROSENTHAL & L. JACOBSON, *Pygmalion in the Classroom*, Holt, Rinehart & Winston, 1968.

R. TAGIURI & L. PETRULLO, *Person Perception and Interpersonal Behaviour*, Stanford University Press, 1958.

4 Roles

A few years ago when I was having a drink in a pub on a hot summer's evening I struck up a conversation with another man at the bar. The conversation turned to a recent case of a murderer who, the court had decided, was insane. The man informed me that in his view all psychologists were outright charlatans, using as evidence the part played by psychologists at this trial. As a psychologist, I was interested in his opinion, especially as up to this point I had regarded him as an intelligent and interesting person. The threat to my self-esteem was simultaneously a threat to my favourable first impression of him.

'The trouble with psychologists,' he said, 'is that they always dehumanize human beings. They don't treat people as people at all. To the psychologists, a person is nothing more than a *type*—a neurotic, a psychopath and that sort of thing. They are so busy putting us all into little boxes that we cease to be responsible and unique individuals.'

I sympathized and we discussed the nature of the human sciences for some time. Several drinks and two hours later we were generally in agreement, though I suspect it was more a function of the drink than of our arguments. Initially the man had taken a very narrow view of human behaviour. Although each individual is indeed unique he does possess certain characteristics which he shares in common with all or many other individuals. Further, human behaviour is, within limits, regulated in a number of ways. Social scientists, whether they be psychologists, sociologists or anthropologists, look for these regularities. One aspect which my drinking companion failed to take into account is that human behaviour takes place within structured situations and it is very doubtful whether we can say much about the unique individual unless we can analyse the structures within which such a person thinks, feels and acts. The individuality of a man is achieved within a social environment, so to explain the

69

individual we must, amongst other things, consider the structure of the social system in which he operates.

All social systems, from the small unit such as the family to the large unit such as a nation, consist of a complex structure of interrelated *positions*. These positions are really categories of persons with certain similar attributes who hold certain structured relationships with members of other positions. To say 'mother' is to describe how a certain position which can be occupied only by females is related in a particular way to another position, 'child'. Some of the positions occupied by people are *ascribed*, that is to say, a person occupies a position quite independently of his wishes or accomplishments. For example, age, sex, family and race are ascribed positions. Most people cannot change their age, sex, family or race, however much they might like to do so. It is very rare for a person to be able to change his sex and it is unknown for a person to change his age or his biological parents. It is equally impossible for a person to change his race, though he may sometimes be able to convince others that his race is not what it actually is, as in the case of the light-skinned Negro who can 'pass as white'. Other positions are *achieved* and depend on the qualities or abilities of the individual. A man's occupation, for example, may be largely a reflection of his own attainments. In practice, of course, it is difficult to make a clear distinction between ascribed and achieved positions, because many positions are the result of a fusion of the two elements. A person's occupation in Western society is related not only to his personal accomplishments but also to the social class of his parents, his age, his sex and his race.

These positions are quite independent of particular individual persons. The term 'doctor' is quite meaningful without thinking of 'Doctor Smith'. The position persists despite particular incumbents. Individuals become old and die, but the position of 'old woman' is not very much affected by them. That is why we speak of people *occupying* a position, as if it were a seat in a waiting room being constantly filled and emptied by widely different individuals over time. Most people occupy a very large number of positions both simultaneously and consecutively. In addition to the basic positions of age, sex, social class and race, we occupy a multiplicity of positions such as neighbour, colleague, friend, committee member, driver, customer, patient and so on. Some of these positions are much more important to the incumbent than others. Age, sex, family and occupational positions are obviously much more central to a person's self-identity than others, such as the positions of neighbour or patient, because they are much more pervasive and permanent in that incumbents spend more of their lives in direct occupation of the positions. One is a woman all the time, but only a small proportion of one's time is likely to be spent in the position of committee-member.

The concept of position is a fundamental one in sociology, where one major concern has been to examine and codify the ways in which these positions are inter-related in various social systems, such as families, organizations, tribes and nations. Another problem of interest to the sociologist has been the *functions* of these positions, or the part they play in, and the contribution they make to, the social system as a whole. The findings and theories of sociologists are of great importance to social psychologists, but as we shall see shortly, the social psychologist has interests of a rather different sort.

The concept which is closely allied to that of position is the concept of *role*, which is perhaps the main concept which offers a potential link between the too frequently estranged disciplines of sociology and psychology. Like many basic concepts in the human sciences, role lacks any clear definition to which the leading writers subscribe. The variety of ways in which this concept is used in different books and articles is very confusing to the novice, and there is little hope that the recent attempt by Biddle & Thomas (1966) to draw together the different strands may help to promote a more universally accepted terminology. In this book we shall use the concept of role in a broad way to refer to behavioural expectations associated with a position. That is to say, attached to any position are a set of expectations about what behaviour is appropriate to the person occupying the position, who is often referred to as the actor. Thus the concepts of position and role are bound together by definition and one might use the term position-role to refer to the whole complex, the position itself and the behavioural expectations attached to it. Consider the position of *mother*. In our society there are a number of expectations which most people share about how a mother should behave. It is expected, for instance, that she will love her children, look after their bodily and emotional needs to the best of her ability, take care of them when they are ill or in need, encourage them to acquire moral values appropriate to the culture and so on. There are also expectations about what she should not do—a mother is not expected to be cruel to her children or to neglect them.

Role, then, refers to prescriptions about the behaviour of a person occupying a given position, a set of guide-lines which direct the behaviour of the role incumbent or the actor. Roles consist of sets of expectations. But what of the actual behaviour of the actor? We shall call the behaviour of an actor in a given position-role his *role-performance* which, as we shall see, may or may not conform to the expectations associated with the position. There are mothers who do indeed neglect their children. It is because some individuals do not conform to the expectations that we must distinguish role from role-performance.

The fact that all positions within a social system are related to other positions has important consequences for the position-role complex. Some positions, like that of mother, are common to large numbers of the population and most of us have direct experience of interacting with a woman in the position of mother. In addition, we have usually interacted with other people's mothers and observed many mothers interacting with their children. Consequently most of us have fairly definite expectations about how mothers are supposed to behave, though we may not all agree on the details of the prescription. Other positions, such as that of Sanitary Inspector, are held by a small number of persons and relatively few of us have interacted with such a person. Consequently very few of us have definite expectations, if any at all, about the sort of behaviour which would be appropriate to that position. Many positions have what may be termed a *complementary role* with which the position is specially related. A mother, for instance, is primarily linked with the complementary positions of sons and daughters. Likewise, husbands have wives, doctors have patients, teachers have pupils. In each case the two form a role partnership; the positions make sense only in terms of the complementary position. Can we conceive of a doctor without the concept of a patient, of a teacher without pupils?

The occupant of a complementary position-role, the role partner, plays a very important part in defining the role associated with a position. It is his expectations which make a highly significant contribution to the content of a role. There are three basic reasons for this. First, it is toward the complementary position-role that an actor principally directs his role performance. A doctor spends most of his time performing his role as a doctor in relation to his patients. Similarly, a teacher directs most of his role performance to his pupils. So the expectations of the main role partner tend to assume a special significance. Second, because much of an actor's role performance is directed to him, the role partner can, to varying degrees, constrain the actor to fulfil his expectations by rewarding him when he conforms to the expectations and punishing him when he deviates. For example, a patient whose doctor fails to fulfil his expectations can seek medical attention elsewhere. Third, the nature of the role relationship between role partners is so structured that the two actors are bound together in a special way. Associated with every position are *duties* (or obligations) and *rights* (or privileges). But the duties of the actor are the rights of his role-partner, and the rights of the actor are the duties of his role partner. For example, a doctor has a duty to his patient to give him the best possible medical attention and advice, and it is the patient's right to expect this. The doctor has a right to expect the patient to follow his advice scrupulously and this is the duty of the patient.

Almost everyone has expectations about the behaviours appropriate to such common roles as mother, teacher and clergyman, because we have extensive experience of interaction with them. Sometimes we have expectations of position-roles because they are unique within the social system, or because they are highly visible to the community at large, or because the occupants of such positions make important contributions to the maintenance of the social system. It is for such reasons that we tend to have definite expectations about the behaviour associated with certain positions, such as that of monarch or prime minister, with whose occupants we are very unlikely to interact. Most position-roles are, however, primarily linked with one particular complementary position-role with a small number of occupants. (Compare mother–children or teacher–pupils with monarch–subjects.) It is with this role partner that the actor performs most of his role. Yet the complementary role partner is not the *only* role partner, nor the only or necessarily the most important source of role expectations. Most roles are linked to a number of other position-roles, referred to by Merton (1957) as the *role-set*. The teacher's complementary role partner is the pupil, but the teacher is linked to other role partners, such as the headteacher, colleagues, pupils' parents, the chief education officer, the school governors, the education welfare officer, the school caretaker. When the teacher interacts with such persons he does not cease to be performing his teacher role. For the teacher, in common with many position-roles, has an extensive role-set which involves him in a wide range of social relationships. With each different role partner he behaves in a readily distinguishable way. In all of them he is performing his teacher role.

Every role partner from the role-set has expectations about how the teacher should behave towards him. Headmasters and pupils do not expect to be treated in the same way by the teacher. But in addition each role partner usually has expectations about how the teacher should behave towards the other role partners. For example, a headteacher has not only expectations about how the teacher should behave towards him, but also expectations about how the teacher is supposed to behave toward the pupils, other members of staff, parents, governors and so on. It is unlikely that all members of the role set will have precisely the same expectations concerning each of the teacher's role relationships. Each member of the role set is likely to have more elaborate expectations for some role relationships than for others. A related phenomenon is that the expectations of some role partners will exert more influence on the teacher's conception of his role performance than will other role partners. (And it does not follow that the role set members with the most elaborate expectations exercise the greatest influence.) Three of the main role relationships that make up the teacher role and their connection with the principal role partners in the teacher's role set are outlined in Figure 4.1. If we

73

charted all the teacher's role relationships in relation to a more complete role set it would be clear that the concept of the 'teacher's role' is remarkably complex.

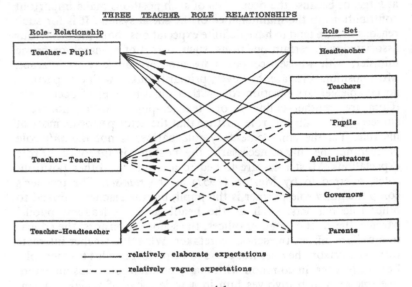

FIG. 4.1

A position-role, then, is linked to a number of other position-roles, the incumbents of which have expectations about the actor's behaviour towards all the other role partners. It is obvious that the whole body of expectations which make up the role is likely to contain some conflicting elements. For this, and other reasons, most actors do not and cannot entirely fulfil the expectations associated with the position they occupy. It is in considering the situations in which an actor experiences difficulty in conforming to the expectations that make up his role that we discover one of the principal values of the concept of role. All these situations can be subsumed under the term *role strain*, or its alternative name of *role conflict*. It is not possible to discuss the innumerable types of known role strain—and there are doubtless many types still to be discovered—but we shall confine ourselves to some basic forms.

One form of role strain arises when an actor simultaneously occupies two positions whose roles are incompatible. A good example is the military chaplain. He is occupying the position of priest and officer at the same time. An officer is expected to be rather distant from the enlisted men with whom his relations must be rather formal. But a priest is expected to be a highly approachable person with whom one can discuss personal problems in a warm and familiar relationship. The military chaplain has the problem of reconciling the incompatibility between the two roles. Another example is the teacher whose own son or daughter is in his class. As a father he is expected to be warm and personal in his relationship to his children, but as a teacher he is usually expected to hold a more formal relationship with his pupils. Another example is the woman who is both a mother and a wife. Both her husband and her children may call on her loyalty and support when there is a disagreement between the father and the children. Many children seem to have learned the art of using this situation to their advantage by provoking a conflict between their parents.

Role strain of a different sort arises when there is a lack of consensus among the expectations that make up a role. Such dissensus takes a variety of forms. One of the most important contributions to the content of a role is the actor himself. Yet it is not uncommon to find a lack of role consensus among the occupants of a position themselves. Among teachers there exists a variety of opinions about certain basic aspects of the teacher's role, such as the nature of the relationship between the teacher and his pupils. 'Traditional' and 'progressive' are examples of the ill-defined labels we attach to different role conceptions within the teaching profession. Musgrove & Taylor (1965, 1969) have shown that there are different role conceptions of teachers working within various segments of the educational system. The teachers involved in this research were asked to rank six objectives as they valued them. These aims were Moral Training, Instruction in Subjects, Social Training, Education for Family Life, Social Advancement, and Education for Citizenship. Grammar school teachers were found to have a much more restricted conception of their role than teachers in modern schools, in that they tended to reject a greater number of the educational objectives as 'none of their business'. Although teachers in all types of schools conceived their role primarily in intellectual and moral terms, placing greatest weight on Instruction in Subjects and Moral Training, there were some interesting differences between grammar and modern school teachers. Eighty-six per cent of the male grammar school teachers ranked Instruction in Subjects first or second compared with 63 per cent of the Modern school teachers. The modern school male teachers were much more likely to include 'social' elements in their role than their grammar school colleagues. Forty-two per cent of the modern school men rated Social Training first or second and 74 per cent put Moral Training first or second. The relevant figures for the grammar school men are 20 and 60 per cent respectively. These differences in role conception among teachers probably reflect differences in the *ethos* of different types of secondary school, differences in the characteristics of the pupils they serve, and differences in the background and training of the teachers.

A different type of role strain emerges when the lack of consensus is not among the actors but among one of the role partners. Musgrove (1961) has shown that parents in different social classes have different conceptions of certain aspects of the teacher's role. One of the areas which Musgrove investigated was parental expectations of the teacher relating to the child's behaviour. Working-class parents are much more likely than middle-class parents to place responsibility on the school for training the child, especially in respect of good behaviour, obedience, respect for elders, manners, honesty, truthfulness and cleanliness. Middle-class parents tend to emphasize the responsibility of the home in such training.

A particular severe form of role strain has its roots in a conflict between an actor's conception of his role and a role partner's conception of that role. In short, A and B have divergent conceptions of A's role. Often this conflict is bilateral: A and B have different conceptions of both A's role and B's role. In such circumstances it becomes very difficult for either actor to sustain a very adequate role performance toward his role partner. A good illustration of such role strain is provided in Leila Berg's (1968) study of the famous Risinghill School affair. The headmaster, Michael Duane, had a 'progressive' conception of the roles of headteacher and teacher which conflicted with the role conceptions of some of the teachers on his staff, and with other role partners.

> Mr. Duane felt his duty was more to his pupils than to his staff. Sometimes a teacher became upset, and wanted to be put first; then the teacher didn't want impartial fairness, but comfort.

> many of the staff did not want discussion—they wanted to be 'told'; and secondly they wanted to be led in a completely different direction from the way Mr Duane was going.

> The administrators wanted someone who would keep a school quiet and orderly, a man who would only have to stroll through the playground and you could hear a pin drop; and the politicians wanted someone who could be manipulated or who at least would show he was grateful to be chosen.

> The inspector cold shouldered the child-centred approach, which since this policy is rarely applied in state schools, needed help, and spent considerable time with the traditionalists, whom he approved of and with whom he sympathized.

How could Inspector Macgowan ever co-operate with Mr. Duane without going against his own convictions? Wasn't there bound to be antagonism—manifest antagonism—between a man who believed that 'six of the best would cure almost any disciplinary problem' and a man who believed corporal punishment was brutal and encouraged brutality, between a man who disapproved of informality between head and staff, and between head and children, and a man who thought life could only be lived on terms of friendliness . . . ?

Sometimes role strain may arise because the expectations of different role partners conflict with one another. The deputy head-teacher is particularly susceptible to this form of role strain. The head may expect the deputy to be loyal to him; to be his executive officer who carries out the head's policies among the staff; to be an informer on staff gossip. The staff, on the other hand, may expect the deputy to be loyal to them; to fight for their views and policies with the head; to be their informer on the head's intentions. Children also frequently suffer this form of role strain since the parents often disagree on the content of the roles of son, daughter or teenager. Children are also very skilled at deflecting this role strain by provoking an inter-parent conflict. When the child finds himself rebuked by one parent, he claims that he is abiding by the expectations of the other parent and then disappears whilst his parents argue it out together.

Another variation of this theme is the role strain which derives from incompatible expectations held by a single role partner. This sounds a rather unlikely situation, in that the role partner must be to some extent inconsistent to make conflicting demands. Yet it is a common occurrence. In the teaching situation the pupils usually expect the teacher to help them to pass the examinations successfully but they also expect to be set minimal amounts of homework. Sometimes this strain takes the form of the 'heads-I-win-tails-you-lose' game, as in the case of teachers who complain that the headmaster is autocratic and domineering, but when he becomes more democratic accuse him of failing to make important decisions or to give a lead or to take responsibility.

In other cases the role partner may hold expectations of an actor which he does not believe are incompatible but which the actor himself perceives as incompatible. For instance, some parents advocate a more child-centred education in schools and believe that such an orientation on the part of the teacher will not hinder the promotion of examination success. Yet some teachers believe that to abandon this more 'traditional' orientation would lead to a decline in academic standards which would be reflected in examination performance.

Similarly some parents support the abolition of corporal punishment in schools and do not see such a change as incompatible with school order. The teachers, by contrast, may believe that corporal punishment is a necessary deterrent for the maintenance of order and discipline in school.

Role strain of a different sort arises when the expectations are unclear. Where there is a lack of clarity in the role expectations the actor is uncertain how he ought to behave. Roles vary in the degree of clarity of the expectations, but even in the case of roles where the expectations seem very explicit and specific the actor is likely to find himself in an unexpected or novel situation where the expectations become unclear, diffuse or ambiguous. Many of the expectations of the teacher's role are of a very general order and it is often unclear which expectations should apply in a given situation and which expectations should be given priority.

Many roles, which may not in themselves be usually lacking in clarity, are perceived as unclear when the actor assumes the position for the first time. When a child first enters school he may lack even a basic knowledge of the expectations of the pupil role, especially if he has received no nursery education and if he comes from a working class home where, relative to the middle class child, he is less likely to have received preparation for this new role. For a time he will call his teacher 'mother' and expect maternal treatment from the teacher. Part of the child's early reluctance to go to school may derive from the strain he experiences in learning and adjusting to this new role. In a similar way role strain may be experienced by a man on retiring from his job, and the problems of learning the husband and wife roles among the newly married are well documented in the literature. (Where the couple lives with one set of in-laws there may be additional conflict between roles—e.g. the roles of son and husband.) Lack of clarity is frequently found in newly created roles. The school counsellor in Britain sometimes has a rather vague and diffuse conception of his own role. His role partners may be ignorant of his qualifications and capacities and the part he can be expected to play in the life of the school. The school counsellor may have to inform and persuade other teachers about the content of his role. The difficulties arising from lack of role clarity are often mitigated by the introduction of special training courses or orientation sessions.

The last type of role strain we shall discuss has its roots in the personal qualities of the actor which make it difficult for him to perform the role. An actor may lack an attribute which is essential to an adequate role performance. A man with little dexterity is likely to be an unhappy and unsuccessful watchmaker. Similarly a teacher who never got beyond 'O' level Latin is unlikely to make a success of teaching classics to the sixth form. As a person with neither ability nor interest in any form of sport, I was horrified to discover on arrival at the school of my first teaching appointment that I had been appointed House Athletics Master. Often the actor's values, attitudes and personality may interfere with role performance. Many agnostic and atheist teachers consent to teach religious education only because they are pressured to do so by the headteacher. A lack of organizing ability or interpersonal skills in a headteacher is likely to produce administrative chaos or frictions with his teachers because of his inability to meet their role expectations of him.

Essentially this form of role strain is a conflict between the self and the role. This is a complex area since we know that roles often contribute to and form part of one's self-image. In general we strive towards maintaining congruence between self and role, so when we move into a new position-role we often change or modify certain aspects of our self-image in order to bring self into line with role. Attitudes are one important link between self and role that can be modified to enhance self-role congruence. Liebermann (1956) has shown that when industrial workers become foremen or shop stewards they change many of their attitudes accordingly. The workers who became foremen became more favourable to management and less favourable to the Union. New shop stewards followed the reverse trend. Interestingly when some of the foremen had to be demoted because of reduction in staff they reverted to their original attitudes.

Sometimes an actor may find that the conflict between his self-image and the role demands are not so easily reconciled. This will obviously arise when the role demands are incompatible with very basic or central aspects of the self. If he abides by the role which is incongruent with his self-image he will feel that he is 'playing a part'. He will achieve social integration at the cost of personal integrity. If he abides by his self-image he will find it difficult to meet the role expectations. Personal integrity will be preserved at the cost of social integration. Either way he will experience psychological discomfort and will find it difficult to perform his role effectively. A square peg cannot fit into a round hole by pretending to be round or by insisting upon being square.

Most teachers teach those subjects which they have chosen to teach and for which they are qualified by their training, and throughout their careers they maintain an expertise in their subject which makes them superior in achievement to their pupils. An exception, however, is the teacher of physical education. As he grows older, he may discover that his physical skills are on the decline and that he is rapidly being surpassed by his pupils. He loses the attributes which he regards as necessary to an adequate role performance. Thus we meet the tragedy of the worn-out games teacher as described by Terson (1967).

I'll tell you how it happens. You're eighteen, you were in the school cricket team, or football team, or this and that team, or you were a genius at every sport except snooker. You were clean living and healthy, you could swim the 100 metres in some time that is good for the 100 metres, or you could bowl spinners or dribble down the left wing at outside left. The world was open. You could have played for Newcastle United Reserves, or been on the ground staff of Hampshire, or gone in for boxing; but no, you had intelligence as well and your father was a pitman or railway clerk and wanted his son secure and 'professional', so you went into teachers' training college. Here you had three blissful years; you jumped about in the gym, you somersaulted on the trampoline, you were watched for style, under water, through the observation glass in the swimming pool, you got drunk, sang Rugby songs, you jeered at the History students or English specialists, you were banned from the college drama festival for jeering at something you did not understand, you brought back pub signs from Bristol, Cheltenham or Wolver-hampton when you played away, you gloried in your strength, you got all the girls. You usually left with a pregnant girl.

. . . Then you were among the kids. They didn't care for your bygone feats, you were an old man with them anyway. You tried to keep up the Saturday afternoon football, but that wore off with your worries and pram-pushing. In two years you refereed 63 major football matches and 184 house matches. You jeered because the school records on Sports Day were not a patch on what you could do at their age.

. . . Then suddenly, last summer, you weren't as fit as you used to be and you won't retire until you're sixty-five and the other blokes in the other subjects are streets ahead. One day soon, that kid in 3A is going to outrun you, or beat you to a tackle, or knock you for six, and jeer 'Come on, Sir?' And you'll say, 'Ah, but I'm an old man, I'm past it.'

It is for this reason that many teachers of physical education like to keep one foot in the 'academic' subject departments, in which they can spend an increasing proportion of their time as their physical skills wane. It may also account for the disproportionate number of physical education teachers who apply for courses leading to an Advanced Diploma or a Master's degree in Education. For to change his expertise is one of the best ways of changing his role, and a further qualification in educational studies is the most obvious means by which he can gain entry to a College of Education as a lecturer in education. Such a role change would also solve the P.E. teacher's restricted career mobility, since it is well known that the P.E. teacher's 'Head of Department allowance' is invariably of a less substantial order than in the case of 'academic' subjects.

To summarize, we have distinguished eight basic types of role strain. An actor may be said to be in a situation of role strain when:

1 he simultaneously occupies two positions whose roles are incompatible.
2 there is a lack of consensus among the occupants of a position about the content of a role. $(A_1A_2A_3 \ldots$ lack of consensus about the role associated with position A.)
3 there is a lack of consensus among the occupants of one of the complementary role-positions. $(B_1B_2B_3 \ldots$ lack of consensus about role A.)
4 his conception of his own role conflicts with the expectations of a role partner. (A and B have different conceptions of A's role.)
5 the various role partners have conflicting expectations. (B, C and D have conflicting conceptions of A's role.)
6 a single role partner has incompatible expectations.
7 the role expectations are unclear.
8 he lacks the qualities required for adequate role performance.

Role strain is obviously a very common occurrence and there are probably very few roles which are not to some degree liable to one or more of the conflicts we have examined. As an actor performs his role he is likely to be involved in several types of role strain, and sometimes to experience several forms simultaneously, though the type and severity of the role strain concerned will vary with the situations in which the role is performed. Why, then, are we not constantly suffering from the dilemmas, confusions, and conflicts which the various types of role strain seem to imply?

One important reason is that the objective conditions of role strain, namely a disagreement between an actor and his partner, do not always produce *subjective* role strain, that is a state of psychological discomfort in the actor. To establish the existence of role strain between an actor and a role partner we need to know the following details.

(a) The actor's conception of his role.
(b) The partner's expectations of the actor.

From these we can assess a discrepancy in expectations between an actor and his role partner. As an example of such discrepancies we can consider some of the findings of *Schools Council Enquiry 1: Young School Leavers* (Morton-Williams & Finch, 1968). Pupils, their parents, teachers and headteachers were asked to rate the importance of certain educational objectives. The percentages of members in each category rating an objective as very important are given in Table 4.1.

TABLE 4.1 *Rating of educational objectives*

	1 Moral	2 Job (a)	3 Job (b)	4 Exams	5 Character
Pupils—boys	66	86	81	66	41
Pupils—girls	76	88	81	67	51
Boys' parents	74	89	79	72	56
Girls' parents	80	87	77	68	58
Teachers	76	47	33	19	92
Headteachers	84	38	14	19	96

From: *Schools Council Enquiry 1* (Morton-Williams & Finch, 1968)
Numbers refer to percentages in each category rating the item as a very important school objective.
1 Teach you about what is right and wrong.
2 Teach you things which will help you to get as good a job or career as possible.
3 Teach you things which will be of direct use to you in your job.
4 Help you to do as well as possible in exams like GCE or CSE.
5 Help you to develop your personality and character.

It can be seen that with respect to the objective 'Teach you about what is right and wrong' (column 1) there was general agreement among the subjects in different categories. However in the case of other objectives some clear discrepancies become apparent. Pupils and their parents put a strong emphasis on the vocational aspects of schooling as measured by the items 'Teach you things which will help you get as good a job or career as possible' (column 2) and 'Teach you things which will be of direct use to you in your job' (column 3). Yet heads and teachers are much less likely to agree that preparation for a job is an important school objective. The same picture emerges with respect to examinations where the item was 'Help you to do as well as possible in exams like GCE or CSE (column 4). In the last column are given the percentages rating the item 'Help you to develop your personality and character.' Here we find the opposite trend; what teachers and headteachers almost unanimously rate as very important receives much less support from parents and even less from the pupils. (It is worth noting that the headteachers diverge from parents and pupils much more than do the teachers.)

These findings might be taken to indicate the existence of role strain, since what the pupils and their parents rate as a very important educational objective receives much less approval from teachers, and what teachers regard as a basic objective is valued less by pupils and their parents. As the authors write:

It is evident . . . that conflict and misunderstanding may arise between the short-term viewpoint of parents and pupils who are concerned with starting work in the immediate future and the long term objectives of teachers who see their responsibility as preparing pupils for the whole of their future lives.

Yet subjective role strain is by no means a necessary concomitant of this objective discrepancy, for none of groups involved is necessarily aware of these discrepancies in the valuation of educational objectives. Pupils and parents might conceivably believe that teachers agree with them and vice-versa. Subjective role strain requires that the actor perceives the discrepancies between himself and his role partner. Thus we require a third factor.

(c) The actor's perception of the partner's expectations of the actor's role.

It is important to note that subjective role conflict can arise in the absence of objective role strain. This happens when an actor incorrectly perceives that a role partner has discrepant expectations. For example, Musgrove & Taylor (1969) show that (i) teachers rate moral training as an important component to their role; (ii) parents agree in that they expect teachers to give a high priority to moral training; but (iii) teachers perceive that parents expect teachers to give moral training a *low* priority. Similarly, teachers and parents regard social advancement (preparing children to 'get on in life') as a low priority in the teacher's role, but teachers believe that parents expect teachers to give it a high priority. In both these cases there is a suggestion of the existence of subjective role strain in the absence of any objective conflict. (It is worth noting the apparent contradictions between the findings of Musgrove & Taylor and those of *Enquiry 1*. Whilst it is possible to attribute these differences to different samples and methods, it also suggests that we must exercise a great deal of caution in interpreting and generalizing from such studies.)

(iv) The actor's personal awareness of feelings of role strain. It is only on the basis of this last evidence that we can verify the existence of subjective role strain. Unfortunately, most studies of role strain rely on evidence derived from (i), (ii) or (iii). Attempts to derive evidence of type (iv) are extremely rare. Yet it is this evidence which we require if the concept of role strain is to tell us much about the problems human beings experience in performing roles. To demonstrate the existence of objective role conflict is simply to report discrepancies of expectation between an actor and a role partner. This may be of very little significance to the actor (or the partner) unless they perceive the discrepancy and its existence provokes some psychological discomfort which encourages the actor to take remedial action.

It may be, of course, that there are important consequences which follow from the existence of objective role conflict (e.g. misunderstanding which affects role performances), but these must be demonstrated, not merely assumed. Even in those cases where the actor perceives the discrepancy between his own expectations and those of his role partner, we cannot assume subjective role strain, because the actor may ignore or be quite unconcerned about the conflicting expectations of a role partner. Teachers can experience subjective role strain in relation to parents only if the expectations of parents in regard to the teacher role matter to the teachers.

The concept of role strain is considerably more complex than it appears to be at first sight, and empirical studies of role strain must be interpreted with great caution. Many researchers give what may be a spurious importance to their findings by demonstrating discrepancies between the actor's and the role partner's expectations which are *statistically* significant. This is not enough. The researcher needs to demonstrate the psychological or sociological significance of these discrepancies for the actor and his partner, for their role performances, or for the functioning of the social systems with which the actor and his partner are associated. I suspect that such discrepancies may need to be quite marked before they provoke a reaction of psychological or sociological significance. Small differences in expectations may have a significance which is only statistical. Sometimes, too, researchers do not ascertain that the areas in which they reveal discrepancies of expectation are in fact central and important aspects of the role involved. The demonstration of discrepancies in expectation concerning peripheral aspects of a role may be without significance for the persons involved or for the social systems of which they are part.

One of the main reasons why researchers have shown more interest in objective role conflict rather than subjective role strain is that the actor often uses various techniques to resolve or reduce his experience of strain even though the discrepancies in expectations persist. The fact that an actor does not currently feel subjective role strain does not mean that he did not do so at an earlier stage, but simply that he has found some means of coping with the strain. It is not easy for the researcher to distinguish between situations where objective role conflict is not, and never has been, accompanied by subjective role strain, and situations where objective role conflict has been accompanied by subjective role strain which has been resolved.

In short, several types of role strain allegedly arise when there are discrepancies between an actor and his role partners in their expectations of the actor's role. This is not an adequate view, unless the discrepancies can be shown to have real consequences for the actor, his role partners or other people. It is likely to be difficult to demonstrate such consequences especially in the case of objective role conflict without subjective role strain. Three minimal conditions seem necessary before subjective role strain could be inferred. First, there must be a discrepancy between the actor's conception of his role and the actor's perception of a role partner's expectations. Second, the discrepancy must be marked and related to aspects of the role which the actor regards as central or important. Third, the role partner with the perceived discrepant attitudes must to some degree be a 'significant other' to the actor. These remarks refer only to the forms of role strain where discrepant expectations are involved. Most other types of role strain also need to be subjected to rigorous analysis before some of the assertions in the literature can be accepted and there is an urgent need for the formulation of more precise conditions under which role strains are likely to arise.

However, the main reason which accounts for our ability to perform roles with little discomfort despite the role strains involved is that there are various mechanisms which an actor can use to resolve or reduce role strain. If an actor is simultaneously occupying two positions with roles which are likely to conflict he can solve such a conflict by giving one of the roles a priority over the other. Thus a working mother will usually give priority to her mother role when her child is sick, and her occupational role takes second place. When a teacher has his own son in his class he may treat his son like any other pupil and thus give the teacher–pupil role relationship priority over the father–son relationship. Even within a single role there is likely to be a hierarchy of obligations which dictate which expectation is to be accorded priority. Thus a teacher will tell his class that he is sorry that he cannot spend the lesson reading a play, which he knows the children want, because he must give them practice in comprehension passages to prepare them for the examination.

When an actor's role partners have conflicting expectations of him, the actor can resolve or reduce the strain involved in various ways. He can, for instance, distort his perception of the role partner's expectations. If he can persuade himself to believe that the role partner's expectations are congruent with his own then there will be no subjective role strain—though the objective role conflict may have important consequences for the actor. To take a theoretical example from the *Enquiry 1* data given in Table 4.1, the teacher might adjust his perception of the pupils' expectations concerning the importance of vocational preparation in education to match his own. This could resolve a possible role strain, but it might have important consequences if the pupils continue to feel that not enough attention is paid by the teacher to vocational aspects and withdraw their interest from the teacher's lessons.

A common solution is to conform to the expectations of the role partner with the greatest power. If the actor were to conform to the least powerful role partner then he would have to forego the relatively larger rewards and incur the relatively greater punishments at the disposal of the most powerful partner. To conform to the expectations of the most powerful role partner is common sense in that it is the path of maximum gain and minimum cost. Thus a teacher is more likely to conform to the expectations of the headteacher rather than the pupils' parents, since the parents have very little power over the teacher, but the headteacher is able, amongst other things, to exercise a great influence over the teacher's career prospects.

Should there be no power differential between two conflicting role partners, the actor can still resolve the strain by regarding one partner's expectations as more legitimate than the other. When an actor regards one role partner as possessing no right to prescribe behaviour for him, then he frees himself from the need to conform to his expectations and from the strain that is induced by nonconformity. But if, of course, the role partner who is regarded as having illegitimate expectations is also a person with power over the actor, the role strain is not so easily resolved, as many teachers who regard the expectations of their headteacher as illegitimate ('unfair' 'unrealistic' 'theoretical' 'autocratic') will testify.

It is very costly to ignore the role partner who can invoke numerous sanctions. One possible solution is to insulate the role performance against (simultaneous) observation by conflicting role partners. This is in fact the mechanism by which the teacher is protected from the potential role strains inherent in his position. None of his major role partners but the pupils normally observe his role performance in the classroom. The headteacher, the inspectors, Local Authority officials, governors, parents, other teachers etc. have limited if any opportunity to see whether or not the teacher is in fact conforming to their expectations in the way he conducts himself in the classroom. But at times these members of the role set may have a right of entry into the classroom. If the teacher is failing to abide by their expectations, then the teacher must develop defensive ploys to conceal this non-conformity. Since Miss Jean Brodie, the heroine of Muriel Spark's (1961) novel, believed in unorthodox methods, by no means approved by the headmistress, she would keep a long division sum permanently on the blackboard or have the pupils prop up history books in front of them. These provided useful cover in the event of unwelcome and unexpected intrusions by the headmistress in the classroom. Indeed, as the novel shows, the headmistress had great difficulty in collecting firm evidence of Miss Brodie's malpractices and in the end was forced to rely on the betrayal of a pupil informer.

All teachers do deviate from the expectations of the headteacher, colleagues, parents etc. to some degree, but the surprising thing is the extent to which they conform to such expectations. This is partly because their general expectations have been codified into a set of professional norms and partly because they are perceived as more legitimate and more powerful exponents of expectations of the teacher's classroom role. They can promote conformity to many of their expectations even when the teacher is not under their surveillance. They can, in the very apt term of Getzels & Thelen (1960), be regarded as 'invisible participants' in the classroom. Should the headteacher or an inspector be present during his lessons for a short period, it is an easy matter for the teacher to create a special temporary role performance which will meet the expectations of the observer. On a 'Parents' Evening' the teacher may present himself to a parent as a charming, friendly and gentle person who has been completely mis-represented by the pupil in his accounts at home, and then return to the classroom the following morning as a bad-tempered autocrat. When a parent complains about a teacher's behaviour, many head-teachers will not on principle allow the disgruntled parent to meet the teacher concerned. The teacher could hardly be more insulated than he is from simultaneous observation by different role partners. His role performance tends to consist of different performances to different role partners, who are carefully segregated from one an-other. A typical reaction of a teacher in a situation where his class-room insulation from observation is invaded is given in Evan Hunter's *The Blackboard Jungle* (1955). Rick Dadier, the teacher, is visited by the Department Chairman, Mr Stanley. The intrusion has interesting effects on both teacher and pupils.

The class filed in, spotting Stanley instantly, and behaving like choir boys before the Christmas Mass. There'd be no trouble today, Rick knew. It was one thing to badger a teacher, but not when it led to a knock-down-dragout with the Department Chairman. No one liked sitting in the English Office under the cold stare of the Stanley man.

The cold stare showed no signs of heating up during the lesson. Rick gave it all he had, glad he'd prepared a good plan the night before, able for the first time actually to follow the plan because the kids kept their peace in Stanley's presence. He called primarily on his best students, throwing in a few of the duller kids to show Stanley he was impartial, but he steered away from Miller and West, not wanting to risk any entangle-ments while Stanley was observing . . .

Nor was this the last visit. Stanley began stopping by frequently, sometimes remaining for the full period, and sometimes visiting for ten- or fifteen-minute stretches, and then departing silently.

In the beginning, Rick resented the intrusions. He would watch Stanley scribbling at the back of the room, and he wondered what Stanley was writing, and he felt something like a bug on the microscope slide of a noted entomologist. Why all this secrecy? What the hell was this, the Gestapo?

With the growth of team teaching and similar developments, perhaps teachers will be less sensitive than at present about the private world of the classroom.

If insulation against observation by role partners is not feasible, an actor may try to reduce the role strain by reaching a compromise between the two conflicting expectations, by partially meeting the expectations of both in his role performance. Alternatively, he may try to bring the conflicting expectations into the open and arrange a conference between the role partners in which some sort of consensus may be achieved. Conferences of various sorts are also means of giving greater clarity to roles for persons assuming the role for the first time. Some schools open a day earlier for new pupils so that they may familiarize themselves with the physical arrangement of the school and receive some preparatory induction into their new roles within a new social organization, largely through talks from the headteacher and his staff. Training courses of all sorts tend to socialize the members into the roles for which they are being trained so that the eventual transition is facilitated. It is probably true that children who attend nursery school find the early days in the infants' school much less disturbing than the child who is attending a school for the first time.

It is generally assumed that role strain is 'bad', a state of affairs which, whether it is objective, subjective, or both, must have negative and undesirable effects for persons, relationships and social systems. We have devoted too little attention to the potential positive effects of role strain. It could be argued that a certain amount of role strain can have beneficial effects under certain conditions. For persons, relationships and social systems, the benefit might reside in a valuable and necessary corrective to acquiescence, conformity, conservatism and resistance to change. Friction and conflict may be essential ingredients for the promotion of health and change. It would not be easy to argue that a 'progressive' staff in a school should resolve their conflicts with an 'authoritarian' headmaster by conforming to his expectations of teacher and pupil roles. It would certainly be sad if the present conflicts in universities and colleges were resolved entirely in favour of the governing bodies of these institutions. In the end, we cannot escape making value judgments about the ways in which role strains are resolved. But let us not make the foolish mistake of claiming that *any* resolution of role strain is preferable to its persistence. A total lack of role strain may be characteristic only of Utopia and Heaven—but for my own part I am not entirely convinced of that. Even Heaven seems to require its Lucifer.

These are not by any means exhaustive lists of the varieties of role strain or of role strain resolution, nor can it be claimed that they include the most basic forms. Lists of different types of role strain or of role strain resolution are simply products of the academic's attempt to elaborate distinctions within the over-all and inclusive concept of role strain. Such attempts on the part of social scientists serve important purposes in the development of the concept, since, it is argued, with each distinction and clarification our understanding of the concept is enhanced. Yet paradoxically the more elaborate our typologies of role strain become, the less easy it is to place a particular example of role strain drawn from 'real life' into a particular category. Role strains taken from 'real life' are often complex and multiple, so that we have to place the role strain concerned into several categories simultaneously or we are not sure how to classify it at all. It is when our formulations become too cumbersome that we recognize the need for new formulations and conceptualizations and this point has probably been reached in role theory.

A good example of what might be termed a 'complex of role strains' is available in the role of the student teacher and his transition to the full teacher role. During his professional training the student teacher formulates his conception of the teacher role. It would not be true, of course, to say that the student *acquires* his role conception during training, for his own lengthy experience as a pupil and his decision to become a teacher suggest that he will already hold a fairly elaborate conception of the teacher's role. The influence of training on his role conception is basically two-fold: the influence of his tutors and lecturers, and the influence of the pupils and teachers he meets on teaching practice. Several strains are likely to arise. For example, the student may have been educated in a traditional grammar school and then find that most of his tutors hold progressive views about education. In such a case the models provided by his own school education may inhibit his acceptance of the tutors' conception of the teacher's role in which they are seeking to train him.

When the student goes on teaching practice, he is likely to find some conflicts between his tutors' and the school's teachers' conception of the teacher's role, a problem which is not infrequently exacerbated by the (albeit veiled) hostility between the tutor and the supervising teachers, each of whom regards the other as holding 'wrong' ('unrealistic', 'reactionary') views on the nature of the teacher's role. Moreover, the student may find that the views he has acquired from his tutors are not easily put into practice because of the pupils as well as the teachers. The student has to face problems which

89

were not considered in the safety of the college seminar and to which he feels he has no answers. And these problems sometimes never allow him the opportunity to try out his new ideas and beliefs. Finlayson & Cohen (1967) have demonstrated the deep differences in role conception between student teacher and headteacher. They report:

> Headteachers are much more strongly in favour of interpreting right and wrong for the child, using punishment, being stricter in their discipline, and insisting on immediate conformity from children than are the students. On the other hand they are more inclined to reject activities in which the children formulate their own rules of behaviour, opportunities for children to learn from their own experiences and for them to discover their personal difficulties with teachers. In all of these items, the head teachers' greater concern for good order and discipline and outward conformity is clearly manifested.

A number of studies have shown that there is a distinct change in the teacher's conception of his role as he moves from being a student, to a student on teaching practice, and then finally to a fully-fledged teacher. Wright & Tuska (1968) have conceived of this as a transition from 'dream', the student's conception of the teacher role before practice, through 'play', the student's conception of the teacher role after teaching practice, to 'life', the full teachers' conception of his role. The same writers have shown that after one year of teaching the woman teacher sees herself in her role very differently from the way she originally thought she would be. She sees herself as less happy, relaxed, perceptive, confident and inspiring but as more blaming, demanding and impulsive than she had imagined at the beginning of training. Similar disillusionments have been reported by others. Rabinowitz & Rosenbaum (1960) tested student teachers during training and after three years' full-time teaching on the M.T.A.I. They report a major change in score (and thus in attitudes), especially in the area of discipline.

> in the three years between testings the teachers became less concerned with pupil freedom and more concerned with establishing a stable orderly classroom, in which academic standards received a prominent position. The change was accompanied by a decline in the tendency to attribute pupil misbehaviour or academic difficulty to the teacher or the school.

Similar findings are reported by Wallberg (1968) Jacobs (1968), Muss (1969) and Hoy (1969). Finlayson & Cohen reported that the peak of radical and liberal educational attitudes is in the second year in the College of Education, followed by a distinct decline in the third year. It seems that the teacher must in the first few years of his career come to terms with the 'harsh realities' of life in school. It is not possible at present to state how much of the changes in attitudes of a teacher after training is attributable to the influence of the attitudes and norms of other teachers (as revealed in the research of Finlayson & Cohen reported above) to which as a member of the school and staff-room he is pressured to conform, or to the practical problems of classroom management which damage the ideals of education which he has acquired in training. But all this research does seem to suggest that there is truth in the charge that tutors in Colleges and Departments of Education are 'unrealistic' either in their views of what teachers should be doing and/or in failure to prepare students for the difficulties of putting their views into practice in school. It seems doubtful whether this situation will be solved until lecturers are more ready to listen to the charges levelled against them and until the supervising teachers in schools are more ready to give a sympathetic ear to new ideas and experiments. This is unlikely to take place until there is much more co-ordination between tutors and supervising teachers and until the tutors are willing to give much more responsibility for training the student to the supervising teachers. Perhaps the best solution would be to make tutorial posts in Colleges and Departments of Education short-term appointments. Without such radical innovations it is difficult to see how a flow of personnel between schools and Colleges, which would be beneficial to both, could be effected.

Some of the student teacher's role strains arise directly within his training institution. Many student teachers see the function of their course as essentially practical; they want to learn to be good classroom practitioners. Thus they like the curriculum and method courses and teaching practice since they directly serve this objective. The staff, however, believe that students should acquire some knowledge of the more theoretical aspects of education and require students to follow courses in the major educational disciplines of history, philosophy, psychology and sociology. The students often feel that these theoretical aspects of the course are irrelevant. The content of these courses is learned reluctantly merely to pass the final examination. It is very doubtful whether we can make a clear distinction between theory and practice in the training of teachers, but there is little doubt that training institutions alienate some students from the more theoretical aspects by making them a compulsory part of the course without making the basic rationale for their inclusion explicit to the students. Conflict seems inevitable unless the staff make the justification both clear and acceptable to the students.

A further strain resides in the student's relationship with his tutor. With respect to teaching practice the tutor tries to enact the role of friendly help and guide to the novice teacher. But at the same time he

91

must evaluate the student's performance by grading it as part of the practical examination. These two functions of helper and evaluator may not be inconsistent to the tutor, but they frequently are to the student. He may, for instance, be reluctant to confide his problems to the tutor for fear that he will give the impression that he is incompetent. When his teaching is being observed by the tutor he will try to give an idealized performance in order to boost his practical teaching grade. The observed lesson will be carefully planned and elaborately staged with a wide variety of fashionable teaching aids. Paradoxically we find that it is the student in the greatest difficulties who is least willing to betray his weaknesses to the tutor and is thus inhibited from gaining the help and advice and support which he may need so desperately at this crucial stage in his career as a teacher.

Recommended reading

B. J. BIDDLE & E. J. THOMAS (eds), *Role Theory: Concepts and Research*, John Wiley, 1966.

N. GROSS, W. S. MASON & A. W. MCEACHERN, *Explorations in Role Analysis*, John Wiley, 1958.

R. RUDDOCK, *Roles and Relationships*, Routledge & Kegan Paul, 1969.

5 Interaction

The concept of role and many of its allied concepts are drawn from the world of the theatre. The language of role theory relies heavily, though not exclusively, on the language of the drama. Of greater importance is the tendency of role theorists to develop a conception of man in terms of a dramaturgical analogy, an analogy which contrasts with psycho-analytical or behaviourist psychologists' conception of man. To speak of a person as an actor performing a role is to see human behaviour as analogous to a character in a play. In Shakespeare's words:

All the world's a stage
And all the men and women merely players;
They have their exits and their entrances;
And one man in his time plays many parts.

This analogy has been extremely useful in providing a set of concepts with which we can describe and analyse the conduct of man, but we must also be prepared to examine the limitation of the analogy and the dangers inherent in this dramaturgical conception of man. For unless we are prepared to admit the shortcomings of role theory we shall be unable to create a more refined set of concepts with which we can describe and analyse human behaviour more adequately. More specifically, we need to show why the concepts of role theory are not in themselves sufficient to provide us with a theory of interaction.

One of the most basic ideas in role theory is that the role is quite independent of a particular actor. The words that Hamlet speaks and the ways in which he speaks them are fairly fixed and survive the various interpretations of the actors who may play the role on stage. Whilst it is true that Sir John Gielgud and Ian McKellen give very different interpretations of the part, what is common to the two

93

versions is greater than the difference between them. Both are clearly recognizable as Hamlet. It can be argued that in the same way a social role, such as that of the teacher, has fairly fixed elements within it and we can recognize the role despite the differing interpretations of Mr Smith and Mr Jones.

Yet when we look at the details of two teachers performing their roles and try to make sense of their behaviour, there seems to be comparatively little that they have in common. This is still true even if we take, say, the role performances of two male teachers of science, with similar qualifications and background, teaching children of the same age in the same school. The differences in role performance are quite as striking as the similarities. But if we contrast the role performances of these teachers with the role performance of a doctor, we are then impressed much more by the similarities of the two teachers. In other words, role theory consists of a valuable set of concepts for distinguishing between role performances associated with *different* position-roles, but these concepts are much less useful for distinguishing different performances of the *same* role.

It is in the nature of roles that they consist of a modicum of prescriptions with which the actor and his principal role partners agree. Yet it is important to recognize that this consensus about how the actor should behave is limited to a relatively small number of aspects of the actor's role performance. There is a high level of agreement about a few things the actor *must* or *must not* do. In many more areas there is a prescription for *preferred* behaviour. Then there is by far the largest area in which there are no prescriptions at all—the actor may do as he pleases because these behaviours are irrelevant to the performance of the role. Thus it is generally agreed that teachers are obliged to have some sort of qualification to teach and are forbidden from propagating extreme political opinions among their pupils. If a teacher is known to deviate from these role requirements, various sanctions will be invoked against him, including dismissal. There is less agreement about whether or not a teacher should have received a professional course of teacher-training, should administer corporal punishment, should give up his spare time to supervise extracurricular activities, or should supervise school meals, though there is probably some consensus about a preferred course of action for teachers in these respects. Teachers may, and do, deviate from the preferred course without much recrimination, because these are preferred 'optional extras' rather than basic role requirements. Finally, there is a wide spectrum of behaviour in which the teacher is totally free to choose his own course of action, for example whether or not he makes jokes in the lessons to amuse the pupils, or marks his books at home or during a free period. This view does rather oversimplify the issue, since many role expectations are related not so much to the behaviour *per se* but to the situation or context in which the behaviour is displayed. For example, it is generally agreed that teachers should not smoke in the classroom in the presence of pupils, but there is less agreement about whether or not the teacher should smoke on a school visit, and there are no expectations about whether or not the teacher should smoke in his own home, even though he may be performing his role, by preparing a lesson for instance.

Further, much of the behaviour that occurs within a role relationship is governed by expectations and norms that are not role-specific at all. Many norms are what may be termed general social norms, expectations about how we should behave in all situations and role relationships. For example, we expect teachers to be punctual and to be honest. But these are not role-specific expectations, for it is expected of most adults that they should be punctual and honest whatever particular occupational role they happen to be performing. Yet a little thought will quickly reveal how much of our behaviour within a role is influenced by these general social norms. They are essential guides to action, considerations we constantly bear in mind in our behaviour and relationships, even though we tend to take them for granted as 'natural'.

A related point is that in real-life role performances, behaviour is influenced by the actor's other roles as well as by general social norms. Roles are separable and distinct only in theoretical analysis. Let us consider a teacher interacting with a headteacher. One is performing the role of teacher, and the other the role of headteacher. Yet it is not as simple as this, for each actor's performance of this particular role is influenced by other position-roles that he occupies, such as age and sex roles. A young male teacher performs his role differently than an old female teacher. Roles spill over and *fuse* into one another, and this nowhere more evident than in interaction. An old female headteacher interacting with a young male teacher is very different from a young male headteacher interacting with an old female teacher.

The general nature of the role expectations, the existence of pervasive general social norms, and the fact of role fusion make it difficult to conceive of role performance as simply a matter of interpretation by an actor parallel to the interpretation of the stage performer. These difficulties are easily reinforced by further consideration of matters we have already discussed. Role expectations do not affect an actor directly. He must perceive them and organize them into something which might be called his conception of the role, which then becomes an important determinant of his role performance. This role conception derives largely from the role partners in the role set. Yet we have seen that a single role partner is likely to conceive some of these expectations in an idiosyncratic way; consensus about the role expectations will be far from perfect among the role partners; the clarity and force with which these expectations are expressed by the role partners will be very variable. Further, the way in which these expectations are *perceived* by the actor is of crucial significance, and this perception will be influenced by many idiosyncratic elements within the actor himself. The analogy between a social role and the Shakespearean text begins to be rather strained.

What we have said so far makes nonsense of a crude sociological view that suggests that human behaviour is essentially role-determined. This view has been superbly characterized by Peter Berger (1963). Society, in this view, provides the script for all the actors and the social play can proceed if all the players will learn and conform to the roles that are assigned to them. The role thus shapes the actor and the action. The actor, unlike the actor on the professional stage of the theatre, actually becomes the part he plays.

> Role theory, when pursued to its logical conclusions, does far
> more than provide us with a convenient shorthand for the des-
> cription of various social activities. It gives us a sociological
> anthropology, that is, a view of man based on his existence in
> society. This view tells us that man plays dramatic parts in the
> grand play of society, and that, speaking sociologically, he *is*
> the masks that he must wear to do so . . . The person is per-
> ceived as a repertoire of roles, each one properly equipped with
> a certain identity. The range of an individual person can be
> measured by the number of roles he is capable of playing. The
> person's biography now appears to us as an uninterrupted
> sequence of stage performances, played to different audiences,
> sometimes involving drastic changes of costume, always
> demanding that the actor *be* what he is playing.

In reality social roles, including the teacher's role, are not Hamlet-like theatrical roles, for none of the words or specific actions are fixed. Social roles require not just an interpretation, according to the actor's personality, of set lines and actions, but rather lay down some very general guide-lines of how the actor is expected to behave. The actor of the social role has, relative to the stage actor, very little to guide his conduct. There is no fixed script, no stage directions, no definite plot, no stage director. There is not enough substance to the social role to give a detailed guide to the actor's behaviour in specific interactional situations. So for the actor of a social role it is not so much a question of interpreting a role as of *improvising* and *constructing* the behaviour we call role performance. The role provides a loose prescriptive framework within which the actor must to a large degree make up the lines and action as he goes along. Thus the role performance becomes unique and idiosyncratic to the actor within a specific interactional context. Performing a role is a highly creative act. As Turner (1962) suggests, we should speak not of role-taking but of role-*making*.

Social behaviour takes place within situations, usually interactional ones. Normally the situation itself imposes constraints which a person must take into account in performing his role. To say the very least the role performance has to be adapted to those situational conditions. As Yablonsky (1953) pointed out:

> Few roles are so explicitly defined for an actor that he can mechanically act out its specifications within a situation. Even when certain roles are most explicit a shift in the structure of the situation has an effect on the actor and requires some measure of spontaneity.

People are not simply actors performing roles. It is true that most of us spend a large proportion of our lives performing various roles, but to see a person as nothing more than a collection and amalgamation of roles is a dangerous oversimplification. Much of our behaviour is indeed *within* various roles, and these roles structure the way we behave. But then to argue from this that our behaviour is role *determined* is an unjustifiable distortion of role theoiy's potential contribution to our understanding of human behaviour. For the truth is that no role can in itself dictate the detailed step-by-step behaviour of the actor. As we have seen, so many of the role expectations are of too general an order to determine the details of behaviour, and those expectations which relate to preferred, not obligatory or forbidden courses of action, require from the actor a creative improvisation within the role framework. Through improvisation the actor brings to his performance his needs, attitudes and personality and much irrelevant role behaviour, so that his actual performance is highly unique.

The limitations of a role analysis are most striking if we consider our relationships with the people we know best. We do not interact with people we know well purely or even largely in terms of roles. We regard them and treat them as persons. With people we do not know as individuals and with whom we have a highly specific and very short relationship (e.g. an encounter between a customer and a sales assistant in a shop) we do interact almost entirely in terms of a role relationship. It is all that we require to satisfy our needs. Indeed, we may actively resist any sort of 'personal' involvement. In this respect it is worth noting that the training for many occupational roles involves strong advice to avoid such personal involvement with clients as inimical to adequate role performance. (Teachers are, of course, particularly prone to such problems.) Yet for the most part we do not regard brief or formal relationships as our most important

97

relationships. In short, I wish to suggest that the more we interact with another person, the less our relationship can be executed or analysed in role terms. As Jones, Davis & Gergen (1961) have shown, it is when we behave out-of-role rather than in-role that we are perceived as individuals.

> The performance of social roles tends to mask information about individual characteristics because the person reveals only that he is responsive to normative requirements . . . The stronger and more unequivocal the role demands, the less information is provided by behaviour appropriate to the role . . . On the other hand, one who rejects or ignores pressures to play a defined role is considered to reflect his true disposition and is perceived with confidence.

Because it is so difficult to sustain a very meaningful or extensive relationship entirely in terms of role, when one person in a long-standing social relationship seeks to insist on behaving in a manner that is principally role-determined or role-conformist, then he is usually so doing in order to avoid the natural development towards a personal encounter with the other that goes far beyond role.

Even when sociologists have managed to avoid a crude deterministic conception of role, they have, with a few notable exceptions, suffered from a tendency to analyse role performance and social behaviour in terms of the formula role-plus-personality. This is epitomized in the famous Getzels & Guba (1957) model (Fig. 5.1).

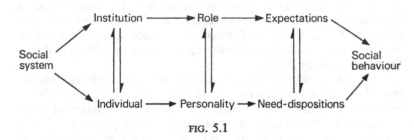

FIG. 5.1

Goffman (1961) has drawn attention to that area which

> falls between role obligations on the one hand and actual role performance on the other. This gap has always caused trouble for sociologists. Often, they try to ignore it. Faced with it, they sometimes despair and turn from their own direction of analysis; they look to the biography of the performer and try to find in his history some particularistic explanation of events, or they rely on psychology, alluding to the fact that in addition to play-

ing the formal themes of his role, the individual always behaves personally and spontaneously, phrasing the standard obligations in a way that has a special psychological fit for him.

To ascribe social behaviour, or even mere role performance, to an interaction between role and personality is either a gross over-simplification or, taking 'role' and 'personality' as synonyms for 'social' and 'individual', merely a statement of the obvious. Attempts have been made to link the concepts of role and personality by such concepts as that of *role style*. This, as we shall see later, offers at best a marginal advance. From a social psychological point of view the main weakness of this approach is that it ignores the interactional nature of role performance and social behaviour. We have learned from G. H. Mead and symbolic interactionism that every actor needs to take account, minute by minute, step by step, of the behaviour of the other as it unfolds by constantly assessing and assigning meaning to the other's behaviour. No theory of interaction can be regarded as adequate unless it builds in this idea.

This is not to say, of course, that role theory is a very poor tool for understanding behaviour. It is one of the most important and funda-mental concepts in the human sciences. It demonstrates how our relationships are deeply embedded in the social structure and helps to bridge the gap between the 'social' and the 'individual'. Yet even where role seems the most obvious concept to use in analysing be-haviour, it is rarely adequate in itself. Let us consider an experiment by Garfinkel (1967), who asked some of his students to behave in their homes as if they were lodgers:

> family members were stupefied. They vigorously sought to make the strange actions intelligible and to restore the situation to normal appearances. Reports were filled with accounts of astonishment, bewilderment, shock, anxiety, embarrassment, and anger, and with charges by various family members that the student was mean, inconsiderate, nasty, or impolite. Family members demanded explanations: What's the matter? What's gotten into you? Did you get fired? Are you sick? What are you being so superior about? Why are you mad? Are you out of your mind or are you just stupid? One student acutely em-barrassed his mother in front of her friend by asking if she minded if he had a snack from the refrigerator. 'Mind if you have a little snack? You've been eating little snacks around here for years without asking me. What's gotten into you?' One mother, infuriated when her daughter spoke to her only when she was spoken to, began to shriek in angry denunciation of the daughter for her disrespect and insubordination and refused to be calmed by the student's sister. A father berated his daughter

for being insufficiently concerned for the welfare of others and for acting like a spoiled child.

Now we can interpret this experiment as demonstrating the fundamental importance of role. Once the students adopt a different role in a particular interaction, then a highly dramatic upset ensues. Thus, it can be argued, the experiment shows how very much of our behaviour is role determined. However another interpretation is possible. It seems to me that the parents may have been so distressed not so much by the *inappropriateness* of their children's behaviour in the lodger role but rather by its *unexpectedness*. In other words, the students not only failed to conform to the *normative expectations* of the son/daughter role but also failed to conform to the *probabilistic expectations* based on how their children had always behaved in the past. One must recall from the previous chapter that in perceiving Other, Person develops a conception of Other which is used as a basis for creating expectations and predictions about Other. It is these expectations, built up through regular and extensive interactions, which are probabilistic rather than normative. In changing roles the students moved from an intimate, informal relationship to a highly formal one and consequently infringed parents' probabilistic expectations. Although it is very difficult in practice to distinguish the two sorts of expectations, it is the breaking of normative expectations which can be used to demonstrate the importance of role as such, independent of other factors.

The normative expectations themselves are not as simple as they appear to be at first sight. We seem to have two sorts of expectations relating to the position. There is a set of expectations about how the actor ought to behave in the *ideal* sense. At the same time we recognize that actors are not perfect so we modify the ideal to what can be reasonably expected of the actor. The expectations are still normative, even though they may fall far short of the ideal. For example, ideally teachers should not lose their tempers with their pupils. Most teachers and most members of the teacher's role set share this ideal. But they also recognize that at times the teacher will lose his temper in the classroom. The ideal becomes scaled down to the expectation that under certain circumstances the teacher will lose his temper. The normative aspect is still strong, for it is only in particular circumstances that a loss of temper will be permitted. Further, the teacher will be expected to express his anger in certain ways, but not in others. This distinction is particularly important in the measurement of role expectations, since if we simply ask, 'Do you expect the teacher to lose his temper?', there are potentially two answers, one referring to an abstract ideal and another to what can reasonably be expected in practice. (Some role partners may, of course, refuse to modify the ideal, which can produce an interesting but unexplored type of role conflict). Finally there is the third answer, tapping probabilistic expectations of a particular actor. Mr Smith is known, on the basis of past behaviour, to lose his temper frequently, whereas Mr Jones never does so. Many role studies fail to make it clear which of the three types of expectation is under scrutiny.

In short, although the concept of role seems to be essential to the analysis of human interaction, especially the early stages of interaction, it is not in itself adequate to complete the whole job. I have already hinted at some of the concepts we shall need. Now we must look much more closely at the detailed dynamics of social interaction, something which classical role theory never did. In doing so, we shall be casting many doubts on the value of the concept of role performance for it will become evident that relatively little of human behaviour in interaction can be with justification so termed. From the social psychological view it is the most dangerous term in role theory, because whilst it is important to recognize that a very large part of our behaviour is deeply influenced by those expectations which make up roles it is equally important to recognize that very little of our behaviour can be adequately described or explained simply in terms of fulfilling role expectations. Role performance is often used in ways which suggest that it can be directly observed, but this rarely is the case. In real life the performance of one role cannot be easily disentangled from the performance of other roles with which the first role is fused, nor is it easy to specify which particular actions are determined by the role and which actions have their origin in other non-role forces. Role performance is not the same thing as social behaviour. It can be used legitimately to describe behaviour which is in-role but without the deeper implications implicit in the usage of this term by many social scientists.

Interaction is a dynamic concept. An interaction involves at least two people each of whose behaviour is orientated towards the other. If we think of two people in interaction, we imagine two people in conversation in the same place and at the same point in time. This is the most common, but not the only form, of interaction. The participants can interact from widely different locations, as by telephone, and even from different points in time, as in written correspondence. Interaction can be non-verbal, as in letter-writing, playing a game of chess, or in making love. How then can we define interaction? Asch (1952) has described interaction well in the following way:

> The paramount fact about social interaction is that the participants stand on common ground, that they turn *toward one another*, that their acts interpenetrate and therefore regulate each other . . . In full interaction each participant refers his action

to the other and the other's action to himself. When two people, A and B, work jointly or converse, each includes in his view, simultaneously and in their relation, the following facts: (1) A perceives the surroundings, which include B and himself; (2) A perceives that B is also oriented to the surroundings, that B includes himself and A in the surroundings; (3) A acts toward B and notes that B is responding to his action; (4) A notes that B in responding to him sets up the expectation that A will grasp the response as an action of B directed toward A. The same ordering must exist in B.

This can be summarized under *the principle of reciprocally contingent communication*. To express this more simply, in every form of interaction the participants must communicate, that is transmit symbols with a shared meaning, in such a way that the behaviour of each is in part a response to the behaviour of the other. Interaction is a process of reciprocal influence and mutual dependence.

This can be clarified by examining different types of interaction (cf. Jones & Thibaut, 1958). Some interactions may be termed *pseudo-contingent*: the contingency principle appears to be in operation, but is not really so. The interaction appears to be contingent, because each actor synchronizes his speech so that it does not overlap with the speech of the other. When we go to the theatre to watch a play, the actors time their lines in such a way that they appear to be interacting, though the lines have been pre-determined. To create the illusion of reality, each actor must wait for his cue, then speak his lines as if he had just had the thoughts which lead him to speak. If the actor does not look as if he is responding to the other actors, if he 'fluffs' his lines so that it is no longer credible that he is reacting to the others, or if he misses his cue and the temporal synchronization goes awry, then the illusion of the play breaks down. A similar pseudo-contingent interaction occurs in many religious rituals. The priest chants, 'O Lord open thou our lips', and the congregation responds, 'And our mouth shall show forth thy praise.' Now the congregation has not given a spontaneous response to the priest. They have no choice in the matter but to give the fixed reply. If they decided to respond with 'O Lord make haste to help us', then the service would soon be reduced to chaos. In such an interaction the only problem for the participants is to know the lines and make the right response at the right time.

Another form of interaction can be described as *asymmetrically contingent*. This arises where the behaviour of one participant is fully contingent on the other, but the behaviour of the second participant is only partially contingent on the other. A good example is the interview situation. The interviewee has to respond to the (unknown) questions of the interviewer, and what he says is very much dependent on the questions. The interviewer's questions, however, do not always depend on the interviewee's replies, though they may do so in a 'follow-up' question. The interviewer is likely to have a predetermined schedule of questions, and when he has received a satisfactory reply to one question, he will non-contingently turn to the next question. Teachers often behave in asymmetrically contingent ways, as when they are asking questions of children in the classroom. All the pupils' answers will be related to the teacher's questions, but not all the teacher's questions will be related to the children's answers.

In truly reciprocally contingent interaction, each participant reacts to the other, and the behaviour of each is in part determined by the other. It is not more than *in part* a function of the behaviour of the other, because each participant has to create a response. The actual content of the response and the manner in which the response is made are both functionally related to a number of factors additional to the immediately preceding behaviour of the other. One of the most important influences on content and manner will be the expectations deriving from the roles of the two participants. In a teacher–pupil interaction, for example, the response of the pupil to a teacher is influenced not only by the behaviour of the teacher but also by the norms governing the appropriate behaviour, both content and manner, on the part of the pupil. Thus in interaction we have to make some sort of compromise between the response we might wish to make, our own personality, needs and goals, and the limitations on our creative response which accrue from social norms, from the role partner's expectations and from the nature of his preceding behaviour.

The definition of the situation

Interactional situations vary along a continuum of the degree to which there is a structure inherent in the interaction. This structure concerns the mutual orientation of the participants and their awareness of how they should behave in the situation. When the structure is high, the participants are in no doubt about the roles each is going to perform; how the actor is going to perform his role, and how his role partner is going to perform his role; the goals of each participant in the interaction; the contribution and timing of each. Examples of high structure interactions would be a bus conductor collecting fares from the passengers, or priest and laymen saying Mass together. In both these illustrations, at least in normal circumstances, the roles of the participants are very clearly defined and there is an awareness by both participants about who should do what, when and for what reason.

When the structure of an interaction is low, we find the reverse situation. There is considerable uncertainty, confusion or ambiguity about the roles the participants are expected to play, the goals of the actors, the contribution each is going to make, and the way in which these contributions will be meshed together. The most common instance of interactions with low structure is the meeting of two strangers in a situation where few clues to a potential structure are available, such as on a beach or in a pub. Another common example is a situation where one actor is unaware of the structure but the other actor is, such as the person who becomes a patient of a psychiatrist for the first time.

When the structure is low, either because it is lacking or because it is not explicit, the interaction will have two outstanding features. The first is a certain psychological distress for each actor, marked by such thoughts as 'Who is this?' 'What does he want?' 'What am I supposed to do now?' 'God, how can I get away?' The second feature is a tendency to what might be called the interactional jerks, namely awkward silences or both participants speaking at the same time. This stage will in most cases be temporary, since the dissatisfactions experienced by the participants exert a pressure towards resolution. Typically, a 'probing session' in which each actor tries to clarify the situation and to discover the roles and purposes of the other ensues. If little progress is made, or if results of the probing session are distasteful to one or both of the participants, withdrawal will occur and the participants will leave each other's company or turn their attention elsewhere (a book, another person) in order to terminate the interaction.

In most teacher–pupil interactions, the structure is high rather than low. When a new teacher takes a class for the first time, or when an experienced teacher meets the new intake of a school for the first time, then the structure will be much weaker. Even in such a situation there will be some structure, for in most cases the teacher will have some previous experience of children in school and the children will have some experience of teachers. The most extreme case of low structure, then, would be that of a quite inexperienced teacher with a class of children who were coming to school for the first time. We shall be concerned with more typical teacher–pupil situations and interactions, but we shall have to consider the process of structural development if we are to have an adequate understanding of teacher–pupil interactions as they occur in an ordinary classroom.

We shall investigate and clarify this structure of an interaction by using the famous phrase of W. I. Thomas (1928, 1931) the *definition of the situation*. Thomas used this phrase or concept to examine the impact of definitions and meanings on the structure of human action. What sociologists call culture can be seen as a set of collective definitions. Thomas emphasized that these collective definitions are not fully shared.

> There is . . . always a rivalry between the spontaneous definitions of the situation made by the member of an organized society and the definitions which society has provided for him . . . There are rival definitions of the situation, and none of them is binding.

In spite of such contradictions that are evident within cultures, social interactions can proceed only when to a large degree the participants have a common definition of the situation. As Asch puts it:

> A conversation can proceed only when (a) the same (or a similar) context is present in the participants, *and* (b) when the context possesses for each the property of being also the context for the other. This reference on the part of each participant to the other is the condition of psychosocial events; action in the social field is steered by phenomenal fields which are structurally similar in these respects. Only individuals who encompass their common situation in this way can produce social-psychological acts . . . It is individuals with this particular capacity to turn toward one another who in concrete action validate and consolidate in each a *mutually shared field*, one that includes both the surroundings and one another's psychological properties as the objective sphere of action. This relation between psychological events in each of the participants makes possible the sharing of actions, feelings, ideas and mutual acknowledgement. We cannot take a step in practice without presupposing this understandability; we cannot take a step in theory before clarifying it.

Alfred Schutz, in defining interaction in terms of a 'reciprocal Thou-orientation' makes a very similar argument to those of Thomas and Asch, but he speaks of a *communicative common environment* rather than a 'mutually shared field'. Schutz points out that the world of one's daily life is by no means a private world of one's own making, but rather an intersubjective world that is shared with one's fellow men. The sociocultural world in which one exists is also the social environment of others, shared, experienced and interpreted by them. A man's fellow men are part of his environment as he is of theirs. A man acts upon others and others act upon him. This mutual relationship implies that all experience the world that is common to all and experience it in a way that is substantially similar. The mutuality of the definition of the situation springs in part from social norms, mores, customs and all those aspects of the cultural pattern which provides by its recipes typical solutions for typical problems available to typical actors. These recipes serve both as a common scheme of action, indicating to an actor how he is to proceed to attain a goal, and as a common scheme of interpretation by which we can make sense of action.

In the chapter on role we considered the reciprocity of rights and duties in a role relationship, the rights of the actor being the duties of the role-partner and vice versa. Later in this chapter we shall consider another form of reciprocity when we deal with exchange theory. Schutz, noting that the participants in the common communicative environment are given to one another not as *objects* but as *counter-subjects*, is led to the concept of the *reciprocity of perspectives* in interaction. Both Person and Other must assume that each is taking broadly the same things for granted. Each participant assumes an interchangeability of standpoints, by which things would be broadly the same if each participant were to change places with the other.

Both Asch and Schutz, recognizing the unique individuality of each participant, stress that the participant's definitions of the situation must be similar or congruent rather than absolutely identical. At any point in time, argues Schutz, a man finds himself in a 'biographically determined situation' with the consequence that no two persons can ever view the situation in quite the same way. The unique biography gives a person a unique 'stock of knowledge at hand' which will serve as a scheme of interpretation for past and present experience and influence his goals and projects for the future.

The definition of the situation with respect to an interaction has shared elements that are common to both participants; these spring largely from various cultural recipes, including social norms and social roles. Yet each actor is unique, with his own biography, schemes of interpretation and purposes which he brings to bear in the interaction. Some elements of the definition of the situation are *given*, and come to constitute the agreed and taken for granted overall definition of the situation. Other elements are unique to the individual participant, who has his own version of the definition of the situation, over-lapping with but not identical with his partner's definition of the situation. Over time these individual definitions of the situation need, especially if they are not highly congruent, to be *negotiated* and agreed if the interaction is to proceed to the satisfaction of the participants. In many of our habitual interactions the situation has already been progressively negotiated in past interactions. We take the over-all definition of the situation for granted because it does not have to be negotiated *de novo* every time. But this should not blind us to the process of progressive negotiation and modification that has taken place. In order to facilitate the analysis of the process whereby the definition of the situation is established we shall consider the major segments separately. Later we shall draw all the segments together into a more coherent and dynamic model of interaction.

Typically each participant enters the interactions with a relatively clear set of goals. Rather than entering the interaction without purposes or objectives, he has some definite notions of what he wants to get from the interaction. Obviously the range and variety of these goals is endless. For example, these goals can vary in their importance to Person. The goals may be very unimportant—passing a few spare minutes in congenial company—or they may be of very fundamental significance for Person's self-image—obtaining a job to which he has been aspiring for many years. Secondly, each participant has a relatively clear conception of the role he expects to perform in the interaction. Again, this role will vary in its importance for Person. This role conception is, as we have seen, derived in complex ways from Person's perceptions of the expectations of his role-partners in the role-set and from his own needs, personality and background. It is Person's goals and roles—and the two are usually inter-connected—which form the major contribution to Person's conception of how the situation should be defined. Sometimes Person may enter an interaction without any clear goals and roles. In this case the roles and goals will be progressively formulated in the early phases of the interaction. In the present analysis we shall be more concerned with the more typical case where each participant has a relatively clear conception of his goals and roles, though it is useful to recognize at this point that each participant is likely to modify his goals and roles in the light of subsequent interaction.

On entering the interaction, and sometimes, prior to it, each participant has the important task of *deriving information about Other*. This involves trying to find out as much as possible about Other that is likely to be relevant to the interaction. Among the most important questions that Person asks himself, though not always explicitly or consciously, are:

What is Other's role?
How does Other perceive his role?
What are Other's goals?
What does Other expect to gain from the interactions?
What are Other's intentions?
How is Other going to behave towards me?

In other words, Person comes to the interaction with his own goals and roles and has the immediate task of assessing Other's roles and goals. This raises once more the problems of interpersonal perception that we have already discussed at length. Each participant will select information from Other's behaviour and make inferences and attributions about Other's roles, goals, intentions, motives, personality

107

and so on. Each participant must do this in relation to the other participant. Each will conclude with a conception of Other which will influence his behaviour towards Other.

In the chapter on person perception we were in the main concerned with Person's perception of Other. It was essentially a one-sided view for we were concerned with what happened in Person and ended up with a model of Person's perceptual processes. When we consider social interaction we are forced to come to terms with the facts of *inter*personal perception. Other himself is also a perceiver of Person. Person's behaviour towards Other is influenced by Person's conception of Other; Other's behaviour towards Person is influenced by Other's conception of Person. It becomes more complex when we realize that Person also has a conception of Other's conception of Person and that Other has a conception of Person's conception of Other. Whenever we get to know someone quite well, we become fascinated by this person's conception of us. When we say 'What do you really think of me?', we are asking for information which will allow us to develop or clarify our conception of the other person's conception of us. In short, Person's behaviour towards Other is influenced not only by Person's conception of Other but also by Person's conception of Other's conception of him. So Person asks himself two sets of questions. The first set, by which he creates his conception of Other, we have already dealt with. The second set, by which he creates his conception of Other's conception of him, includes such questions as:

What does Other think my role is?
What does Other think my goals are?
How does Other think I am going to behave towards him?
What motives and intentions is Other going to attribute to me?

One of the most important writers in this area is R. D. Laing (1966, 1967). Laing deals with three elements, the direct perspective (my view of me), the metaperspective (what I think you think of me) and the metametaperspective (what I think you think I think of me). He shows that the third level of the metametaperspective can be very important in human interaction.

Let us consider one of Laing's illustrations.

Direct perspective

John does not love Mary	Mary does not love John.

Metaperspective

John thinks Mary loves him	Mary thinks John loves her.

Metametaperspective

John thinks that Mary thinks he loves her	Mary thinks that John thinks she loves him.

Although neither is in love with the other they might get married because neither is willing to tell the truth and thus clarify the misunderstanding because neither wants to hurt the other. It is easy to set up a hypothetical classroom equivalent.

Direct perspective

The teacher likes the pupil	The pupil likes the teacher.

Metaperspective

The teacher thinks the pupil dislikes him.	The pupil thinks the teacher dislikes him.

Both teacher and pupil are now likely to interpret the other's behaviour in the light of their false beliefs about the other's feelings. The teacher decides to tease the pupil gently, because this is a natural mark of favour towards a pupil he likes and because he hopes it will persuade the pupil that he really does like him. The pupil, however, interprets the teasing in the light of his false belief that the teacher dislikes him and thus regards the teasing as 'making fun of him'. So he reacts with hostility and his belief that the teacher dislikes him seems to be confirmed. The teacher interprets the pupil's hostility as a confirmation of his belief that the child dislikes him. We thus find that even though teacher and pupil actually like one another, one person's attempted expression of the liking goes awry and reinforces the false metaperspectives.

We can now introduce the third level.

Metametaperspective

The teacher thinks the pupil thinks the teacher likes him	The pupil thinks the teacher thinks the pupil likes him

Here each has got the other's metaperspective wrong. Because the teacher likes the pupil and thinks that the pupil knows this, the hostile reaction of the pupil strengthens the teacher's belief that the pupil must dislike him. Similarly, because the pupil likes the teacher and believes the teacher knows this, the teacher's behaviour is interpreted as a positive attack on him. This metametaperspective has the effect of adding insult to injury and giving further confirmation to the false belief that he is disliked by the other. There is the alternative third level.

Metametaperspective 2

The teacher thinks the pupil thinks the teacher dislikes him.	The pupil thinks the teacher thinks the pupil dislikes him.

Here the teacher recognizes that the pupil's metaperspective is false and so he may be led to realize that the pupil's hostile reaction is the result of a misinterpretation based on the pupil's belief that the

109

teacher dislikes him. He can thus take corrective action. With this metametaperspective the pupil may still react badly since he still believes the teacher dislikes him. But he may also realize that his hostility may be interpreted by the teacher as confirmation of the teacher's belief that the pupil dislikes him, which the pupil knows is not true. This second form of the metametaperspective is important for although it is not certain that it will improve relations between teacher and pupil it certainly will not make matters worse, as was the case with the first form of the metametaperspective.

These Laingian spirals are obviously very complicated, though Laing himself has shown that they can be systematized into a relatively small number of types. They demonstrate in a very convincing way that we cannot always make sense of human interactions simply by considering the direct perspectives of the participants. All these ideas are of great potential interest when applied to interactions in educational contexts and we shall return to them later. For the present we must return to our central task of developing a model of interaction.

In addition to deriving information about Other, Person has, on entering his interaction with Other, to face another important question: what sort of person do I want Other to think I am? Since Other has the task of deriving information about Person, Person is in a position to control to some degree the release of information which Other is seeking. Person may be able to influence Other's conception of him and Other's conception of Person's conception of Other by giving away information in his own behaviour towards Other. This is the task of *self-presentation*. Thus deriving information about Other and self-presentation are opposite sides of the same coin. Each derives information about the other through the other's self-preservation.

The writer who has contributed most to this field is the American Erving Goffman, especially in his fascinating book, *The Presentation of Self in Everyday Life* (1959). In our discussion we shall draw heavily upon the concepts and ideas of Goffman. In presenting himself each actor has to behave, intentionally or unintentionally, in such a way that he *expresses* himself by words, gestures, facial manipulations and so on and that the other receives an *impression* of him. This recalls our work on person perception. My expressions are the raw data on which, through selection, inference and attribution, you build up an impression of me. This means that I shall have to control my expressions in order to make sure you get the right rather than the wrong impression of me. For you to have the right impression usually means that you have interpreted my expressions accurately, that is, you have received an impression which is very close to the impression I wanted you to form. A wrong impression arises when you make inferences and attributions which are an incorrect reflection of my (actual or projected) roles, goals and expectations. In this case, you have 'misunderstood' me and I will attempt to alter your impression if I become aware of the impression I have made. Sometimes, of course, I may try to express myself in a way that is intended to mislead you. A pupil, for example, by keeping his eyes riveted on the teacher who is explaining some problem to the class may wish to give the impression that he is paying attention, when in reality he is thinking about his girl-friend. If the teacher suspects that he has the wrong impression, he may ask the pupil a question as a check on whether or not the pupil's expression is a genuine representation of his mental state. If it becomes clear that the pupil was not paying attention, then the pupil's misleading expression is 'seen through'.

Goffman demonstrates that the techniques required in self-presentation are extremely elaborate. There is a subtle art in managing our expressions to create impressions on others. Many of these techniques are so common-place that they are regarded as 'natural', though in fact considerable learning has taken place. One of the most important clarifications made by Goffman is the distinction between an expression that is *given* and one that is *given off*. The first concerns mainly verbal communication to convey information. The second includes a wide range of cues, often not easily controlled, that may or may not be in support of the expression given. For example, the pupil who is lying to the teacher may blush, and the teacher may regard the blush as an indication that the truth is not being told. Similarly a teacher who is having a 'mock rage' in order to bring the class to order may reveal his deception by the twinkle in his eye. So a participant in an interaction must make sure that the expression given off is consonant with the expression given if he is to make the right impression on the other.

A related distinction made by Schutz (1932) is that between the 'expressive act' and the 'expressive movement'. In an expressive act an actor intentionally communicates or seeks to communicate his subjective experience, whereas in the expressive movement there is no communicative intent on the part of the actor, even though it may be taken by Other as an indication of the actor's subjective experience. Other may have difficulty in distinguishing an expressive movement from an expressive act, especially when an actor seeks to control his self-presentation by turning what would normally be an expressive movement into an expressive act yet at the same time seeking to leave Other with the impression that it is an expressive movement not an expressive act.

Goffman calls the expressive equipment employed by the individual during his performance 'front'. This is the part of the individual's

111

behaviour which serves to define the situation for the other. In creating the appropriate front the actor may use props and scenery in addition to insignia, clothing, speech, posture and facial expressions. The use of the academic gown by teachers in English grammar schools is a good example of a prop which indicates that the wearer is both a member of the teaching staff and a superior, qualified expert in a particular discipline. It is often possible to detect the headmaster from among a group of teachers by his special front— a dark, formal suit, an impeccable appearance, a haughty, self-confident manner. In creating a front an actor helps to define the situation by giving the other information about his role and goals and the way he intends to behave during the interaction.

Frequently actors try to give others the impression that their performance is *idealized*, that their behaviour is an exemplification of what is expected by others. Thus on Open Days in schools the premises are made especially clean and tidy and both staff and pupils are better dressed than usual. Teachers may, in the presence of parents, refer to pupils by their first names rather than by their surnames as is customary in the classroom. I know one headmaster who, in his 'briefing' before Speech Day, gave detailed advice on the idealization of performance in the presence of parents and visitors. 'It doesn't matter if you are bored by speeches', he would say, 'but try to *look* as if you are interested'. During a general inspection of a school by Her Majesty's Inspectors, the teachers may use an unprecedented number of audio-visual aids and unheard of amounts of coloured chalk. In all these instances the actors are presenting themselves as closer to the ideal than is really the case. Practising teachers may try to idealize their performances in the presence of student teachers. As one of my own students reports:

> A student teacher watching a teacher in action is often given
> an idealized performance. Several members of staff in the schools
> where I spent my teaching practices admitted that they had
> prepared careful lesson notes for the first time in years for the
> lesson I was to observe.

Often we try to avoid being given an idealized performance. The same student continues:

> Many students, including myself, found the strain of expecting
> a tutor's arrival to observe a lesson was considerable. Although
> we were assured that it was only because of the pressing time-
> table arrangements that we could not be forewarned of a visit,
> we felt that it was because the tutors themselves were afraid
> that otherwise we would be able to prepare and give an ela-
> borately staged and idealized performance.

Inherent in an idealized performance is a degree of misrepresentation. The extreme case is the impersonator or the confidence trickster, who must be extremely skilled in masking their misrepresentation. However, misrepresentation is by no means confined to the criminal, because most of us engage in activities that are incompatible with our front. The teacher will behave as if he is scandalized by the pupil who is found swearing, or reading books of nudes, or playing at poker—but he may do all these things himself in the staffroom. Indeed, one of the reasons why in many schools pupils are not allowed to enter the staffroom is that this place acts as a 'backstage' where the front can be dropped. If the pupils were able to observe teachers with a dropped front, then the performance would be discredited.

In order to sustain a front or performance, especially an idealized one, the actor must ensure that in his self-presentation no action or cue is expressed that might discredit the performance. If we wish to give a boring friend the impression that we are interested in his stories, we must stifle our yawns. Similarly, the teacher who wishes to maintain his front as an 'expert' may be led to cover up behaviour which is discrepant with this front. The teacher who makes an error which is recognized by the pupils may try to retain his front by saying, 'I'm glad you spotted the deliberate mistake. I put that in specially to see if you were all wide awake this morning.' In face of an awkward question, the expert front can be maintained by such diversionary tactics as, 'That's a good point. We'll discuss it next week', or 'Now why don't you look up the meaning of that word in the dictionary?' or 'Can we have questions at the end of the lesson?'

In our discussion so far it has been implied that problems of self-presentation are a matter for the individual actor. This is not so, since actors are often members of teams who have a common front. The teacher is part of a team which by mutual agreement, co-operation and collusion fosters and sustains a front against the pupils. Thus teachers in conversation in the presence of pupils will often refer to each other by formal surnames. A teacher enters another teacher's classroom and smiling asks, 'Can I have a word with you, Miss Smith?' and in the privacy of the corridor demands, 'What the hell have you done with my register?' Most school staff frown upon a teacher who talks to the children about other teachers. To do so is 'unprofessional'; it threatens the image as well as the unity of the team front. For a similar reason many schools will not allow student-teachers to wear mini-skirts on teaching practice.

Within the team there is considerable trust between members. The front can be openly discussed—'I just don't seem to be able to look angry when I really want to laugh at the pupils' misdemeanours'—and confidences can be exchanged without fear of the content being reported back to the other team. Teachers and pupils are alike in their intense dislike of team members who 'tell tales' or give 'leaks' to members of the other team. Sometimes teams which are typically in opposition may combine against a third team. On Speech Day the teachers and pupils may co-operate in presenting an idealized performance for parents and visitors. To refuse to join in this co-operative venture is regarded as 'letting the *school* down'.

In creating an idealized performance, there is a dishonesty of intent. Most of us are ready to acknowledge that we frequently use a variety of techniques to misrepresent ourselves to others. But we also feel that in most of our performances our self-presentation is 'genuine'. We are, we might say, 'being our true selves'. Yet the art of self-presentation does *not* refer merely to the skills that are necessary to the creation of a contrived, essentially false performance. It refers just as much to being 'true' as to being 'false'. In Goffman's words:

> Socialization may not so much involve a learning of the many
> specific details of a single concrete part—often there could
> not be enough time and energy for this. What does seem to be
> required of the individual is that he learn enough pieces of
> expression to be able to 'fill in' and manage, more or less, any
> part that he is likely to be given. The legitimate performances
> of everyday life are not 'acted' or 'put on' in the sense that
> the performer knows in advance just what he is going to do,
> and does this solely because of the effect it is likely to have . . .
> But as in the case of less legitimate performers, the incapacity
> of the ordinary individual to formulate in advance the move-
> ments of his eyes and body does not mean that he will not
> express himself through these devices in a way that is dramatized
> and performed in his repertoire of actions. In short, we all act
> better than we know how.

We have to learn, in other words, to be sincere, because sincerity—like insincerity—has to be expressed in such a way that the other is appropriately impressed. A person who pretends to be sincere tries to use precisely the same self-presentation techniques as the person who is really being sincere. The very fact that one can pretend to be something one is not demonstrates that certain self-presentation techniques are essential to being genuine. Simply because much of the material of the writing on self-presentation is drawn from contrived performances should not mislead us into regarding self-presentation as the art of pretence. If such were the case, it would be of only peripheral relevance to the study of interaction. In fact self-presentation is at the very roots of an understanding of interaction.

The point becomes clear when we remember that the professional stage actor has the difficult task of acquiring mastery of those self-presentation techniques which, whilst forming a natural and effortless part of his everyday life, need to be brought under conscious control if a stage character is to be portrayed convincingly. As Stanislavski (1926) puts it:

the very best that can happen is to have the actor completely carried away by the play. Then regardless of his own will he lives the part, not noticing *how* he feels, not thinking about *what* he does, and it all moves of its own accord, subconsciously and intuitively . . .

In ordinary life you walk and sit and talk and look, but on the stage you lose all these faculties. You feel the closeness of the public and you say to yourself, 'Why are they looking at me?' And you have to be taught all over again how to do all these things in public.

Inspiration . . . needs no technique; it operates strictly according to the laws of our art because they were laid down by Nature herself . . .

'Every bit of that was artistically true,' exclaimed the Director with feeling. 'You could believe in it all because it was based on carefully selected elements taken from real life. She took nothing wholesale. She took just what was necessary. No more, no less.'

Goffman's writings are rich in ideas on self-presentation. One review of *The Presentation of Self in Everyday Life* described it as 'one of the most trenchant contributions to social psychology in this generation'. This is not an unjustified claim. It is a book to be read. The concept of self-presentation is fundamental to a study of interaction, because it extends the dramaturgical perspective that is implicit, but undeveloped, in 'classical' role theory. Moreover, it allows role theory to become dynamic and thus more relevant to the study of interaction.

The further aspect of defining the situation concerns *situational proprieties*. This is a term coined by Goffman in another of his books, *Behaviour in Public Places* (1963b). One of the difficulties with role theory from an interactional point of view is that it does not sufficiently allow for situational factors in an interaction. The concept of situational proprieties is a useful element for considering the impact of situational variables upon interaction. The basic idea can be best expressed in Goffman's words, that the rule of behaviour which is common to all situations is the one which obliges the participants to 'fit in'. They must not, in other words, make a scene or cause a disturbance or be out of place. They must not be too withdrawn from what is going on, but neither must they thrust themselves on the other people who are present. Many situational proprieties are really norms or rules which govern the participants' involvement in the interaction. *Autoinvolvements*, the tendency to be too much involved with oneself rather than with the other, are improper. Picking one's nose, cleaning one's fingernails are autoinvolvements that imply too little attention to and respect for the other is being paid. To be 'away' in a daydream or to doodle are similarly frowned upon. Goffman suggests that in interaction both the participants must show *civil inattention* to the other. By this term he means that we must not stare with constant eye-contact at the other. This would be 'rude'. At the same time, it would be equally impolite to make no eye-contact at all with the other. As Goffman puts it:

> one gives to another enough visual notice to demonstrate that one appreciates that the other is present . . . while at the next moment withdrawing one's attention from him so as to express that he does not constitute a target of special curiosity or design.

In the last decade some experimental social psychologists, notably Exline and Argyle, have undertaken research on eye contact in interaction. We know that in conversation most people look at one another for between 30 and 60 per cent of the time; that there is more eye contact while listening than while speaking; that women engage in more eye contact than men. Argyle (1967) has summarized the main functions of eye contact as a social technique. It may be used as a device to try to initiate interaction with another. Teachers, like waiters, often prevent their clients from 'catching their eye' since to do so would be to create an obligation to respond to the other's desire to interact. A speaker tends to look at the other at the end of his speech. This can be a signal to the other that the speaker has finished, for if he is merely pausing in mid-flight he tends not to look at the other. Eye contact is used as a signal for saying 'Your turn now.' The speaker may also look at the other at the end of an utterance in order to get feedback on the other's reaction, to check whether or not the other is still listening, or has understood or agreed. Eye contact can also be used to convey emotion, especially the feelings of friendliness or sexual interest.

Goffman also draws our attention to many of the proprieties of everyday life, such as those of leave-taking. In interactions between superiors and subordinates the superior may give many cues that the interview is over, by standing up and moving towards the door or by saying directly, 'Thank you for coming in to see me.' In interactions, especially informal ones, between equals the matter may prove to be more difficult as every host, weary of his guests at a late hour,

116

knows to his cost. The host's hints that he would like to go to bed may have to be made very overt before the guests realize they are no longer welcome. Many of these proprieties have to be learned. For instance, a teacher who has called a young pupil to the desk to examine an exercise book may, at the end of the examination, turn the pupil round and propel him back to his seat to terminate the interview.

Another contributor to this area, whose writing is as startling as Goffman's, is Eric Berne (1966) with his theory of games. Berne initially considers some universal rituals, such as the rituals of greetings between acquaintances, where little information is exchanged or business transacted. These rituals spring essentially from socially approved codes of politeness. Games, in contrast to rituals or pastimes, are essentially serious, consequential and dishonest.

A common party game is called Rapo. The woman signals that she is available and gets her pleasure from the mild flirtation and the man's pursuit. But once the man commits himself, the game is over. She may thank her suitor for his compliments and then move on to her next conquest, or if she is less polite he may be rebuffed more sharply. As Berne says:

> A skillful player can make this game last for a long time at a social gathering by moving around frequently, so that the man has to carry out complicated manoeuvers in order to follow her without being too obvious.

The marital game Frigid Woman is played as follows. The husband makes advances to his wife and is repulsed on the grounds that he is only interested in sex and does not really love her. The husband makes no further advances but the wife becomes increasingly provocative in her behaviour towards him. When the husband responds to these provocations his first tentative moves are accepted but at the critical point the wife resorts to her original move by claiming that all men are beasts and the game can begin again.

Consensus and modification

Each participant in an interaction is trying to establish a definition of the situation. If then the interaction is to proceed smoothly, there must be some agreement between the participants in the definitions of the situation they are trying to project. Such consensus about the definition of the situation involves a recognition and acceptance by each of the roles and goals of the other, an agreement about how each will treat the other, and the formulation of rules that will regulate conduct. For the consensus to be high, the definitions of situation projected and intended by each participant must be similar or compatible. In Goffman's (1959) view, absolute consensus is rare.

> Together the participants contribute to a single over-all definition of the situation which involves not so much a real agreement as to what exists, but rather a real agreement as to whose claims concerning what issues will be temporarily honoured. Real agreement will also exist concerning the desirability of avoiding an open conflict of definitions of the situation.

Typically, then, the consensus is a *working* consensus. Where the consensus about the definition of the situation is low, the interaction is subject to conflict and strain, which can be removed by the withdrawal of one or both participants in order to terminate the interaction, or by modification and compromise. In everyday life most interactions involve some degree of modification of the definition of the situation by one or both participants.

In reaching a consensus about the definition of the situation, each participant is constrained towards compromise by the demands of the other. Each can perform his role in the way he wishes, fulfil his own goals, and satisfy his needs only in so far as the other will allow him to do so. Unless Person has great power over Other, Person cannot force Other to accept his definition without demur. Both participants must be satisfied by the interaction; the satisfaction of each depends on the satisfaction of the other. If Person behaves in a way which conforms to Other's expectations and allows Other to realize his own goals, then Other will be gratified and disposed to reward Person in a variety of ways. If Person behaves in a way which conflicts with Other's expectations and inhibits Other's realization of his goals, then Other will be angry and disposed to withhold rewards and support from Person. There are very good reasons, then, why Person should try to please Other, for pleasing Other is instrumental to the attainment of his own goals. The problem is that in pleasing Other, Person may be led to behave in ways which prevent him from attaining his own goals. Thus it is that both Person and Other must reach some consensus about the definition of the situation, which is a compromise between 'having it all my way' and 'having it all your way'. The consensus is approached in terms of maximizing one's own satisfactions; the closer one can get to 'having it all my way' the better. Neither participant is usually prepared to settle for less. Thus the consensus concerning the definition of the situation is *negotiated* by the participants, and an obvious aspect of this is the negotiation of roles. In other words, often roles are not simply a *given* aspect of the situation but have to be agreed upon by the participants. It is difficult for me to perform the role of leader (benefactor, rebellious adolescent son etc.) unless you are willing to perform the role of follower (beneficiary, unbendingly authoritarian father etc.).

As yet there has been relatively little systematic analysis by social psychologists of the negotiation skills and tactics of social interaction. One common technique has been called altercasting (Weinstein & Deutschberger, 1963). In altercasting, Person not only clarifies his own role but casts Other in a role that is congruent with Person's

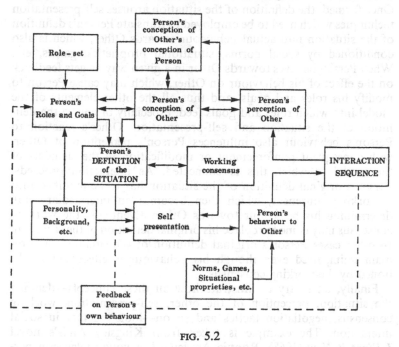

FIG. 5.2

goals. Person seizes the initiative in defining the situation and assigns the roles of both participants. For example, if I say 'As a good friend of mine, will you lend me ten pounds?' I not only cast myself in the role of borrower but cast you in the role of friend, on whom there are obligations to lend. You can escape this obligation to lend only if you can manage to resist being trapped in the (probably undesired) friend role. Similarly, one can altercast with approaches such as 'I'm told you're an authority on ...' or 'Excuse me, sir ...' which define the role of the other as expert or superior. The onus is then on the other to refute this unilateral projection of a definition of the situation, often to the annoyance or the embarrassment of the other.

We are now in a position to construct a rough model of the process of interaction (Fig. 5.2). The interaction sequence is made up of the contingent behaviour of each participant towards the other. For simplicity we shall deal with the interaction from a single participant's point of view. Person's behaviour towards Other is determined by two elements. The first concerns his roles and goals, which are influenced by his role-set and by his personality, needs and background. The second element is his conception of Other and his conception of Other's conception of him. These two elements influence one another and both contribute to Person's definition of the situation. Once formed, the definition of the situation arouses self-presentation techniques which need to be employed to translate Person's definition of the situation into actual behaviour towards Other, which is also conditioned by social norms, situational proprieties, games, etc. When Person behaves towards Other in a given way he gets feedback on the effect of his behaviour on Other, which may cause Person to modify his roles and goals, and thus all the other elements of the model into which roles and goals feed, especially of course the definition of the situation and self-presentation. Other's reaction to Person's behaviour also influences Person's perception of Other, with consequent reinforcement or modification in the elements of the model to which this is connected. As interaction proceeds, Person's original definition of the situation has to be modified into a working consensus, which then becomes an intermediary link determining his behaviour towards Other. Eventually the working consensus may come to replace his original definition of the situation. In other cases Person's original definition of the situation may remain unimpaired even though his behaviour is effectively conditioned by the working consensus.

Finally, let us try to illustrate how all these concepts—defining the situation, perception of the other, self-presentation, working consensus, negotiation tactics and so on—look together in social interaction. The example is taken from Kingsley Amis's novel *I Want it Now* (1968). Ronnie Appleyard, a young television personality, is invited to a society party. When he arrives, the hostess, Mrs Reichenberger, is talking to a man in boots. When she sees Ronnie, she calls over to him.

> 'Mister-Heart-Throb in person. TV's Young Lochinvar. Nice of you to come, Mr. Appleyard.'

> The possible satirical edge to these words was absent from her tone and manner. Actually she could have done with a bit of it in her manner; it was disturbing to think that all that breastwork might be for real. But, of course, Ronnie was overjoyed at this reception. Now that the need to gain her attention and fill her with a proper sense of his importance had been met in advance, all he need bother about was showing her that he was

much too amusing and unspoiled by success not to be asked again very soon. Ten minutes' work at the most, after which he could get down to the more challenging tasks of (1) closing with and exploiting some of the other significant people here and (2) homing in on some unattended or incompetently escorted bird.

In the event it took nearer twenty minutes than ten, but promised proportionate gains. After the man in boots had been sent packing—not, Ronnie noted approvingly, without just the right amount of how fascinating it had been and how a proper get-together simply must be fixed up when everybody was back after the holidays—there was suddenly talk of an art-student son whose difficulties Ronnie might be the very man to understand. As described, the difficulties turned out to be nothing that being less of a talentless loafer would not cure, but he gave a full demonstration of himself being the said very man without having to offer the smallest help. As prelude, he had done some hypocritical surprise about any child of Mrs. Reichenberger's being of an age to have difficulties. Now, by way of epilogue, he did some eyes and tone of voice to suggest that he would not mind going to bed with Mrs. Reichenberger. The risk of being taken up on this total falsehood was as nothing compared with the certainty of being asked to more of the bag's parties.

Ronnie took himself off when a tiny painter and his rather large bird moved in among the pampas. The chap was up-and-coming all right, but had been at it for a decade. It was time he either came or went down again; anyway, he was not for Ronnie. Nor was his bird, who wore rococo spectacles and appeared to have white-washed her face before coming out.

. . . Ronnie went in the other direction, past an actor in a shiny green suit, a woman who was absolutely terrible at talking to kiddies on TV, a disc-jockey carrying in his arms a wriggling toy poodle, some young crap or other in Victorian army Officer's uniform, some little bitch or other dressed as a Spanish Lady, a food-and-drink pundit who was vigorously keeping up with both his subjects on the spot. None of these were any use to Ronnie. He left them behind . . .

Finally, Ronnie finds a girl who interests him.

As she stepped forward Ronnie noticed rather muzzily that her feet were bare and streaked with dirt. She spoke in a husky undertone.

'Hello, will you get me a drink?'

121

'What? Uh . . . what would you like?'
'Scotch and water. No ice.'
'Right.'
While he saw to this, Ronnie was wondering who the hell she
was. And what she was. Not that it really mattered. He would
forgive somebody who looked like that anything in the world.
Even if she turned out to be a folk singer he was going to screw
her.

He forced the whisky out of the bottle as quickly as was consistent with good manners, but by the time he got back to the
girl with the face two other men had zeroed in on her: a television don and a little bastard in a four-button denim jacket and
a very-queer-film-producer's trousers.

'Sorry chaps,' said Ronnie putting his arm through the girl's
and walking her away, 'I'm afraid something's come up . . .
nothing I can do about it . . . sorry . . . pity it's turned out like
this . . .'

They go through the ritual of introductions and discover that they are
both unmarried. Then Ronnie senses rivals. The girl shows interest
in a colleague of Ronnie's who is also at the party. Ronnie decides
to strike at once.

Ronnie put sincerity, plus the merest dash of intimacy, into his
gaze at Simon Whatsername. The . . . snuff-coloured? . . . eyes
gazed opaquely back at him.
'What about skipping out of here? Going and having dinner
somewhere?'
'I'd rather go to bed,' she said in her habitual monotone.
'If you're tired some food'll perk you up.'
'I don't mean that. I don't feel tired. I mean sex.'
This was exactly the sort of thing that Ronnie, in his role as a
graduate student of Britain's youth was supposed to know all
about. But, for the moment, his reaction was a simple, though
uncomfortable mixture of lust and alarm, with alarm slightly
to the fore. 'Okay,' he said reliably. 'Fine. Nothing I'd like better,
love. We'll grab a taxi and go to my flat.' 'I can't wait,' droned
the girl, 'I want it now.'

This is what is known as being hoisted by your own attempt to define
the situation.

Interactions require the making of profits

This proposition is based on the contributions to our understanding
of human behaviour made by the so-called 'exchange theorists',

namely writers who draw on the concepts of elementary economics to provide an analogy of social interaction. In this present discussion we shall be concerned with the work of George C. Homans (1961), rather than other important exchange theorists such as Thibaut & Kelley (1959) and Blau (1964), largely because Homans's book is as readable as it is brilliant.

There are three basic concepts in Homans's exchange model of human interaction. The first, *reward*, refers to needs which are satisfied or punishments which are avoided. Food is a reward when it satisfies my hunger needs, and assistance received is a reward when it satisfies my need for help. To avoid a punishment is also rewarding, since punishment is usually undesirable. *Costs*, the second basic concept, refers to the efforts required to obtain the reward and to the alternative rewards which must be forgone in pursuing one particular reward. Suppose I decide to go to the cinema. The reward consists in seeing the film. In getting this reward, however, I incur certain costs, which would include the price of the journey and the seat in the cinema, the time I spend in seeing the film, and the alternative activities (and their potential rewards) which I must forgo in order to see the film. *Profit*, the third basic concept, is defined as reward less cost. If the rewards I obtain in seeing the film are greater than the costs I incur by so doing, then I have made a profit. It is possible, of course, that at the end of the evening I come to the conclusion that the film was very boring and that I have wasted my time in going to see it. In this case, the costs are greater than the rewards, and I may be said to have made a *loss*—though Homans himself does not use this term.

We can illustrate the application of these concepts to an interactional situation with an example which Homans draws from the study of life in an office (Blau, 1955). According to the rules of the office each man must do his own work. If he needs help in solving his problems, he should consult the supervisor. Some of the men find it easier to do the work alone than others, and it is the latter group which needs the help of the supervisor. They are reluctant to seek the advice of supervisor, however, for to do so is to bring their lack of ability to the supervisor's attention and thus to reduce both the supervisor's evaluation of them and their chances of promotion. A preferable alternative is for the less skilled men to seek the informal aid of the more skilled. Let us make Person the man who seeks help from a colleague, and Other the colleague whose help is sought. Person is not very skilled and cannot complete his work without help. Other is highly skilled at his job and can complete his work allocation with time to spare. Person seeks help from Other and in return for the help is prepared to give Other his gratitude and esteem. Other gives his assistance to Person and in return for these services receives gratitude from Person. They have exchanged help for gratitude.

Now there are really two forms of exchange, homeomorphic and heteromorphic (Gouldner, 1960). Homeomorphic exchange occurs when the commodities that are exchanged are alike or identical. Typically homeomorphic exchange is in the negative form of the *lex talionis*,—an eye for an eye and a tooth for a tooth. But it can also involve the exchange of benefits. For example, two pupils in a school do not like doing their homework on English grammar. On Monday evenings John does the work and lets Tom copy from him; on Thursday evenings, Tom completes the exercise and John copies from Tom's book. In heteromorphic exchange the commodities exchanged are often very different but they are equivalent in value to the persons involved in the exchange. For instance, John may be good at mathematics and allows Tom to copy his work and in exchange John is allowed to copy from Tom's French book, for Tom is better at French than John. Clearly, our example of Person and Other exchanging help for gratitude is a heteromorphic form of exchange. Help and gratitude must have some equivalence of value.

Homans's exchange theory assumes that for an interaction to arise or persist both participants must make a profit, that is, the rewards of the interaction must exceed the costs involved. How can *both* make a profit? The answer basically rests on the fact that the commodities exchanged vary in value between Person and Other. Person very much needs help for without it he cannot complete his work adequately. So to Person help has a high value. In order to get help Person must be prepared to give Other gratitude for any assistance Other will give him, and the expression of gratitude also implies that Person is somewhat inferior to Other. But to Person the reward of gaining help is far greater than the cost of giving gratitude, so he can make a profit in making the exchange. In Other's case the value he sets on the commodities reverses. Because Other is highly skilled he can complete his work in good time. The cost of giving help to Person is relatively low. At the same time the reward of getting gratitude, esteem and the implied superiority from Person for the assistance rendered is relatively high. So Other too can make a profit on the exchange. In short, both Person and Other can make a profit because Person values the gaining of help more than the giving of gratitude, and Other values the gaining of gratitude more than the giving of help. As Homans himself puts it:

> The open secret of human exchange is to give the other man behaviour that is more valuable to him than it is costly to you and to get from him behaviour that is more valuable to you than it is costly to him.

In this illustration we have dealt with only a single exchange. What about future interactions between Person and Other? Homans seeks to account for the on-going interactions between Person and Other by means of three propositions.

1 The more often within a given period of time a man's activity rewards the activity of another, the more often will the other emit the activity.

2 The more valuable to a man a unit of the activity another gives him, the more often he will emit activity rewarded by the activity of the other.

3 The more often a man has in the recent past received a rewarding activity from another, the less valuable any further unit of that activity becomes to him.

From Proposition 1, the more often Person gives Other gratitude, the more often will Other offer his help; and the less often Person gives Other gratitude, the less often will Other offer his help. Similarly for Other, the more often he offers his help, the more often will Person express his gratitude. From Proposition 2, the more Person needs help, the more he will ask Other for it and the more thanks he will give Other for it. And if Person values help more than Other values gratitude, the more gratitude Person will offer to Other for his help in the exchange. Note that these two propositions suggest that the frequency of interaction is a function of the values that Person and Other place on the activities to be exchanged and the frequency with which they reward each other. Proposition 3 suggests that the more often Person receives help from Other in a short time sequence, the less valuable any more help becomes to him, and thus, from Proposition 2, the less often he will seek help.

In the single interaction we studied, both Person and Other made a profit. But, from Proposition 1, we should expect them to continue the exchange. Perhaps they do so in real life, but a limitation should be imposed by Proposition 3, which suggests that with further exchanges the value of help declines for Person and the more costly to his self-esteem the giving of gratitude becomes, and similarly, the value of gratitude declines for Other and the more costly in time or effort the giving of help becomes. They will not spend the whole day in the exchange of help and gratitude. The exchange between Person and Other will cease when the exchange no longer offers a bargain, or a more attractive bargain presents itself. No profit, no exchange. (This does not necessarily mean that Person and Other will cease to interact for they may continue to interact on the basis of a different exchange. For example, they may move to homeomorphic exchange of joke swapping.) If the exchange of help and gratitude between Person and Other is to persist, the 'rate of exchange' must alter. Suppose Other has given Person three units of help and received three units of gratitude in three successive exchanges. With each exchange

Other's costs are rising, for there are increasing demands on his efforts and time. Thus his profits are declining and he will soon be ready to cease the exchange. If, however, Person still needs more help from Other, he can persuade Other to give him this help only by increasing Other's rewards, that is, by offering him more gratitude per unit help. Thus Other can continue to make a profit because Person increases the amount of gratitude for each bit of help. But to give Other more gratitude is to increase Person's costs. In the end, the exchanges will cease because Person's costs begin to exceed the value of the help he receives, or because however much gratitude Person offers him, Other finds it more profitable to be doing something else—his own work, or another exchange.

These simple concepts can be used to make sense of a surprisingly large number of our interactions. We are not, of course, always *conscious* of the fact that we are calculating profits or negotiating bargains with others, but our lack of conscious awareness of the process does not make the analysis any less valid. Moreover these economic concepts make us sound very self-seeking, materialistic people. This is not necessarily the case at all, and the unfortunate connotations of exchange theory terminology should not put us off. A further study of exchange theory will prove—if I dare use the phrase—to be worth the effort.

One of the problems arising from our analysis is the amount or extent of the profits made by Person and Other in a particular interaction. For it is possible that Person could make a very much larger profit than Other, a situation which would probably lead Other to regard the transaction as unfair. This is what Homans calls the problem of *distributive justice*, that is, the justice in the distribution of rewards and costs between persons. The rule of distributive justice states that each man should receive rewards that are proportional to his costs. The greater a man's costs, the greater his rewards should be. If Other is extremely busy when Person asks him for help, then the act of giving help will be more costly to Other than would be the case if Other had nothing to do for the rest of the afternoon. In consequence Person is likely to offer more gratitude in the first case than in the second, for the extra costs must be compensated with extra rewards. If Other demands a large amount of gratitude when he is not busy, then Person is likely to feel that the exchange is unfair, because Other is then getting a reward that is out of proportion to his costs, and thus making an excessive profit. Person may have to pay this extortionate price for the help he needs, but if he does so he is likely to feel angry or disgruntled or resentful. Similarly, if Person gives Other a small amount of gratitude for help which is costly (because he is very busy), Other will be upset and regard the transaction as unfair, because Other is then making a profit which is disproportionately

126

small (small rewards but high costs) whilst Person has made a disproportionately high profit (small costs but high rewards). Distributive justice is realized when the profits of Person and Other are equal, in the sense that the rewards are proportional to the costs.

Whether or not a man regards an exchange as fair depends not only on the proportionality of the rewards and costs, but also on what he believes he has a right to expect from the interaction, and this depends on many factors in the lives of Person and Other which Homans calls *investments*. Common examples of investments are age and experience. Because Other is more senior in the office than Person, it is possible that Other will get higher pay than Person. Person and Other are quite likely to accept this as fair because Other has served the firm for a longer period, and is perhaps making a superior contribution to the work. Some teachers are paid 'responsibility allowances' and 'qualification allowances' which are regarded by many teachers as fair, (though predominantly by those teachers who receive them), since the extra pay recompenses them for their greater investments (e.g. qualifications, years of service) as well as for their greater costs (e.g. organizing and administering the department, or teaching special groups of pupils). Distributive justice requires, then that each man's rewards are proportional to his costs and that his profits are proportional to his investments.

In practice, of course, there is often a striking lack of agreement about what are considered legitimate investments or fair rewards and profits. Certainly it seems unlikely that people who introduced the allowance system into the teaching profession were aware of the unfortunate side-effects (dissatisfaction, rivalry, competition, sycophancy and malice, to name but a few) that would follow their innovation. The profession—and the columns of *The Times Educational Supplement*—continues to be plagued with arguments about whether or not degree status is an investment meriting higher pay than College of Education status, and similar controversies rage over the financial advantages which accrue to the 'good honours' graduate, and the heavy allowances which are normally given to the heads of the science and English departments, but almost never to the head of the music department.

But let us return to our interactional situation. Suppose that Other in our office is indeed older and more experienced in the job than Person. Then by the rule of distributive justice Other should make rewards that are proportional to his costs and profits that are proportional to his investments. In consequence it would be fair for Other to make a larger profit than Person because his investmensts are greater than Person's. Thus Person is likely to give greater gratitude to Other, for one unit of help than he would give to another person, X, who joined the firm at the same time as Person and who thus has the same investments. Our experience confirms this. We tend to give more gratitude to our seniors for their help or advice than we do our peers. More gratitude is due to them because they need more gratitude to make a greater profit which is required by their greater investments.

The elaborate details of Homans's exchange theory and the ingenuity of its application to a variety of theoretical and empirical areas cannot be documented here. Such a good book should be read, not summarized. But I hope I have made it clear how this theory brings to light some of the essential ingredients of the interaction process. Firstly it stresses the importance of the participant's *goals* in interaction. We interact with others in order to realize certain purposes, which in the theory are viewed as the making of profits or the cutting of losses. Secondly, in interaction each participant has to discover and take into account the goals of the other, and each will have to consider the effects of his own behaviour on the other. This is essential if the allocation of rewards, costs and profits is to be negotiated between the participants, and this negotiation of profits is a crucial part of reaching a working consensus about the definition of the situation.

Exchange theory abounds with many exciting possibilities for subsuming under its concepts a wide range of human experiences. In particular, it can comprehend much more complex human situations than those illustrations we have used so far. In everyday life, for example, the choice we often face is not one between profit and loss, but between two losses. Let us take the case of a wife who is beaten regularly by her husband. She may see her choice as on the one hand leaving her husband (reward) but jeopardizing her financial position (cost), or on the other hand staying with her husband and being beaten (cost) but retaining her financial security (reward). If both alternatives involve a net loss, then she must choose the lesser of the two losses. Teachers may put pupils into a similar position. For example, when the teacher gives a recalcitrant pupil the choice between making an apology or receiving a punishment, he is asking the pupil to choose the lesser of two losses.

Another interesting problem for exchange theory is the case of the altruist. Such a person appears to be willing to give more than he receives. In an altruistic exchange, Person gives much to Other, but Other is not required to give anything in return. Person appears to be incurring all cost without reward (a pure loss) and Other obtains a reward with no cost (a pure profit). Is not Person, then, an exception to exchange theory? The theory requires that if the exchange is to be satisfying to both participants, each must make a profit. For Person to make a profit, he must have some rewards which exceed the loss of his gift to Other. In the case of altruist, Person's rewards are not ones which involve a loss for Other. This can arise in a variety of ways. The pleasure experienced by Other in receiving the gift may serve as the necessary reward for Person. Other sources of reward to Person might be a relieved conscience, a sense of moral superiority, a social duty fulfilled, the feelings of obligation induced in Other, or the attainment of a better relationship with God. Christian ethics may be regarded as a transformation of an ethic based on human exchange through the introduction of a divine third party who alters the normal accounting process between human beings. The love of God creates a new dimension to the exchange, which in purely human terms seems to involve a definite loss.

Ye have heard that it hath been said, Thou shalt love thy neighbour and hate thine enemy. But I say unto you. Love your enemies, bless them that curse you, do good to them that hate you, and pray for them which despitefully use you, and persecute you; that you may be the children of your Father which is in heaven . . . For if you love them which love you, what reward have ye? (Matthew, chapter 5)

The main weakness of exchange theory lies in its great strength, namely its applicability to a very wide range of social phenomena. Because it is so general it has relatively little predictive power, though it can be used to explain *ex post facto* as Homans himself so ingeniously demonstrates. It could be argued that exchange theory is an expansion of reinforcement theory in psychology—Homans begins his work on the basis of the work of B. F. Skinner—working on the general assumption that human behaviour is motivated towards the gaining of satisfactions and the avoidance of dissatisfactions. In the hands of a theorist of the stature of Homans it is this old idea which is given such imaginative new clothing. Exchange theory does suggest an important human motivational component in social behaviour which can easily be built into our more comprehensive model given earlier. It is worth re-reading the extract from the Kingsley Amis novel where Ronnie's calculations of rewards and costs, profits and losses, are often quite explicit.

Recommended reading

M. ARGYLE, *Social Interaction*, Methuen, 1966.

E. GOFFMAN, *The Presentation of Self in Everyday Life*, Doubleday Anchor, 1959.

G. C. HOMANS, *Social Behaviour: Its Elementary Forms*, Routledge & Kegan Paul, 1961.

R. D. LAING, *The Politics of Experience*, Penguin, 1967.

6 Teacher—pupil interaction

In the last chapter we examined some of the concepts which seem to be essential to a study of the dynamic processes that we call human interaction. Many of these concepts were illustrated from the behaviour of teachers and pupils, but no systematic attempt to understand *classroom* interaction was made. In this chapter we shall be concerned with the interactions between teachers and pupils in the classroom, for it is in the classroom where the major portion of teacher–pupil interaction takes place, and it is here where we find the very heart of the process of education in its interpersonal aspects. So we must now put our theories and concepts of interaction to the test and consider to what extent they are of use in the analysis of interactions in classrooms.

Interaction analysis

What actually goes on in classrooms? Almost everyone has spent many years—thousands of hours—in classrooms as a pupil, and the average teacher will spend more than forty thousand hours in the classroom during his professional career. It should not be too difficult for us to say what goes on. In fact, it is an incredibly difficult task. Let us begin by *observing* the behaviour of teachers and pupils in a classroom. What do we see? What we observe is extremely complex, because many people are involved. The average classroom contains one teacher and between twenty and fifty pupils. During any one lesson we are likely to observe *talking*, most of it by the teacher directed to the pupils, though with some talk from the pupils to the teacher and between pupils; *movement* by both teacher and pupil around the room; *activities* such as reading, writing, hand-raising, distributing equipment, chalking on the blackboard and so on. Finally we observe, but not directly, a great deal of mental activity and *thinking*. This is a far cry from the model of interaction we have tended to use so far, namely a simple verbal interaction between two persons, Person and Other. How then can we devise a method whereby we can simplify this highly complex behaviour into a few concepts which will allow us to make sense of the more elaborate whole? How can we single out the dominant and pervasive elements of behaviour and interaction that will retain the essentials of the situation yet free us from the details? We have no choice but to select certain features of classroom interaction for our analysis. Not even a film of the class in action will preserve every part of every person's behaviour sufficiently well for us to regard such a documentation as complete.

One technique would be to divide all the behaviour that we can observe into a relatively small number of categories, and then categorize each 'bit' of behaviour as we observe it. We can count up the number of 'bits' in each category and use these as a means of drawing a picture of 'what went on'. This has been a popular method of interaction analysis. Since we cannot categorize everything that occurs, the central problem is to decide which particular behaviours we are most concerned to categorize. For reasons of simplicity, many researchers have confined their observations and categories to purely *verbal* behaviour. Such an approach has several advantages. It can be argued that what people say is the most significant single aspect of human behaviour and interaction; it should not prove difficult to devise categories of verbal behaviour that are sufficiently clear-cut to allow different observers to agree on what verbal element should be coded into what particular category; we can always make a cheap recording of what people say for re-checking our categorizations.

One of the earliest attempts to observe and categorize interaction in classrooms by such a method is that of John Withall (1949). He categorized the behaviours into seven categories.

1 *Learner-supportive statements* that have the intent of reassuring or commending the pupil.

2 *Acceptance and clarifying statements* having an intent to convey to the pupil the feeling that he was understood and help him to elucidate his ideas and feelings.

3 *Problem-structuring statements* or questions which proffer information or raise questions about the problem in an objective manner with intent to facilitate the learner's problem-solving.

4 *Neutral statements* which comprise polite formalities, administrative comments, verbatim repetition of something that has already been said. No intent inferable.

5 *Directive or hortative statements* with intent to have the pupil follow a recommended course of action.

6 *Reproving or deprecating remarks* intended to deter the pupil from continued indulgence in present 'unacceptable' behaviour.

131

7 *Teacher self-supporting remarks* intended to sustain or justify the teacher's position or present course of action.

Categories 1–3 are called 'learner-centred' and categories 5–7 are called 'teacher-centred'. A profile of any teacher can be drawn by checking the number of statements falling into each of the categories. A simpler way of reporting the profile is to check the proportion of statements falling into the first three against the proportion falling into the last three categories. If the first three contain more than the last three then the teacher is called 'learner-centred', and if the situation is reversed, then the teacher is referred to as 'teacher-centred'.

One of the most striking things about this method of interaction analysis is the amount of interpretation which is required from the observer, who has to infer the intentions of the teacher, presumably from the tone of voice, facial expression and the immediate context of words spoken. This is not a major problem, however, as long as we can show that different observers make the same inferences and thus code the statements into the same category. Even more striking is how little of the total behaviour of people in classrooms is used, for attention is focused exclusively on the verbal behaviour of the teacher: what the pupils do and say is completely omitted. In the final analysis, our coding of classroom events has reduced what actually happens into a statement which tells us whether the teacher is learner- or teacher-centred. Thus the method barely deserves to be called a form of interaction analysis at all.

Some later investigators have included the pupils' verbal behaviour as well as the teacher's. Ned Flanders (1960), for example, has two categories of pupil talk.

1 Talk by students in response to talk initiated by the teacher.

2 Talk by students initiated by the students themselves.

Flanders's system makes only a very small advance on the method of Withall, for his categories of teacher verbal behaviour are very similar to those of Withall, though he changed the terms learner- and teacher-centred to 'indirect' and 'direct' influence, and his dichotomy of pupil verbal behaviour is extremely crude.

Other methods of analysing classroom interaction have been less restrictive in the range of behaviour that is categorized. For example, Amidon & Hunter (1966) in their Verbal Interaction Category System (VICS) have distinguished between the sorts of questions teachers may ask their pupils. Questions can be 'narrow', where the general nature of the pupil response is predictable ('What is the capital of France?'), or questions can be 'broad', which are thought-provoking and more open-ended ('Why do you think Paris became the capital of France?'). Similarly, the categories of pupil talk are more refined, including 'predictable' and 'unpredictable' responses to the teacher's questions, and the division of Flanders's category of 'pupil-initiated talk' in two categories, initiation to the teacher and initiation to another pupil.

Yet the more behaviour we try to include in a system, the more refined and exclusive the categories become. Thus the number of categories increases and the consequent profile of classroom interaction becomes more elaborate as it becomes more comprehensive. The drawback is that as the system of interaction analysis gets more complex, the further away from the classroom-interaction-as-it-really-is we seem to go. As Cogan (1963), one of the leading interaction analysis researchers, puts it:

> If one reads carefully the work that has been done by the men and women who are attempting to make sense of what teachers do in classrooms, one must ultimately conclude that the underlying weakness that permeates the whole endeavour is a weakness of the primary data the researchers are dealing with. Most of the data amounts to superficial, rootless verbalisms about the events of classrooms. The truth is that these data are so attenuated, they are so remote from the sights, sounds, the smell, the feel, and the sense of the classroom that the reality escapes us. Whatever order we do find is thereby transmuted to something pallid, alien to the real events of the schoolroom. With all our questionnaires and our interviews and schemes for scoring classroom interaction, we are like mineralogists without specimens—our data have escaped us. Our slices and sections of reality are so thin and fragmentary that even when we first examine our specimens and our samples we are already miles removed from the phenomena we are dealing with—as though we had elected to study the moon by way of its reflection in a puddle of water.

We cannot complain, of course, about losing the fine detail of teacher–pupil interactions, for in creating the categories it was the fine detail from which we wished to be free. The complaint is rather that in having lost the detail we have failed to preserve its essential qualities. In the rest of this chapter we shall be trying to recapture much of what interaction analysts have let slip through their fingers—mainly without appearing to have noticed. Here I wish to make the point that the interaction analysts have been compelled by their method to ignore much of the elementary nature and structure of teacher–pupil interactions. The basic unit of an interaction between a teacher and pupil, or between any two persons, consists of one bit of behaviour emitted by one person (the teacher) followed by a bit of behaviour emitted by the second person (the pupil) which is contingent on the first person's behaviour. The first bit of behaviour can be called the *pro*action, and the second contingent behaviour the *re*action. For example, consider the following interchange:

Teacher (proaction): How much is five and five, John?
Pupil (reaction): Ten, sir.

This basic unit of interaction, a proaction followed by a reaction, we might call an interaction *link*. Diagrammatically, we can represent it as in Figure 6.1. Yet in classrooms many interactions which look

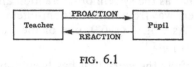

FIG. 6.1

like simple links are not so. This is because when the teacher speaks to a child the teacher's remark is often heard by many if not all of the pupils in the classroom and they may also react to the teacher's remark. This is what Kounin & Gump (1958) have called the 'ripple effect'. The teacher's behaviour is like a stone dropped in a pond. It sends out ripples which affect areas that were not part of the original target. So in the above example, the situation is influenced by the fact that the teacher's question is heard by the other pupils present who also react, by thinking of the sum, by raising their hands and so on. The diagram should more accurately be drawn as in Figure 6.2:

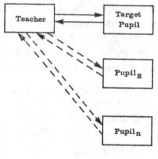

FIG. 6.2

In other words, what appear to be simple interactional links are really *multiple links* affecting the behaviour of many people.

Interactional links join up and become *chains*. A chain consists of several links which together make up an interactional episode. An example of chain would be the following interaction.

Teacher: What are five and five, John?
Pupil: Nine.
Teacher: No, that's not right. Try again.

Pupil: I don't know.
Teacher: I think you do.
Pupil: Ten.
Teacher: That's right, well done.

After the teacher's initial proaction, each participant continues to react to the other to make the chain. The chain may be lengthened by a further verbal reaction such as 'Can I try another sum?' from the pupil. Or the teacher may begin a new chain by a proaction such as 'Are you paying attention to the lesson, Margaret?' Just as links are often multiple, affecting more pupils than the immediate participant, so also are chains. The essence of teacher–pupil interactions is contained within these links or chains. If we abandon them, we shall soon lose the flavour of teachers and pupils in classrooms. The interaction analysts have had to sacrifice the links and chains for categories, and in so doing they have betrayed a major flaw of their method of analysis. The significance of a chain or episode disappears when it is broken up into individual verbal acts. The episode as a whole has a meaning for teacher and pupil(s) which is greater than the sum of its parts.

Because Interaction Analysis takes a 'scientific' approach to classroom behaviour, where teachers and pupils are treated essentially as 'objects' observed from without, no account is taken of the meanings which the participants give to their interactions. The assumptions and perspectives of the teachers and pupils, which are often covert and implicit, are not explored. We discover little of the over-all teacher–pupil relationship as it is experienced by the teacher or by individual pupils. Yet it is this relationship which may not only influence the meaning assigned to particular verbal statements or acts but also exercise a pervasive influence which is not immediately obvious or directly open to measurement by traditional methods.

This is not to say that many interesting findings have not been produced by the interaction analysts. One fundamental contribution has been the demonstration of a range of classroom behaviours which differentiate teachers. We all know that different teachers behave differently in classrooms. Interaction analysis is one method of revealing these differences. At the simplest level we can compare teachers on the different proportions of their behaviour falling into particular categories. At a higher level we can construct an 'interaction matrix' from which we can derive information on the sequence of teacher behaviours—what follows what. Teachers vary not only in the amounts of time they spend in behaviour of different categories but also in the way these behaviours are related in a time sequence. Thus we can compare teachers with different personal characteristics, teachers with pupils of different age grades, and teachers of different

135

subjects—though it must be admitted that no *general* picture has emerged in relation to these factors.

Attempts have been made to use methods of interaction analysis to specify the behaviour of 'the good teacher'. Amidon & Giammatteo (1967) have shown that superior teachers score more highly than average teachers in the categories of acceptance of feeling and use of students' ideas. In the classrooms of superior teachers student-initiated statements occur twice as often as in regular classrooms. On the other hand, average teachers spend twice as much time as superior teachers in the categories of giving criticism and directions. Both types spend about 40 per cent of their time in lecturing, though the superior teachers tend to lecture in shorter 'bursts'. The problem with such studies as this is the nature of the criteria by which we designate a superior teacher. Amidon & Giammatteo used the ratings of supervisors and administrators, but it is not clear what precise criteria were used in these ratings. If fellow teachers, or pupils, or parents had been asked to nominate the superior teachers, a different group of teachers might have been selected, since these groups might not use the same criteria as administrators to define the good teacher. In consequence a different picture of the behaviour of the superior teacher might have emerged.

Another obvious potential application for interaction analysis is that of the relationship between teacher behaviour and pupil attitudes and achievement. Flanders (1964) predicted that teachers whose behaviour is indirect (learner-centred) should produce pupils with better attitudes to learning and higher attainment than would teachers whose behaviour is direct (teacher-centred). These predictions were confirmed. Pupils taught by indirect teachers had significantly higher attainment and significantly better attitudes. In the design of this research the behaviours of a number of teachers were first analysed and from these two groups of teachers, direct and indirect, were selected. Then the pupils of these two groups of teachers were tested in the areas of attitude and attainment. The explanation of the results could be the reverse of the one suggested. Teachers with better pupils (higher attainment and attitude scores) might have adopted an indirect teaching style in response to their superior children, whereas the teachers with more difficult children (lower attainment and attitude scores) are pressured to be more direct in their behaviour.

Much of the work on interaction analysis in the classroom, especially in the early days of the research, represents an attempt to discover the *climate* of different classrooms, following the pioneering work of Lewin, Lippitt & White which we shall be discussing shortly. This term climate is intended to characterize the dominant and pervasive quality of interpersonal relations in the classroom. The best definition is probably that of Ned Flanders (1967).

The words *classroom climate* refer to the generalized attitudes toward the teacher and the class that the pupils share in common in spite of individual differences. The development of these attitudes is an outgrowth of classroom interaction. As a result of participating in classroom activities, pupils soon develop shared expectations about how the teacher will act, what kind of person he is, and how they like their class. These expectations colour all aspects of classroom behaviour, creating a social atmosphere or climate that appears to be fairly stable, once established. Thus the word *climate* is merely a short-hand reference to those qualities that consistently predominate in most teacher–pupil contacts and contacts between pupils in the presence or absence of the teacher

It is clear from this definition that the term *climate* approximates fairly closely to our term *definition of the situation*. Some progress has been made in specifying this climate. It is significant that the vast majority of researchers are united in conceiving it basically in terms of a dichotomy, though they use different names for the two principal forms of classroom climate. On the one hand the climate can be 'teacher-centred', 'direct', 'traditional', 'dominative' or 'autocratic', and on the other it can be the opposite, 'learner-centred', 'indirect', 'progressive', 'integrative' or 'democratic'. Unfortunately, however, the work of the interaction analysts on classroom climate is of limited value to us in our study of the definition of the situation in classrooms. It tells us little about how the definition of the situation is negotiated and established, and even less about how it is maintained. In addition, the contribution of the pupils to the definition of the situation tends to play a very small part in studies of classroom climate. In short, studies of classroom climate may help us to *describe* the definition of the situation, but they do not offer much assistance in the analysis of the definition of the situation as a *dynamic process*.

Teacher–pupil interaction

If we are to analyse the ways in which teachers and pupils define the situation and reach a working consensus, we must first of all consider those characteristics which are peculiar to interactions between teachers and pupils. Our model of interaction must, in other words, take account of the distinctive quality of teacher–pupil relations in classrooms. Perhaps the most striking feature of the world of the classroom is that the pupils are compelled by law to be present in school. Most interactions are entered freely by the participants. If, for whatever reason, the interaction is unattractive or unsatisfying, the participants can withdraw. Pupils at school have no choice: they are required to enter into interaction with the teacher. This feature is not unique to schools. In mental hospitals, in prisons, and in the armed services, the patients, prisoners and enlisted men are in a similar situation. In practice, of course, pupils are usually by no means reluctant participants, but this does not alter the involuntary basis to their presence in school, for there are times in the lives of all pupils when they would prefer not to be at school. If they do *want* to be present, so much the better. If they do not want to be present, they will be compelled to attend. A pupil who does not wish to go to school has two means of escape. He can play truant, which is at best a short-term measure, since the system employs agents who will seek him out and require him to return. Alternatively he can withdraw from school not physically but *psychologically*, for instance by day-dreaming, or by committing himself only minimally to the require-ments imposed on him when he is present. (Psychological withdrawal is more frequent and a more complex reaction by the pupil to his involuntary presence in the classroom and we shall consider it in more detail later). Even 'normal' children, who presumably do not complain excessively about having to go to school, and who enjoy much of their lives in school, tend to associate absence from school, especially holidays, as a time of freedom and blessed release.

Teachers also recognize the involuntary nature of the child's presence in school in many subtle ways. For example, it is not un-common for teachers to reward 'good' children by letting them out of school a little earlier than the rest of the pupils. As Philip Jackson has pointed out, this is reminiscent of the remission system in prisons —time off for good behaviour. Parents, on the other hand, tend to see schools as places of unalloyed pleasure for their children and for the most part they expect their children to be enthusiastic about going to school. This springs perhaps from the parental dread of long school holidays and from their tendency to romanticize schooldays to which E. M. Forster (1920) has referred. The distorted nature of this romanti-cization of schooldays by adults is clear as soon as they verbalize it. They recall the exciting football matches and school plays, the eccentric teachers, the tricks they played on teachers or other pupils. They do not lovingly muse over the work aspect of school; that is not remembered as having been 'fun'.

The second distinctive feature of teacher–pupil interaction is the enormous power differential between the two participants. This again is not unique to classrooms. Most men and women enter situations with a similar power differential when they interact with their super-iors at work. What is notable about the teacher–pupil relationship is the fact that the pupil spends so much of his time directly in such a relationship and that the power differential between teacher and pupil is so great. In classrooms, teachers are permitted to and fre-

quently do make almost all the decisions affecting the child's behaviour. What the teacher says goes. It is the duty of the pupil to
accept and obey—preferably without question. The teacher's power
derives from several sources: from his status as an adult; from his
traditional authority as a teacher; from his legal authority; from his
expertise in the subject matter he is teaching. The power differential
shows itself in the asymmetrical rights possessed by the teacher,
rights which extend far beyond the right to invoke formal sanctions.
For example, the teacher has a generally accepted right to intrude
into or interrupt the child's activities at will. A child in school can
have no legitimate privacy and no legitimate secrets. 'Show me what
you have in your desk' or 'Empty out your pockets onto the table' are
permissible requests from teachers—even though they may be resented by the pupils. On the other hand, the child must learn to respect
the teacher's privacy, for the teacher cannot be intruded upon at the
child's whim. If he wishes to speak to the teacher, he must signal
the teacher with a raised hand and wait patiently for the teacher to
catch his attention and give permission to speak. A teacher's action
to the pupil which is defined by the teacher as 'being helpful' could,
if it were reciprocated by the child to the teacher, be easily defined as
'being rude'.

The outcome of these two distinctive characteristics of teacher-
pupil interaction is the great inequality of the two participants in the
process of defining the situation. The dice are loaded in the teacher's
favour. He can take the initiative in defining the situation and possesses the power to enforce it on the pupils. In terms of exchange
theory, he is in a position to ensure that he makes a good profit from
the interaction (more rewards than costs), whereas the pupils may
be left in a situation of a net loss (costs exceed rewards). In short, the
teacher is in a position where if he wishes he can *exploit* the pupils
at will.

Since the teacher is in a position to determine and enforce his own
definition of the situation on the pupils, then the behaviour of the
pupils will be highly dependent on the teacher's behaviour. This is to
be expected since, in terms introduced in the last chapter, teacher–
pupil interactions are typically *asymmetrically contingent*: the pupils'
behaviour is much more contingent on the teacher's behaviour than
the teacher's behaviour is contingent on the pupils' behaviour. That is
to say, pupils' classroom behaviour is a product of, and a response
to, the teacher's interpretations of his role and his teaching style.
Whilst it is true that the pupils' behaviour is influenced by many other
factors than the teacher's behaviour, and whilst it may be that the
teacher adapts his behaviour in response to the special characteristics
of the pupils in his class, we would expect the pupils to adapt to the
teacher to a much greater degree than the teacher to the pupils.

139

Interaction analyses of classrooms offer considerable support for this contention. The classic research is that of Lewin, Lippitt & White (1939). This famous study was *not* an investigation of teachers and pupils, though students of education constantly misrepresent it as such. The subjects were ten-year-old boys in a club situation where they were engaged in craft work. The boys worked in groups of five under the supervision of an adult leader. These adults supervised the club under three different leadership styles and each club was given experience of each leadership style. During every meeting the behaviour of leaders and boys was very closely observed, recorded and categorized.

The three leadership styles are designated by the investigators as 'authoritarian', 'democratic' and 'laissez-faire'. For simplicity we shall consider the first two styles only. Under the authoritarian regime the leader determined all policy; dictated the assigned tasks, work companions and work methods; was arbitrary and 'personal' in his criticism and praise; remained aloof from participation with the club members. Under the democratic regime, the leader encouraged and assisted club members, but left policy to group discussion and decision; gave members freedom to choose their work and work companions; gave advice and outlined alternative procedures from which members could choose their work methods; participated in a friendly manner and was 'objective' in his criticism and praise.

How did the boys respond to these different group climates and patterns of supervision? The majority of clubs under an authoritarian leader responded with a general *apathy*. As compared with the democratic clubs, the members were very submissive and dependent. There was little smiling and joking and much less conversation than in the democratic situation. A smaller number of the authoritarian groups produced an *aggressive* response, marked by high levels of hostility, hostile criticism, demanding attention, and competitiveness and scape-goating. On the other hand, the authoritarian groups showed greater absorption in the work than did the democratic groups, though when the leader was temporarily absent the level of work involvement fell much more sharply in the authoritarian groups. From this it is inferred that the democratic groups have a higher interest in the work. As compared with the authoritarian groups, the members of the democratic clubs were more friendly, more co-operative and more group-minded; showed greater concern for others and greater mutual approval; were more ready to share group property; displayed greater originality in the work; maintained a steadier work level; were less critical, less hostile and less discontent.

The political connotations of the leadership styles in this study are unfortunate, for they predispose us to think of the democratic as 'good' and the authoritarian as 'bad'. (Many later interaction analysts use a similarly loaded nomenclature.) The significance of these names makes more sense when we remember that Kurt Lewin, who inspired all these studies, was an exile from Nazi Germany. His predisposition to favour the American ('democratic') system and to disfavour the German ('authoritarian') is clear elsewhere in his writing (1948).

To one who comes from Germany, the degree of freedom and independence of children and adolescents in the United States is very impressive. Especially the lack of servility of the young child towards adults or of the student towards his professor is striking. The adults, too, treat the child much more on an equal footing, whereas in Germany it seems to be the natural right of the adult to rule and the duty of the child to obey. The parents seem to treat the children with more respect. Generally, they will be careful when requesting the child to bring some object, to ask them in a polite way . . . [whereas] the German parent is much more likely to give short orders . . . The same difference in the basic relationship between the child and the educating adult is found in the schools . . . In fact, the good American nursery school is in my judgement pedagogically better than that of any country.

Lewin's philosophy played an important part in the collection and presentation of the data (Harding, 1953). For example, suppose that Lewin had been a Nazi. Could he not then have emphasized the greater work output, the greater obedience and submission to authority, the great work-minded conversation and the lower levels of aggression (in the apathetic groups) of the boys under an authoritarian leader? Would he not have been more anxious to stress the less attractive features of the democratic groups? More fundamentally, if he had performed these experiments in Nazi Germany, is it not likely that he would have produced quite different results? For all the boys preferred the democratic leadership to the autocratic, to which they were probably not accustomed. Only one boy preferred the authoritarian climate and he, significantly, was the son of a military officer.

The later studies of Anderson & Brewer (1945) are of more direct relevance to us, because these were experimental investigations of the behaviour of teachers and pupils in classrooms. From their observations they suggest two basic extreme teacher types, the *dominative* and the *integrative*. The dominative teacher, who is rather like Lewin's authoritarian leader, can be characterized as working *against* the pupils. He thinks he knows best; issues orders and imposes his decisions; wishes the pupils to obey and conform; dislikes discussion or criticism; tends to threaten and blame. The integrative teacher, in

141

contrast, can be said to work *with* others. He requests rather than orders; consults the pupils and invites their co-operation; shares the control and responsibility; encourages the pupils' ideas and initiative. The effects of these different teaching styles on pupil behaviour follow the Lewin, Lippitt & White study, as might be expected. Under integrative teachers the pupils make greater contributions to the lesson; show great appreciation of others; are more friendly and co-operative; are less inattentive, aggressive and resistant to instruction.

It is this research which offers the greatest support for our contention that it is the pupils who adapt to the teacher rather than vice versa. The teachers remain consistent in their teaching style even with different classes of children. But the pupils, like the boys in Lewin's clubs, change their behaviour as a function of the teaching style. When pupils move from an integrative to a dominative teacher (or vice versa) marked changes in their behaviour occur. It is the teacher, then, who is the principal creator of the climate that prevails in the classroom; the pupils' response is largely determined by the teacher's behaviour.

The teacher's classroom role

Given that the teacher has greater initiative and power in defining the situation, how does he indeed define it? Obviously his first step is to define the situation in such a way that he can perform his role as teacher in a way he regards as adequate. His definition of the situation must be congruent with his conception of his classroom role. Such a role conception will in the last analysis be unique to the individual teacher. Our present task is to consider the range of role conceptions that teachers might assume in the classrooms. Many writers have devised sets of teacher *sub-roles* that are relevant to classroom interaction. These sub-roles really refer to the *tasks* which fall upon the teacher in the performance of the teacher role in the classroom. Sorenson (1963) suggests six principal sub-roles for teachers.

Adviser—recommending courses of action for student.
Counsellor—helping the student to discover for himself.
Disciplinarian—adhering to rules and administering punishment.
Information giver—directing learning and lecturing.
Motivator—using rewards to stimulate conformist activity.
Referrer—securing help of outside agencies.

Another set of six sub-roles, with special reference to the Primary school, has been suggested by Blyth (1965). There seems to be relatively little overlap with those of Sorenson.

Instructor—direction of learning and lecturing.
Parent-substitute

Organizer—including discipline.
Value-bearer—transmission of society's dominant values.
Classifier—evaluation of academic and other behaviour.
Welfare-worker

Trow (1960) suggests eight classroom sub-roles under two headings.

Administrative and executive roles
  Disciplinarian (Policeman)
  Measurer and record keeper (Clerk)
  Learning-aids officer (Librarian)
  Programme director (Planner)

Instructional roles
  Motivator
  Resource Person
  Evaluator
  Adapter

Many other similar lists of sub-roles, of varying lengths, have been offered. Redl & Wattenberg (1951) suggest fifteen basic subroles for the teacher, including the roles of *referee* and *detective*. Unfortunately there seems to be no easy way of reconciling these different compilations of teacher sub-roles. Nor does there seem any logical limit to the number of sub-roles that we could postulate. A more serious difficulty is that almost none of these lists applies to *all* teachers. Indeed some lists, such as that of Blyth, are intended to apply to teachers of one particular sector of the teaching profession. In my own view it would be simpler to try to construct a set of sub-roles which apply to *all* teachers, i.e. a set of sub-rules which comprehends the teacher's tasks that are common to all teaching situations. The additional sub-roles suggested in the literature would then be subsumed under the more basic sub-roles.

All teachers have two basic sub-roles which they cannot escape. These are the roles of *instructor* and *disciplinarian*. The task of the teacher as disciplinarian is the establishment and maintenance of discipline and order in the classroom. It is the task of who shall do what, when and how. It is the creation of rules of conduct and rules of procedure. This includes the teacher's task in organizing the grouping of the pupils, the distribution of equipment, the timing, form and extent of movements by pupils within or in and out of the classroom. Many of these rules define how the pupil is expected to treat and respond to the teacher, as well as how the pupils must treat one another. Also included in the disciplinarian role are the means of maintaining the rules, including the fixing of rewards and punishments for adherence to or deviance from the rules.

This is a wider interpretation of the term 'discipline' than is usually the case. Moreover it is not a value-laden concept, for teachers resolve the discipline task in a wide variety of ways. For example, the teacher may interpret his role autocratically by making all the rules himself, imposing them on the pupils and requiring the pupils to show unquestioning obedience to the rules. Or he might take a much more democratic approach, by which the rules are established as a joint decision of teacher and pupils after a full and free discussion. How the task is resolved will be unique to each classroom. The point I wish to stress here is that any situation which requires teaching and learning between persons has a discipline problem, from the primary school to the universities, though there is an enormous variety in the extent to which this problem is problematic.

In schools most teachers are concerned to establish or maintain themselves as masters of the situation. The teacher feels that he must be *in control*. To be in control means that the teacher must be able to make the rules of conduct and obtain conformity to these rules by the pupils. When the teacher is either unable to impose rules and/or attain obedience to them, he is said to have failed in his disciplinarian role, for the pupils are out of control or undisciplined—masters of the situation.

How then do teachers ensure that they are in control? The root of being master of the situation is thought by most teachers to consist in the child's obedience to the teacher's orders, because if the pupils do not obey him the social order is threatened and the teacher's prerogative of rule making will be eroded. It is for this reason that discipline is in the popular mind linked with punishment, for punishments are one of the standard ways in which the teacher tries to enforce obedience. Another element in maintaining control is for the teacher to create a certain formality in his relations with the pupils. As Waller (1932) puts it: 'Formality arises in the teacher–pupil relationship as a means of maintaining social distance, which in its turn is a means to discipline.' This is the doctrine of 'non-involvement' with the pupils, which seeks to preserve the teacher's emotional independence and inhibit familiarity between teachers and pupils. To abandon formality is to risk loss of what teachers call 'respect' from the pupils. The 'respect' is a concept very dear to many teachers, though they do not usually find it easy to define. I suspect that by this term teachers are referring to the response of the pupils to a teacher who is in control. It refers, in other words, to a blend of formality, politeness and obedience which constitute an acceptance of the teacher's rule-making function.

Because the disciplinarian sub-role is so central a task in classrooms, behaviour on the part of pupils which threatens this role must be defined as 'bad', and be discouraged or punished. Thus impoliteness, insolence, impertinence and disobedience are among

144

the cardinal sins of schoolchildren. Student teachers are often surprised at the apparent intolerance of teachers on such matters. In the students' eyes, a little insolence does not merit a major punishment. Such students have not accepted the not uncommon view among teachers that threats to the discipline role cannot be accepted and that the offenders must be punished *pour encourager les autres*.

The second basic sub-role that is common to all teachers is what we shall call the instructor role. As *instructor* the teacher must get the pupils to learn *and* show evidence of their learning. There are two basic aspects to the instructional task. The first concerns *what* shall be learned and refers to the content of the curriculum. The second concerns *how* this shall be learned and refers to the teacher's teaching methods. As part of his instructor sub-role the teacher must exercise his expertise in his subject; he must evaluate the pupils' progress; he must motivate them to want to learn and persist in making efforts towards that goal.

As in the case of the disciplinarian role, we must insist on a distinction between the role itself and the way in which the role is interpreted and performed by a particular teacher. It would be quite erroneous to regard the term instructor as implying a particular means of getting the pupils to learn. Teachers differ about the content of the curriculum, though here some uniformity can be imposed by the requirements of the syllabus of an external examination, as in most secondary schools. In the primary schools, where such external constraints are more rare, much greater variability in curriculum content is to be found. It is in the area of teaching method that the greatest variability is to be found. Teachers may fulfil their instructional sub-role by lecturing and dictating notes, by organizing group projects, or by promoting learning by each pupil on an individual basis. They may examine frequently or hardly at all. They may motivate the pupils to work by threat of punishment or by praise and encouragement, or by a blend of both.

In performing the instructor role, most teachers make efforts towards keeping the interaction in classrooms highly *task-related*. Talk must be instrumental to the furtherance of learning. Most teachers succeed in this aim. Ned Flanders has shown that between 85 and 95 per cent of classroom talk is task-related. Teachers tend to be suspicious of talk between pupils during lessons, because they are aware the pupils are likely to engage in talk which is not instrumental to learning. When non-task-related talk arises, it must be on the teacher's initiative, with the teacher's consent and on the teacher's terms. Otherwise it will be seen as a threat to learning process (wasting time, a red herring, and so on). Thus it is only the teacher's jokes that are funny. Many of the crimes that pupils commit in classrooms are indicative of the pupils' failure to abide by the expectations which spring from the teacher's instructional task—laziness, copying, talking. The fact that the single word 'talking' can appear on a detention register or misbehaviour report is significant. Such 'talking' is by definition illegitimate, that is, either talking when silence has been called for or talking about matters that are not task-related. In their communications to one another teachers do not need to specify more than the single word 'talking'; the rest is part of the teacher's shared assumptions and perspectives.

The distinction between the disciplinarian and instructor sub-roles of the teacher is a useful analytical one, but in practice they tend to fuse. As *organizer* the teacher may group the pupils to serve disciplinary objectives ('You two boys come and sit at the front where I can keep an eye on you'), or instructional objectives ('I will allow the bright children to work together so that they aren't held back by the slower ones'). Similarly the teacher's *moral* functions can be expressed within his instructional role, as in religious education, or within his disciplinary role, as when the teacher punishes a boy for being impolite. Frequently the two sub-roles blend together in the everyday behaviour of teachers. The value of the distinction between the two sub-roles rests on two points; that it represents a simpler and more comprehensive view of the teacher's classroom behaviour than lengthy lists of such roles; and that it is not possible to conceive of a teacher who does not *in some form* have to concern himself with the establishment of discipline and the promotion of learning.

It has already been suggested that there are great variations in the ways in which teachers interpret and perform the two basic sub-roles. What are these factors which influence how the teacher performs his role in the classroom? The most obvious factor is the uniqueness of any particular teacher—his background, training, attitudes, needs, personality. All these become major influences on the way in which the teacher perceives and performs his role. The second factor is a situational one, the role-set of 'significant others' with whom the teacher is in close contact in the performance of his duties. The pupils are the principal role-partner of the teacher in his classroom role, and it is with them that the teacher must directly negotiate his role. We shall consider the teacher's relations with the pupils in more detail shortly. At this point it is useful to recall that the pupils are not the only important influence on the teacher's conception of his classroom role. Within the school the expectations of the head-teacher and the other members of staff can be expected to exert considerable influence on the teacher's conception of his classroom role. Other persons, such as administrators, inspectors and parents may also be influential, but because they have much less contact with the teacher than do his colleagues, their influence is likely to be much weaker. The point is that in many respects it is the headteacher and colleagues, rather than the pupils themselves, who have the greater influence on the teacher's conception of his classroom role. As Waller (1932) put it:

The significant people for a school teacher are other teachers, and by comparison with good standing in that fraternity the good opinion of students is a small thing and of little price. A landmark in one's assimilation to the profession is that moment when he decides that only teachers are important.

This means that the teacher enters into negotiation with the pupils with a predetermined intention of living up to the expectations of his colleagues.

The influence of headteacher and staff on the teacher's classroom roles is of special interest because usually such persons are not present during the performance of his *classroom* role. The teacher in the classroom is typically insulated from their observations of his behaviour. One main reason why they exert so much influence in spite of their inability to observe directly is that they are a major source of the teacher's self-evaluation. The teacher does not judge his abilities as a teacher on the sole basis of the pupils' response. (If he did, how could a teacher who fails to make much progress with a 'difficult' class retain any self esteem?) The teacher tends to use the attitudes of his colleagues to him as a measure of his worth as a teacher. These colleagues estimate the teacher's abilities on the basis of the information they can glean about the teacher's classroom behaviour, which they cannot directly observe in detail. Sources of such information are what they see through the classroom windows as they pass by; what the pupils say about him; how they see the pupils acting towards him in the more 'public' places such as in the corridor, at lunch, on the games field etc.

The teacher knows that he will be judged by his colleagues in terms of his ability to master the two basic sub-roles of maintaining discipline and promoting learning. A teacher must therefore make sure that the information which is available to colleagues is congruent with successful classroom performance. Thus the teacher will strive to be treated by pupils in 'public' places in a manner that implies he is in control in his own classroom. For example, if pupils are polite and not insolent to him on the corridor, his colleagues will infer that this is the way he is treated in his classroom. Should the pupils be rude and disobedient to the teacher in public, the teacher will be embarrassed by this loss of face which implies a lack of competence in the classroom. It is because teachers cannot avoid such encounters with pupils in public and in the presence of colleagues that it is difficult for the teacher to maintain a false 'front' that he is more competent than public evidence supports.

This concern by the teacher to produce evidence that he is succeeding and to suppress evidence to the contrary, greatly affects what the teacher does in the classroom. He will try, notably in secondary rather than in primary schools, to make sure that children are sitting in a neat and orderly way, rather than roaming in apparent chaos around the room, not only because he might think this is a personal test of his own discipline or in his view the best means of promoting hard work by the pupils, but also because he knows that if the head-teacher or a colleague observes this, either through the windows or by an entry into the classroom, this situation will be used as evidence for his success in maintaining discipline and promoting learning. Similarly, he will try to keep the noise level of his pupils to a minimum, not only because it might disturb the teacher in the next room, but also because he feels that the teacher next door might infer from the noise that the class is inadequately disciplined and therefore not learning. Again, teachers in secondary schools must produce evidence that their children are learning, and the most 'objective' or 'concrete' form of such evidence consists of examination results. So the teacher is concerned that his children should achieve good results not only for the children's sake but also because he knows that the pass rate and grade levels of the pupils' examination performance will be taken by his colleagues as an index of his ability to promote learning.

Thus the expectations and evaluations of colleagues are a major influence on the way in which a teacher interprets and performs his classroom role. In my own view, this reliance of teachers on the esti-mations of their colleagues—which I might add, they are very anxious to deny—represents the greatest conservative force, the greatest in-hibitor of educational change and experiment in our secondary schools. For a teacher to introduce innovations into his classroom, he has to take risks and live dangerously. Yet the teacher is unlikely to take such risks unless he is prepared to reduce, but not totally reject, his dependence on colleagues' evaluations, which are usually rooted in the traditional indices of discipline and learning, namely noise and examination results. Noise is a relatively poor index of either disci-pline or learning. Certain types of teaching method, such as project work, may involve a level of noise which is by no means inconsistent with a disciplined learning. And as long as teachers rely on examina-tion results as the major proof of the ability to promote learning, a tendency which is encouraged by headteachers, colleagues, parents and pupils alike, preaching about developing the child's *interests* will have as much impact on the teacher's perspective as does a light breeze on the Tower of London.

Role style

It has several times been stressed that every teacher interprets and performs his role in a unique way. Indeed every lesson a teacher conducts can never be precisely repeated. At the same time, every teacher behaves with some degree of consistency on different occasions and may thus be said to develop a *style* of performing his classroom role. It should be possible to distinguish different styles which are common to many teachers and construct a list of teacher types with recognizable role styles. One example we have already discussed is the distinction made by the 'interaction analysis' researchers between 'direct' or 'teacher-centred' and 'indirect' or 'student-centred' teachers. As in the case of classroom sub-roles, many writers have suggested long lists of different role styles (Waller, 1932; Redl, 1951; Bush, 1954; Thelen, 1954; Soles, 1964; Hoyle, 1969). We shall consider a selection of the suggested role styles, which clearly embody different interpretations of the classroom role, different conceptions of the teacher–pupil relationship, and different educational philosophies. Willard Waller, whose book written in 1932 was the pioneering study of interpersonal relations in schools, suggested the following list:

1 the parent substitute.
2 the cultural or social ideal.
3 the officer and gentleman.
4 the patriarch.
5 the kindly adult.
6 the love object.
7 the easy mark.
8 the nincompoop.
9 the tyrant.
10 the weakling.
11 the flirt.
12 the bully.

Waller's brilliant and insightful cameos of these teacher types, though now somewhat dated, remain unsurpassed. Take, for example, the case of the *patriarch*. Waller notes that this elderly teacher, who takes a fatherly interest in his pupils, is treated with great affection. His discipline is not firm and he rules by personal influence rather than by means of formal sanctions. In some respects he may appear to be incompetent in academic matters, but the pupils will happily defend him and insist that he has taught them a great deal. His idiosyncracies, his readiness to fall for 'red herrings', his anecdotes, his absentmindedness, all these are fully approved by the pupils. Waller rightly notes that age alone does not make one into a patriarch and that there is rarely room in a school for more than one such teacher.

Clearly these different role styles described by Waller represent different definitions of the situation, different conceptions of teacher and pupil roles, different educational philosophies and goals. They are variations on a theme; many represent deviations from an implicit 'normal' or 'ideal' teacher. Whilst they are not treated in any systematic way, they all have the flavour of 'real' teachers and an experienced teacher, on reading Waller's descriptions, instantly and effortlessly finds that present or former colleagues spring to mind.

Herbert Thelen (1954) takes a different approach to teacher role styles, which are conceived in terms of *models*.

> It is as if the teacher had a model in mind and operated consistently to make the classroom conform to this model; it represents the teacher's idea of what the classroom should be like. When the classroom situation deviates from this image, the teacher then tries to rectify matters by taking action . . . . The teacher's model summarizes for him the principles of learning; his action is taken to maintain the model, using principles of educational method as his guide.

Thelen suggests seven basic models, which can be briefly summarized.

Model 1: Socratic Discussion: After the pupils have learned some facts, the teacher introduces a challenging discussion to clarify concepts and values and to test conclusions.

Model 2: The Town Meeting: The teacher and pupils discuss, in an open, friendly and co-operative manner, how to organize and carry out specified learning activities.

Model 3: Apprenticeship: The pupil identifies with the teacher and learns many attitudes in imitation of the teacher, who actively nurtures the pupil in his own image.

Model 4: The Army Model: The teacher tells the pupils what to do and how to do it, sees that it gets done and finally evaluates how good a job he thinks it is.

Model 5: The Business Deal: The teacher makes the best deal he can with individual pupils, discussing specifications for each piece of work and being available for consultation as the work progresses.

Model 6: The Good Old Team: The teacher's objective is to achieve a high standard of learning and almost any device of persuasion, cajolery, promises and threats that will produce high-level performance are regarded as legitimate.

Model 7: The Guided Tour: The teacher conducts the children through the field of study, giving information, stories and opinions, calling attention to objects of special interest, and answering questions.

In many respects these models are an advance on Waller's teacher types. They are more directly related to the teacher's instructional role and in addition they are less concerned with the teacher's personality and more truly interactional in specifying roles for both teachers and pupils.

Role, role style and the 'good' teacher

At this point I want to interrupt our general analysis of teacher–pupil interaction in order to digress briefly into the fascinating question of the 'good' teacher. It is a necessary digression because a study of different role styles leads us to ask the question whether or not some role conceptions and role styles are better than others. In answering the question we may find that the digression leads us to uncover some important aspects of the central theme of teacher–pupil interaction.

Most teachers have some ideal, however vague, of what constitutes the good teacher and any other teacher who approximates to this ideal will be regarded as good. So if a tutor in a Department or College of Education reports that a student is a good teacher, he may be telling us little more than that the student has a role conception and style of which the tutor approves. Our notion of the good teacher can obviously be more adequately conceptualized than this. (Here the analysis will be confined to the teacher's instructional role, since the idea of 'good discipline' will be dealt with in the next chapter.) Two things seem to be implied by the term 'good'. Firstly, a *moral* judgment seems to be involved. Some teaching styles and methods may be held to be morally superior to others. It might be claimed, for example, that 'democratic' teaching styles are morally superior to 'autocratic' ones or that discovery methods are more moral than indoctrination. Such assertions raise complex philosophical problems which cannot be dealt with here. The second meaning of 'good', which is more relevant to the present discussion, is the implication that the good teacher is effective or successful.

On this view the good teacher is one whose pupils learn what the teacher intended them to learn as a direct result of the teacher's teaching. This seems to be a very simple proposition to test empirically. A set of pupils knows nothing about logarithms; Mr Smith teaches them logarithms; finally we give the pupils a logarithms test. On the basis of the test performance of the pupils we can place Mr Smith along the continuum of highly effective to totally ineffective— in other words, find out how 'good' he is. In practice the problem cannot be solved so easily. First, we often lack satisfactory attainment tests. Second, attainment is probably a very crude measure of teacher effectiveness, in that presumably the good teacher is also one whose teaching results in understanding and interest. Third, even if we confine ourselves to an attainment measure, the number of variables potentially affecting the attainment is very large. Among the most elementary would be:

(a) The teacher—his role style, personality, values, etc.

(b) The content of what is taught—the subject or aspect of the subject.

(c) The teaching method—lecture, demonstration, group work, private study, etc.

(d) The pupil—his intelligence, social background, motivation, attitudes, intentions, etc.

(e) The situation—the time and location of the class, the general teacher–class relationship, immediately prior events, etc.

It is difficult to assess which of these variables is influencing the learning reflected in the attainment test. It may be one single variable, or a combination of two or more variables, that is the principal cause of the learning. Because there are so many variables, each one of which is exceedingly complex, it is difficult to apply the standard experimental design of calculating the effect of one or more variables whilst holding the others constant. Fourth, this method ignores the important long-term effects of teaching, especially on the developing interests of pupils. How can one with certainty ascribe the interests and attitudes of the young adult to particular classroom events many years previously, when so many other factors have intervened in the interim? Also, is it not possible that a whole set of classroom events, each inconsequential in itself, might have a cumulative effect over an extended period? Fifth, what constitutes an effective teaching style for one pupil may not be so for another pupil in the same lesson. Tom may enjoy Mr Smith's lectures and learn from them, whilst Dick is bored during such a lesson. Even more worrying to the researcher—and to Mr Smith—is the fact that Dick may learn nevertheless *in spite of* Mr Smith's lectures. Sixth, we cannot evaluate role styles or teaching methods *in vacuo*, but only in relation to specific teachers, pupils and situations. It is absurd to try to test the effectiveness of lecturing pupils unless we consider the relative abilities of different teachers in their lecturing behaviour. There are 'good' lectures and 'bad' lectures.

For these reasons the question 'What role style should I adopt to be a good teacher?' does not permit an easy answer. We do not have clear evidence on the point and I doubt that we ever will. The answer consists in suggesting that the question cannot be asked in that form, since it ignores the realities of the nature of teaching, learning and classroom interaction. Here the social psychologist has a definite contribution to make. As Thelen (1954) puts it:

The most fundamental thing about classroom experience is that it is social; it is a continual set of interactions with other people. I call this the most fundamental thing because there is no escape;

the demands are there, and they must be met. . . These inter-actions are most fundamental for another reason: they make a difference in the learning process . . . Social interactions set the conditions under which learning occurs.

Whilst the psychologist stresses the characteristics of individual pupils affecting learning, such as intelligence, and the sociologist emphasizes factors such as home background, the social psychologist is more concerned with the interactional here-and-now of classroom relation-ships between teacher and pupils and between pupils. The social psychological perspective on the 'good' teacher is concerned with such factors as the overall relationship between teacher and pupils, their mutual attraction or hostility, the ways in which they perceive, evaluate and react to one another, the ways in which the teacher's behaviour creates, sustains and changes these relationships.

It is impossible to prescribe a particular role style since the effec-tiveness of a particular role style depends on its appropriateness to the teacher, the pupils and the situation. Each teacher has to consider the uniqueness of every teaching situation in which he finds himself and choose the role style that (a) he can execute well, and (b) is the most appropriate to the pupils, the nature of the task and the general classroom situation. In Thelen's words:

> The teacher controls the learning situation by controlling his own role, and his role is different in different types of activity . . . The appropriate activity, then, is what results from the teacher playing the proper role at the proper time; its creation is a natural process of interaction, of living together. The only be-haviour the teacher can control directly is his own, and he does this through diagnosing the class need and then shifting himself into the type of role needed from him to enable the class to meet its need. The members of the class must then shift their roles, to accommodate the teacher, but if his diagnosis and operation is correct, the class will have high motivation and involvement in making the shifts in their own roles.

So there are no simple answers to the question 'What is the best way to do x with a class?' The question can be more properly phrased as 'What is the best way for me to do x with these particular pupils in this particular situation?' This defies a general answer; a specific question requires a specific answer. Moreover there may not be a single answer, for there may be several teaching styles or techniques that are functionally equivalent or equally successful in attaining the objective.

Thelen's words, quoted above, disclose the central interest of the social psychologist in the 'good' teacher, namely the social skills involved. These have not yet been adequately conceptualized or analysed. If a characteristic of the 'good' teacher is the ability to adopt a role most appropriate to the activity, i.e. a role which will facilitate the pupils' learning, then it is probably true to say that, at the very least, the teacher must (a) have a wide range of roles or role styles on which he can draw when needed, (b) have the ability to diagnose accurately the situation 'out there' among the pupils, by a system of continuous monitoring, so that he knows when and in what way to shift roles, and (c) have the ability to get accurate feedback on the effect of his own behaviour on the pupils so that he can gauge the appropriateness and success of his own role changes and modifications. He needs, in short, considerable flexibility and sensitivity. These terms do not in themselves, of course, specify the nature of the social skills involved or the way in which they are acquired. In my view research into these two areas is urgently required and is likely to be more profitable than much current research on teacher effectiveness. For the moment we must, through ignorance, let the matter rest, though we shall return to this area again in the next chapter.

The teacher's definition of the situation

Up to this point we have considered how, in defining the situation in the classroom, the teacher makes clear to the pupils his own conception of his classroom role and the specific ways or style in which he intends to perform his role. Implicit in the teacher's definition of his own role is a definition of the pupil role. The teacher cannot specify how he intends to behave without at the same time specifying how he intends the pupils to behave. Certainly he cannot succeed in realizing his own role conception unless the pupils' response is in accord with that conception. Thus the teacher's definition of his own role and his definition of the pupil role are opposite sides of the same coin. The teacher's expectations of the pupils will derive from, and be congruent with, his conception of his own role. So when the teacher explains to the pupils how he is going to behave towards them, he is simultaneously clarifying his expectations of them in their pupil role; and when he prescribes behaviour for the pupils, he is simultaneously disclosing his conception of his teacher role.

If the teacher's two basic-roles in the classroom are those of disciplinarian and instructor, then the teacher's expectations of the pupils will hinge upon the way in which the teacher interprets these two sub-roles. This means that it is as difficult to specify teachers' expectations of pupils as it is to specify how teachers conceive their own role. There will be as much variation in the one as in the other. So the teacher's expectations of his pupils cannot be characterized except in the most general terms, subject to modification in the case of a particular teacher. I shall take the case of the teacher with a fairly traditional conception of his role. As disciplinarian, the teacher tends to expect the pupils to be obedient, respectful, polite, formal and quiet. As instructor, he will expect the pupils to pay attention, work hard, not copy, and show interest and enthusiasm. It is such behaviour on the part of the pupils which is complementary to, and congruent with, the teacher's conception of his own classroom role. If the pupils do not behave in such ways, he cannot live up to his self-conception as disciplinarian or instructor—except, of course, by altering his view of the nature of the sub-roles.

The pupils will be evaluated by the teacher on the basis of the degree to which they conform to his expectations. In conforming, the pupils offer support to the teacher's role, and thus make the teacher's role performance highly satisfying and (in exchange theory terms) highly profitable to him. Hence he approves of such pupils and regards them as 'good'. If the pupils deviate from the teacher's expectations, they are withdrawing support, showing antagonism, and making it difficult for the teacher to perform his role. In consequence, the teacher's interactions with the pupils involve a loss and are experienced as dissatisfying; he disapproves of such pupils and regards them as 'bad'. Because teachers can have very different definitions of the situations, including very different conceptions of teacher and pupil roles, the same pupil behaviour can be perceived and evaluated very differently. A pupil can be seen as 'well-behaved' by one teacher but as 'lacking in life and initiative' by another teacher.

The teacher perceives and evaluates his pupils on the basis of the two aspects or dimensions of his own classroom role (cf. Parsons, 1959). See Figure 6.3.

		Discipline	
		Conforms	Deviates
Instruction	Conforms	'Good' pupil	
	Deviates		'Bad' pupil

FIG. 6.3

The first dimension concerns the instructional role. The pupil is judged traditionally on the degree to which he pays attention, works hard, and shows enthusiasm and interest. The second dimension

155

consists of discipline. The pupil is perceived and judged on the extent to which he conforms to the teacher's expectations as disciplinarian—traditionally the extent to which he accepts the teacher's authority, is obedient, polite and quiet. The pupil who conforms to the teacher's expectations on both dimensions is evaluated as a 'good' pupil; the pupil who deviates from the teacher's expectations on both is a 'bad' pupil.

This leaves two residual categories, in which the pupil conforms along one dimension but deviates along the other. On the one hand we have the hard-working-but-badly-behaved pupil and on the other we have the well-behaved-but-lazy pupil. I suspect that teachers differ in their preference for, and more favourable evaluation of, one of these two types of pupil. But in all teachers' minds they represent an 'impure' type, fitting neither into the extreme 'good'/'bad' categories nor into 'average' category containing those pupils who conform only moderately on both dimensions.

It is very important to realize that conformity to the disciplinary role plays as important a part in the teacher's evaluation of the pupil as does conformity to the instructional role. Teachers tend to think of the term *evaluation* as referring to academic behaviour only, linking the concept in their minds with tests and grades. But deviance from the disciplinary expectations meets with teacher disapproval just as does deviance from the instructional expectations. 'Bad' pupils arise from disciplinary failure ('weakness of character', 'the wrong attitudes', 'defective values') as well as from instructional failure ('laziness', 'stupidity').

There is considerable evidence which supports this analysis of the teacher's perception and evaluation of his pupils. Wickman (1928) has shown that teachers regard pupil moral offences, threats to the teacher's authority and violations of classroom order as more serious than shyness or unsociability.

> Those problems which transgress the teachers' moral sensitive-ness and authority or which frustrate their immediate teaching purposes are regarded as relatively more serious than problems which for the most part affect only the welfare of the individual pupil.

Hollins (1955), in an improved design, substantially confirmed this work on a British sample and found no differences between men and women teachers or the types of school in which they taught.

From our analysis we would predict that teacher would show greater liking for 'good' pupils than for 'bad' pupils. Williams & Knecht (1962) have demonstrated that the pupils who are most liked by the teachers tend to be the high achievers, that is those who conform to the instructional expectations. Instruction is a more intricate dimension than discipline since its central element—hard work—is confounded with attainment and ability. It is understandable that teachers should in practice experience certain difficulties in disentangling the complex of hard work, ability and attainment. Teachers tend to use the work the child produces, his attainments, as a measure of his hard work, for effort independent of attainment is not easy to observe and measure. I know of only one case where a teacher wrote on a boy's report, 'He has worked hard, but made no progress.' Yet where the attainment is high, it may be a product of high ability rather than high effort. Teachers are aware of this, as when they write on the report of a pupil with low attainment, 'She has ability, but makes little effort.' However, I suspect that although teachers can distinguish effort, ability and attainment from one another where the disparities are reasonably great, they cannot do so either as a matter of course or as much as they think they can. There are two reasons for this. Firstly the teacher is unlikely to have in his possession a full picture of a child's abilities. Secondly both attainment and effort are more open to regular inspection by the teacher in the everyday life of the classroom. Thus we would expect the teacher to give a more favourable evaluation to the pupil with high attainment, which implies that high effort has been expended, than to the pupil with high ability, which has no implication with respect to effort. This is indeed what is suggested by the findings of Williams & Knecht, who report that the correlation between teacher liking and attainment is 0·668, whereas the correlation between teacher liking and ability is 0·478. The importance of the pupil's ability and achievement to the teacher's perception of him is made clear in a study of Jackson & Lahaderne (1967). Whilst the teacher can predict the level of a pupil's satisfaction with school better than chance, the teacher's estimate of a pupil's satisfaction correlates with the pupil's achievement and intelligence at a much higher level than with the pupil's actual satisfaction. It seems that the teacher infers that if a pupil is achieving well he will be satisfied with school.

Other researches support the view that the pupils who give the teacher role support are perceived more favourably. Lambert (1963) has shown that pupils who are rated by the teachers as *successful* are also rated as more able, curious, imaginative, clearthinking, capable, alert, friendly, enthusiastic and cheerful. Ausubel, Schiff & Zeleny (1954) found that pupils rated by the teacher as more *adjusted* were also perceived as being more persistent and competitive and as having a better scholastic record. It seems that teachers fall for the *halo effect* in perceiving and evaluating their pupils. Those who conform to teacher expectations are seen to possess a wide range of characteristics whilst those who deviate possess all the vices. The best demonstration of this is the work of Bush (1954) who found that teacher liking is related to every single characteristic on which he asked the teachers to rate the pupils—intelligence, attainment, class conduct, quality of thinking, emotional balance, and probable college success.

In a small scale British study of one secondary school classroom and one primary school classroom, Pickles (1970) failed to show a significant relationship between teacher liking and pupil achievement, though he does give evidence of other perceptual distortions in the teacher. (The relevant correlation co-efficients are given in parentheses, the secondary classroom ones being given first.) Teachers like those pupils whom they perceive like them (0·66, 0·53), though to a greater degree than the pupils actually like them (0·56, 0·13). Teachers like those pupils whom they perceive as conformist to their role expectations (0·89, 0·69), and they believe that such pupils reciprocate the liking (0·72, 0·62) to a greater extent than such pupils do indeed like the teacher (0·50, 0·26). These distortions may have very important consequences for relationships, though as yet they have not been adequately investigated.

These more favourable perceptions and evaluations of the conformist pupils are expressed in the teacher's treatment of such pupils. We all tend to give preferential treatment, perhaps unconsciously, to those we like and of whom we approve. Thus in the Lumley study (Hargreaves, 1967) the teachers favoured the pupils of the higher streams who were more conformist to teacher expectations than low stream pupils. High stream pupils were more likely to be taken by teachers on visits to local places of interest during school time, and were most likely to obtain places on the oversubscribed holidays abroad which were organized by the school. It has been shown by Toogood (1967) that the children who are given responsibility by teachers (e.g. made monitors and prefects, given supervisory jobs) are the pupils who are also perceived by the teacher as being more likeable, more co-operative, better behaved, higher attainers. Clearly all this work on the teacher's evaluation and perception of pupils is closely connected with the concept of the self-fulfilling prophecy discussed in an earlier chapter. If the teacher's behaviour towards a pupil is conditioned by his perception and evaluation of the pupil, then we need to know very much more than we do at present about the basis on which the teacher makes his perceptions, evaluations and categorizations.

Whilst it is true that conformity to his expectations of discipline and instruction is the principal basis on which the teacher perceives and evaluates the pupil, it is not the only basis. Hallworth (1961, 1962) had teachers rate the personality of their pupils. The results show that teachers tend to perceive pupils in roughly the same way. In the factor analysis of the ratings, two main factors emerged. The first factor, which is based on the teacher's implicit question 'How does this child get on with me?', is called by Hallworth the factor of conscientiousness and reliability. It involves such traits as emotional stability, trustworthiness, persistence, co-operation with the teacher, and maturity. This suggests that the first factor represents the teacher's estimate of the pupil's conformity to his instructional and disciplinary expectations. (That these combine into a single factor confirms our earlier suggestion that the disciplinary and instructional sub-roles are distinct only in analysis but not in real life.) The second factor seems to arise from the teacher's implicit question, 'How does this pupil get on with other pupils?' This factor involves such traits as cheerfulness, sense of humour, spontaneity, sociability; it is called by Hallworth the factor of extraversion. Now these factors are independent of each other; there is relatively little overlap between the first factor—the teacher's pupil—and the second factor—the pupil's pupil. The research showed that there was more overlap between the two factors in secondary modern than in grammar schools. The reader might like to guess at the reasons for this: several clues to the answer have been laid earlier in the book. But this research does show quite clearly that the teacher perceives the pupil not only in terms of his relationship with the pupil, but also in terms of the pupil's relationship with his peers.

The teacher's perceptions, interpretations and evaluations of pupil behaviours are made evident in the common terms and labels which teachers use to describe particular pupil types or pupil characteristics. Some of these labels may be used in public (for example to parents) whilst others are restricted to communications, oral or written, between teachers. Almost all are used regularly by teachers in their verbal interactions with pupils.

	Positive label	Negative label
General	Good lad	Nuisance
	Sound	Pain-in-the-neck
	Promising	Fool
	Nice	Trouble-maker
	Making progress	Going to the dogs
Instructional	Hard worker	Idler
	Bright	Thickhead
	Neat	Untidy
Disciplinary	Quiet	Chatterbox
	Polite	Cheeky
Peer	Leader	Ring-leader
	Friendly	Bully
	Popular	Lone-wolf

Many other factors can be expected to affect the teacher's perception and evaluation of his pupils. For example, we might expect teachers to perceive children from different social backgrounds in different ways. Or we might expect older teachers to perceive pupils differently than young teachers. Both these ideas have been the subject of research. McIntyre, Morrison & Sutherland (1966), in one of the most fascinating researches of recent years, have shown that the teacher's perception of his pupils varies according to the social class background of the pupils. They achieved this by examining the content of the first factor of reliability and conscientiousness (the teacher's implicit question, 'How does this child get on with me?') of teachers with different sorts of pupil. For teachers with middle class children, the factor has high loadings on trustworthiness, co-operation with the teacher, and courtesy. The IQ loading is a very low one of 0·20. For teachers with working class pupils, the factor has the highest loadings on attainment, attentiveness, and the pupil is worth the trouble. The IQ loading is high—for suburban working class children 0·50, and for urban working class children 0·68. It seems that where the classroom consists of middle class children, all the children will be fairly bright, hardworking and enthusiastic. All, in short, conform to the instructional expectations. The children are more easily differentiated along the disciplinary dimension—the degree to which they can be trusted, are polite, and so on. In a working class school, on the other hand, most of the pupils are resistant to discipline, so they can be more easily differentiated along the instructional dimension—the degree to which they are attentive, bright, and hardworking. This same research also showed that young teachers tend to stress good behaviour from the boys, whereas older teachers put greater emphasis on attainment. This suggests that the younger teachers' perceptions are influenced by the problems they experience with discipline. The perceptions of older teachers, who have mastered the discipline problem, are more influenced by instructional problems.

The impact of the pupil's social class and the teacher's perception and treatment of him have been of particular interest to researchers in recent years, especially in relation to streaming. Jackson (1964), Douglas (1964) and Barker Lunn (1970) have all shown that in the primary school the working-class pupils are disproportionately represented in the lower streams and the middle-class pupils in the upper streams. This in itself may simply indicate that middle class pupils are on average of higher intelligence or have higher attainment levels than working-class pupils. However, the data cannot be fully explained in this way. Jackson has shown that the most common criterion for the allocation of pupils to different streams is the teacher's recommendation rather than an objective intelligence or attainment test. Clearly such subjective estimates may be open to considerable bias. This seems to be confirmed by Douglas, who noted the overlap in test scores of pupils in adjacent streams. If we consider the pupils of the *same* measured ability, it is the middle-class child who tends to be assigned to the higher stream and the working-class child to the lower stream. Barker Lunn made a direct comparison between the teacher's ratings of a pupil's ability and his actual performance on an English test. In most cases the two measures agreed. But where they did not agree, working-class children were more likely to be underestimated and middle-class children to be overestimated. We do not know how such perceptual bias, and its consequent discrimination, with respect to the pupil's social class operates. I am inclined to think that it is unintentional on the teacher's part, though some writers seem to suggest otherwise. It seems likely that the differences between middle-class and working-class pupils in speech, appearance, values, attitudes and behaviour, which can soon become very apparent in the classroom, are used by the teacher as a part of the basis for estimating the pupil's intelligence, ability and future potential. It is clear that teachers need to be warned of the potential bias in their perception of pupils' abilities and of the way in which the perception can affect their treatment, including their expectations, of the pupils.

In summary, we can say that the teacher defines the situation in terms of his own roles and goals, especially as they relate to his instructional and disciplinary objectives, and assigns to the pupils roles and goals that are congruent with his own. He selectively perceives and interprets pupil behaviour in the light of his definition of the situation. On the basis of further interaction with the pupils and repeated perceptions of them, he develops a conception of individual pupils (and classes) who are evaluated, categorized and labelled according to the degree to which they support his definition of the situation. He then responds to pupils in the light of these evaluative labels.

The pupil's definition of the situation

We have considered in great detail the ways in which teachers seek to define the situation in the classroom. But what of the pupils? How do they wish to define the situation? In fact it is very difficult to discover how pupils in school wish to define the situation. We have to bear in mind that generally speaking, in their interactions with adults, children are used to having the situation defined for them. Indeed adults, especially parents, actively train their children to accept the situation as they, the parents, define it. The result is that most children in most of their interactions with adults assume a passive position of acceptance. They know who is boss. Life in school seems to be no exception. Children first go to school when they are very young—by the age of six at the latest. As the children are immature and typically somewhat bewildered in this new and strange world of school, they

161

see their task largely in terms of discovering what is expected of them by the teachers. Should the six-year-old try to define the situation in ways that conflict with the teacher's, then the teacher will try to persuade the child into changing his notions of what school and being a pupil are about, and if necessary the teacher can strengthen his argument with formal sanctions. On the whole parents encourage the child to accept the teacher's definition of the situation—at school the child must do as he is told. By the age of ten the vast majority of pupils seem to have accepted the teacher's definition of the situation. School is an inevitable and inescapable and natural part of their lives. It is there and they accept it.

Researches show that most pupils seem to do more than accept the teacher's definition of the situation in school: they *like* it. This is so presumably because we as adults, parents and teachers, have succeeded in persuading children that school is a pleasant experience ('You're going to start school next week, Wendy? Won't that be lovely!' 'You don't want to go to school, Peter? Don't be silly. Schooldays are the best time in your life.') and also because what happens to children in school is indeed pleasant for the most part. All the studies of pupils' attitudes to school suggest that the vast majority like school. Even in a down-town secondary modern school like Lumley, 74 per cent of the boys in their final year say that they like school on the whole. The qualification of *on the whole* is an important one. As Jackson (1968) has reminded us, the pupils' feelings about life in school are probably very complex and somewhat ambivalent. Probably very few pupils are ecstatically happy in school and probably very few experience their schooldays in perpetual abject misery. For the average pupil, school is a mixture of good and bad, pleasant and unpleasant. When pressed to a simple choice between like and dislike in general, with no regard to specific aspects, he will usually choose the favourable rather than the unfavourable response.

Perhaps a more fruitful approach in discovering how pupils feel about their lot would be to study the pupil's reactions to the teacher himself. It is the teacher who defines the situation, who defines the roles of both teacher and pupil in the classroom. If we can find out the sort of teacher that pupils like, from this we might infer those definitions of the situation of which pupils approve and also the roles they prefer to perform, since both of these are strongly dependent on the way the teacher conceives his own role.

Studies of pupils' attitudes to teachers, both in Great Britain and in the United States, have produced a remarkably unified and consistent picture. It seems that Western school children, of all ages and in different types of school, are in very high agreement about the teacher behaviours which they like and dislike. From the researches (e.g.

Bush, 1942; Tiedeman, 1942; Michael, 1951; Allen, 1961; Taylor, 1962) we can draw this picture of pupil attitudes to teachers.

Like	*Dislike*
A teacher who . . .	A teacher who . . .

(a) Discipline

—keeps good control.	—is too strict; is too lax.
—is fair; has no favourites.	—has favourites; 'picks on' pupils.
—gives no extreme or immoderate punishments.	—punishes and threatens excessively/arbitrarily.

(b) Instruction

—explains and helps.	—does not explain; gives little help.
—gives interesting lessons.	—does not know subject well; gives dull or boring lessons.

(c) Personality

—is cheerful, friendly, patient, understanding etc.	—nags, ridicules, is sarcastic, bad tempered, unkind, etc.
—has a good sense of humour.	—has no sense of humour.
—takes an interest in pupils as individuals.	—ignores individual differences.

Each area of the three listed is not given the same weight of importance by the children. In Taylor's (1962) study of British children of different ages and in different kinds of school, the instructional area was perceived as the most important, bearing about 40 per cent of the weight, with discipline next (33 per cent) and personality last (25 per cent). The pupils, like their teachers, see the instructional area, the process of learning, as the primary task of their life in the classroom. They approve of the teacher who can 'put over' his subject in an interesting way. They agree with the teachers that discipline is important as a prerequisite to adequate learning, a necessary means to the end of learning. Finally, they approve of teachers whose pleasant disposition creates a warm, relaxed friendly climate of personal relationships within which the learning process can proceed.

At the same time we should perhaps be a little cautious in accepting these findings at their face value. The picture which emerges is basically a stereotype of the good teacher. It may be that the questionnaires, by conceiving the ideal teacher in abstract terms or as a bundle of discrete characteristics that make satisfactory test items, tend to elicit such a stereotype. In real life the teachers that the pupils like or

163

perceive as good may not have a close resemblance to the 'identi-kit' good teacher of the researchers. Perhaps if they had asked pupils to nominate their favourite teacher and then investigated the characteristics and behaviours of such teachers, a somewhat different picture would have emerged. Perhaps, too, if we took into account the pupils' own views, and examined the terms they use to describe and differentiate teachers (including their labels for teachers), then other dimensions of teacher behaviours would be brought to light.

The working consensus

The implication of our review of researches on pupils' attitudes to teachers is that the majority of pupils accept the teacher's definition of the situation and are relatively content to conform to the teacher's role expectations of them, providing that the teacher does not behave in ways which meet with their disapproval. Only teachers who maintain an excessively strict or lax discipline, who are boring and unhelpful in their teaching, and who have cold, humourless and unpleasant personalities should experience any difficulty in reaching a consensus with the pupils. Assuming that such teachers are in a minority, or behave in such ways for only a small proportion of the time, the definition of the situation in the classroom is readily agreed between teachers and pupils.

Three kinds of outcome in reaching a consensus about the definition of the situation are possible. The first can be called *concord*, which arises when consensus is high. Here the definitions of the situation by teacher and by pupils are congruent and compatible. The situation is pleasant for both participants; both can make profits. The second type of outcome is *discord*. In this case the definitions of the situations are incompatible, and consensus is low. The situation is unpleasant for the participants; each makes a loss. Between these two forms stands a third type, *pseudo-concord*. Here the definitions of the situation are congruent and compatible only in part. The situation is partly pleasant, partly unpleasant. It represents a doubtful bargain: sometimes the participants make profits and at other times losses.

The profound effects of the extreme categories of concord and discord on teacher–pupil relations in the classroom can be examined at a deeper level if we introduce Goffman's (1961) concept of *role distance*. By this term Goffman wishes to consider the fact that in actual performance of a role many actors seem to express an apparent separation or detachment from the role as determined by the expectations of others. That is, the actor adheres less strictly to the role prescriptions and introduces idiosyncratic elements which are irrelevant to the role expectations. The new teacher, inexperienced in enacting his role, tends to restrict his role performance to behaviour which is demanded by the role prescriptions. Typically, he does not find this either easy or satisfying. With greater experience in playing the role, the teacher acquires greater confidence in playing the role and is able to express role distance.

In a situation of concord the teacher is able to show role distance because there is a basic consensus between teacher and pupil about the roles they expect one another to play. Since the teacher's disciplinary and instructional roles are not under threat, he is able to come out-of-role and transcend his role. He becomes a *person*, not just a *teacher*. Given that the general consensus about the definitions of the situation is high, when on occasions the pupils do try to assert a definition of the situation which conflicts with that of the teacher, the reassertion by the teacher of his own definition of the situation is easily effected. Discipline can be restored by the joking invocation of a sanction. Rather than threatening the offenders with a formal punishment, he can recall to the pupil's mind the basic negotiated consensus with a statement such as, 'If you keep on chatting, I'll have to arrange a meeting between my cane and your bottom, won't I, Tommy?' The use of the pupil's first name and the pleasant manner in which the threat is veiled are sufficient to restore concord or pseudo-concord. When the teacher can express role distance, he gains in personal satisfaction, and becomes highly attracted to the teacher role which supports his self-identity. Although Goffman reminds us that role distance tends to be the prerogative of the more powerful participant in an interaction, where concord is achieved the pupils are also permitted some role distance. Because the teacher does not feel under threat he grants the pupils greater autonomy and familiarity; they too are given freedom to become persons rather than mere role-performers.

In the case of discord, the majority of the pupils reject the teacher's definition of the situation and the teacher is unwilling to accept the pupils' definition of the situation. The teacher thus has to enforce his definition of the situation against the pupils' will. Every lesson becomes a battle to assert his definition of the situation in a relationship where negotiation has become impossible. Because the teacher is under threat, he is forced to adhere very strictly by his formal role requirements. To express role distance would be to exacerbate the threat for it would be to imply that his stern manner and his demands for work or order need not be taken seriously by the pupils. Because the teacher becomes a role-performer rather than a person, he finds that he derives few satisfactions from his job and his attraction to it declines.

It seems to me that the extreme types of concord and discord are relatively rare. Much more typical is pseudo-concord, or an outcome that lies between concord and pseudo-concord, where the consensus is far from perfect and where the profits made by the participants are intermittent and relatively small. Only in exceptional circumstances do teachers and pupils in classrooms reach concord or discord. In other words, it is rare for the definitions of the situation between teacher and pupils to be so incongruent that negotiation of a working consensus becomes virtually impossible, just as it is rare for the two definitions to be so compatible that negotiation of a working consensus becomes unnecessary. With respect to the instructional and disciplinary roles the pupils do not seem to be entirely content with the teacher's definition of the situation in the classroom. In the event most of the pupils do conform to a fair degree to the teacher's definition of the situation (which is why pseudo-concord often *looks* very much like concord), partly because the teacher is in a position to impose his definition of the situation with his much greater power. Negotiation becomes, at least in theory, unnecessary where one participant has the power to enforce his definition on the other. But the pupils do have some power, so when the teacher tries to replace negotiation with imposition he finds that he activates resistance, subversion and interpersonal antagonism that effectively promote discord. The absolute imposition of the teacher's definition of the situation is really impossible and the side effects of attempts to do so make such a course of action inadvisable. The teacher has to balance his own personal satisfactions with the need to impose a definition of the situation that is expected by the headteacher, colleagues and other role partners. So in practice the teacher does not always enforce his definition of the situation where he has the power to do so and where it seems to be demanded by his role partners. He aims instead at a negotiated settlement whereby teacher and pupils each go half way with respect to some demands and whereby in other areas the teacher withdraws or moderates his demands on the pupils in return for conformity to other teacher demands. This negotiated settlement may fall short of the teacher's ideal definition of the situation but it is realistic in that it averts discord and ensures that a fair number of his demands are met and that teacher–pupil relationships are generally good. The pupils, realizing that their position is not a strong one from which to bargain, are usually content with the concessions made by the teacher. Fortunately for most teachers the pupils underestimate the strength of their position especially in its collective form. The fears of teachers and headteachers with respect to 'pupil power' groups such as the Schools Action Union stem from the recognition that once pupil power is collectively mobilized it may be used effectively against the teachers' wishes. These teacher fears also indicate that teachers are aware that many school policies and classroom definitions of the situation are not entirely acceptable to the pupils. Like employers in an earlier age, the teachers are unwilling to concede to the pupils the rights which they demand. Since most of the pupil power leaders have been among the older pupils in grammar and comprehensive schools, headteachers can cope with such problems in the short term simply by expelling the leaders. The pupils, like the early trade unionists, are producing their martyrs and sending their organizations underground.

Whilst it is a hazardous and formidable task for the single pupil to persist in rejecting the teacher's classroom definition of the situation, it is much easier for the pupils *as a group* to succeed in such a challenge. For when the pupils unite and make a concerted attack on the teacher's definition, the situation becomes much more threatening and it is much more difficult for the teacher to exert sanctions against them. The teacher is highly vulnerable to a united opposition. It is for this reason that many teachers, faced with a difficult class, follow the principle of 'divide and rule'—single out the main troublemaker and 'make an example of him'. A good illustration of a class where the pupils unite against the teacher is given in *The Blackboard Jungle* (Hunter, 1955).

'All right,' Rick said, 'will you take the first one, Miller?'

He had chosen Miller purposely, hoping the boy would start things off right, especially after his chat with him the other day. . . .

Miller made himself comfortable in his seat again, and then studied the first sentence. Rick wasn't really anticipating much difficulty with the test. This was a fifth-term class, and they'd had most of this material pounded into their heads since they were freshmen. The first sentence read: *Henry hasn't written (no, any) answer to my letter.*

Rick read the sentence, and then looked at Miller, 'Well, Miller, what do you say?'

Miller hesitated for just a moment, 'Henry hasn't written no answer to my letter,' he said.

Rick stared at Miller and then looked out at the class. Something had come alive in their eyes, but there was still no sound. The silence was intense, pressurized almost. 'No,' Rick said, 'It should be "Henry hasn't written any answer." Well, that's all right. I want to learn your mistakes. Will you take the next one, Carter?'

Carter, a big red-headed boy, looked at the second sentence in the test.

If I were (he, him), I wouldn't say that.

'If I were him,' he said rapidly, 'I wouldn't say that.'

Rick smiled. 'Well,' he said, 'if I were you, I wouldn't say that, either. "He" is correct.'

Something was happening out there in the class, but Rick didn't know what it was yet. There was excitement showing in the eyes of the boys, an excitement they could hardly contain. Miller's face was impassive, expressionless.

'Antoro, will you take the next one, please?' Rick said. He had been making notes in his own book as he went along, truly intending to use this test as a guide for future grammar lessons. He looked at the third sentence now.

It was none other than (her, she).

'It was none other than her,' Antoro said quickly.

'No,' Rick said. 'The answer is "she". Take the next one, Levy.' Levy spoke almost as soon as his name was called.

'George throwed the ball fast,' he said.

'Throwed the ball?' Rick said, lifting his eyebrows. '*Throwed*? Come now, Levy. Surely you knew "threw" is correct.'

Levy said nothing. He studied Rick with cold eyes.

'Belazi,' Rick spoke tightly, 'take the next one.'

'It is them who spoke,' he said.

He knew the game now. He knew the game and he was powerless to combat it. Miller had started it, of course, and the other kids had picked it up with an uncanny instinct for following his improvisation.

Situations such as this are exceptional, for it is usually only under the guiding hand of a talented leader, such as Miller in the above example, that the pupils are able to mobilize themselves against the teacher. More typically the pupil is, with respect to the teacher, very much on his own.

It can be said that the state of pseudo-concord is the most typical expression of the working consensus in classrooms. It is pseudo-concord which most accurately conveys the flavour of classroom life, with its ups-and-downs, its good days and bad days, its moments of joy and delight alternating with boredom, frustration and depression. As R. F. Mackenzie (1967) puts it:

> [Difficult children] are still in the minority. The majority we can cope with; or rather, they are nice children and suffer their education patiently. They become moderately interested, like a group on a seaside holiday who are prepared to put up with charades until the rain stops. It's not what they would choose to do, but it's all right. But behind the half-hearted play-acting there is already the murmur of mutiny. . . .

The process by which pseudo-concord is stabilized into a generalized working consensus is exceedingly complex. It is hinted at by those novelists who have written about the experiences of young teachers in their first teaching post. It is an exciting area for future research. In the next chapter we shall be concerned in greater depth with this negotiation process with respect to classroom discipline. For the present, it can be said that pseudo-concord consists of four main elements.

1. *Consensus*—those aspects of the definition of the situation which are acceptable to both teachers and pupils. Sometimes consensus arises from the fact that teachers and pupils share the same goals and role conceptions, as in the case of the research which shows the kinds of teacher the pupils like. Consensus may arise from different goals, however. For example, pupils may agree to work hard in school simply in order to pass the examination and obtain a qualification. The teacher's goal would probably not be so simple.

2. *Compromise*—those aspects where teacher and pupil definitions of the situation conflict but where an agreed compromise is reached. This will arise where the power differential between teacher and pupil is small; where the teacher does not feel that a compromise would involve a fundamental neglect of his duties; where the teacher knows that an attempted use of power to impose his own definition of the situation would be perceived by the pupils as an illegitimate use of power, for this would excite resentment or opposition. Thus the teacher will happily agree to read a play in an English lesson as long as the pupils will agree to do grammar in the next lesson.

3. *Imposition*—those aspects where teacher and pupil definitions of the situation conflict but where the teacher imposes his own definition of the situation. The conditions conducive to imposition are the opposite of those conducive to compromise. It will commonly arise where the teacher is in possession of much more power than the pupils; where compromise would involve a neglect of his duties; where the pupils perceive that the teacher has a right to exercise his power. Many teachers prefer imposition to compromise, since they regard compromise as the 'thin end of the wedge'. For this reason such teachers will attempt imposition even when the conditions are favourable to compromise. Here the teacher's power to enforce a particular definition of the situation is small, but he tries to bluff the pupils into accepting his imposition. He often succeeds, since for reasons already outlined the pupils tend not to call his bluff.

4. *Counter-imposition*—those aspects where teacher and pupil definitions of the situation conflict but where the pupil definition prevails. An alternative name, taking the teacher's perspective, might be *submission*. This arises when the pupils are in possession of the dominant power. Teachers fear this much more than compromise, for they know that if counter-imposition comes to form a major part of the total definition of the situation they will be regarded as incompetent or failed teachers. For this reason teachers often yield in situations

where the pupils have greater power but they yield as if they were granting privileges rather than rights to the pupils. If the teacher, in submitting, can persuade the pupils that he is graciously declining to exert his power to impose his own definition of the situation, he will stimulate the pupils' gratitude and successfully distract the pupils from their own power which need not now be exerted.

If pseudo-concord is indeed typical of most ordinary classrooms, then we must seek the causes which lie behind the fact that the definitions of the situation sought by teachers and pupils are only partly compatible. Part of the answer to this imperfect consensus has already been indicated. Children like school and their teachers *on the whole*. Presumably we can say the same for teachers: they also like school and their pupils *on the whole*. In other words, part of the total experience of school, for both teachers and pupils, consists of certain aspects or experiences which are disliked. For each participant the behaviour of the other is not ideal—it falls short of what is desired. From the teacher's point of view the pupils are not always obedient, hardworking and enthusiastic. From the pupils' point of view, the teacher is not always fair, interesting and friendly. In short, neither participant ever fully lives up to the expectations of the other. The reasons for this are obviously complex and numerous. The pupil may be lazy because he is bored or because he feels afraid of failing; disobedient because he is acting out with the teacher a faulty relationship with his father; unenthusiastic because the teacher makes the subject seem dull. The teacher may be unfair because he likes one pupil more than the others; uninteresting in what he teaches because this is a part of the syllabus that he finds tedious; unfriendly because he had a row with his wife before he came to school. The reasons are endless.

From the pupil's point of view a certain level of dissatisfaction seems inevitable because of the structure inherent in the classroom. There is but one teacher but there are many pupils. The amount of time that any teacher can spend in interaction with an individual pupil is severely limited. Thus the teacher is constrained to interact with his pupils as a group, directing his communications to the class as a whole. This means that, unless the teacher is willing to use group methods, the teacher will do most of the talking and the pupils will spend a very high proportion of their interaction time in *listening*. Let us consider a 'discussion session' lasting 45 minutes in a class of thirty pupils. Even if we make the unlikely assumption that the teacher is silent, each pupil will have an average talking time of only one and a half minutes: 95 per cent of the time must be spent in listening. In practice, of course, many of the pupils will remain totally silent during a discussion period, and it is probable that the teacher will talk very much more than any single pupil. It is hardly surprising then that children in school should be bored and dissatisfied by this rather passive listening—just as they are in church, where a similar restriction on talking prevails. Nor is it surprising that children should need to be rebuked so often by teachers for talking and for inattention, for who enjoys and is attentive in a situation where one can say so little? How many teachers commit the same offences (whispering, lapses of attention) in a staff meeting where the head-teacher does most of the talking?

The plurality of pupils creates a further problem. The one teacher is very unlikely to be able to satisfy all of the children for all of the time, for what pleases one child may displease another. Jane likes History lessons, but Freda hates them. Frank wants to do more project work, but Brian favours a return to more conventional classwork. Such inevitable differences among the pupils in their interests and preferences limit the degree to which any teacher can please all the pupils in what he does with them. In view of these facts it makes sense that teachers and pupils are united in their desire for smaller classes, where some of these dissatisfactions would decline, though not disappear.

Sometimes the source of the pupils' dissatisfactions spring from the school's curriculum and time-tabling of subjects into 'periods'. English is the only subject that Mary really enjoys, and she would be happy to do it all day instead of changing every hour to other subjects she finds boring. Sometimes the examination requirements are to blame. The pupils are made to study a topic they (and perhaps the teacher) find dull simply because it is on the examination syllabus and the pupils cannot afford to neglect it.

In all these ways the pupils may be dissatisfied by their life in school. Moreover many of the factors responsible are beyond the control of either the teacher or the pupils. At best the teacher can try to ensure that all of the children are as happy as possible for much of the time. Yet many of the pupils' dissatisfactions are a direct product of the teacher's behaviour which is within his control. Usually the teacher can improve his discipline; be more impartial; explain his subject more clearly; be more friendly in his approach. Often he fails to do so not because he is unable, but because he is not sufficiently sensitive to the pupils' reactions to recognize the need for change. This is specially important when one remembers that the children are not normally allowed to inform the teacher directly of his failings. To do so would be to engage in the reprehensible act of impertinence. It is because the teacher gets such inadequate feedback on his own behaviour that he is often led to locate the fault in the pupils rather than in himself. Bored and dissatisfied pupils do not want to work. Part (but not all) of this boredom may be a result of the teacher's own behaviour. The teacher correctly blames the poor work on the pupils' lack of interest, but wrongly sees no relationship between this and his own behaviour. In his view it is simply that the pupils do not want to learn; this is the 'cause'. Too often the teacher is reluctant to dig deeper in a quest for the reasons behind the children's indifference. If he did he might discover that more pupils than he imagines are very keen to learn, but that this spirit is progressively killed from the primary school onwards by the nature of what occurs in school. Some of the factors may be beyond his control (e.g. class size), but others which concern his own behaviour may be open to change. Even the factors which are apparently uncontrollable may not be beyond remedy. The effects of a large class may be mitigated by greater use of group work; an unattractive syllabus can sometimes be removed by changing to a different examining body.

Of the teacher's dissatisfactions, some are 'internal' and some are 'external' to life in the classroom. The research of Rudd & Wiseman (1962) finds that the two major sources of dissatisfaction among the teaching profession concern salaries and poor staff relations. But some teachers did mention the attitudes and behaviour of pupils as a source of dissatisfaction. Some of the classroom dissatisfactions experienced by the teacher may have unusual origins. They may spring, for instance, directly out of his attempts to make the situation more rewarding for the children. He may dislike organizing project work because he finds it demanding and exhausting, but he does it because he knows that the pupils enjoy it. In many ways, the teacher can reduce his own satisfactions in trying to please the children.

We need labour the point no further. Teachers and pupils are dissatisfied by numerous aspects of school life. What we must consider in more detail is the effect of the pseudo-concord on teachers and pupils. It is interesting, for instance, to look at the ways in which teachers and pupils 'explain' the dissatisfactions they experience. As has already been suggested each participant frequently locates the cause of his dissatisfactions in the other person. The teachers blame the pupils and the pupils blame the teacher. The teacher sighs, 'How wonderful it would be if only the pupils would. . . .' and the pupils echo this attitude with, 'How happy I would be if only the teacher would. . . .' For each to lay the blame at the other's door is a natural human tendency—how many married couples trump up the same charges against each other. The popularity of this response doubtless rests on the fact that it does not require an admission of one's own faults and transfers the responsibility and initiative for changing onto the other. Yet because this reaction is founded in deception, the potentiality for improvement in the relationship is paralysed. It sets up an equilibrium of mutual dissatisfaction and resentment which cannot generate the energy to break it. It is probably true that with greater effort pupils and teachers could be more rewarding to each other. But there is no panacea here, for the causes of dissatisfaction are complex. Blaming the other is a distorted oversimplification of the facts.

The dissatisfactions experienced in the classroom are not shared by teachers and pupils. It is the pupils who bear the greater burden. This must be so for several reasons. Among the most important is the fact of the plurality of pupils, to which we have already referred. Each individual pupil must subordinate his needs to the needs of the majority. Even under the supervision of the 'child-centred' and 'democratic' teacher, the individual will be regularly sacrificed to the needs and desires of the majority of the class. The principal reason why the teacher should experience fewer dissatisfactions than the pupils consists in the teacher's great power. It is the teacher who can and does take the initiative in defining the situation and he has the power to enforce it on those pupils, whether large or small in number, who are reluctant to accept it. If the teacher feels that the prevailing definition of the situation, whether it is initiated by him or by the pupils, is too costly, then he is at liberty to change it—though it may not always be too easy to do so. The pupil, on the other hand, possesses no right to change the definition of the situation. If he tries to do so, it will be by subtle means, as we shall see. Often the children are probably content to accept the teacher's definition of the situation. In those cases where they are not, the pupils may win the occasional battle, but it is almost always the teacher who wins the war. There is relatively little the pupil can do about it, for the more he resists the teacher's definition, the more heavily the teacher will bring to bear the sanctions at his disposal. The average pupil shrugs his shoulders with resignation and complies. Giving in may not boost his profits, but it certainly cuts his losses.

The pupil, then, is caught up in a situation which is not of his own making, but in which he is compelled to be present and from which he is not allowed to withdraw. It is natural that the pupil—like anyone else in an interaction—should seek to maximize his rewards and minimize his costs in order to attain the greatest profit or the least loss. How does the pupil do this? To answer this question requires us to take a somewhat different approach to the study of classroom interaction. The interaction analysts' approach and the role approach give, as we have seen, by far the greatest emphasis to the teacher's classroom behaviour. When the pupil is considered it is essentially from the teacher's perspective. In part this is because pupil behaviour and pupil roles are heavily contingent upon teacher behaviour and teacher roles, but in part this unequal emphasis stems from researchers' greater interest in teachers. A different approach is required to achieve a deeper and more subtle appreciation of pupil behaviour and the pupil's definition of the situation, an approach in which we must try to capture in a more effective way the pupil's perspective on life in the classroom. We need a more phenomenological approach in which we attempt to understand the meaning of classroom events as apprehended by the pupil. The distinctive nature of this approach has been well expressed by David Matza (1969). Although he is writing in the context of deviance, his words have a more general relevance.

> The decision to appreciate . . . delivers the analyst into the arms of the subject who renders the phenomenon, and commits him, though not without regrets or qualifications, to the subject's definition of the situation. This does not mean the analyst always concurs with the subject's definition of the situation; rather, that his aim to comprehend and to illuminate the subject's view and to interpret the world *as it appears to him*. The view of the phenomena yielded by this perspective is *interior*, in contrast to the external view yielded by a more objective perspective. The . . . phenomenon is seen from the inside. Consequently, many of the categories having their origin in evaluations made from the outside become difficult to maintain since they achieve little prominence in the interpretations and definitions of . . . subjects.

In educational writings it is the child-centred theorists who have come closest to such an approach, largely because of their concern to treat pupils as *persons*. It is no coincidence that one of the most outstanding current contributors to understanding the pupil's perspective, John Holt, ends his book *How Children Fail* with a passionate plea for a more child-centred education. Such writers have in general been unsympathetic to educational research where children are treated as objects in the traditional scientific manner. To the educational researchers, the child-centred writers have often appeared as mere romantic theorists. This divorce has produced a most important omission in our study of classroom interaction. The contrast between a phenomenological and a traditional research approach has been stated explicitly by Herbert Blumer (1966).

> On the methodological or research side the study of action would have to be made from the position of the actor. Since action is forged by the actor out of what he perceives, interprets and judges, one would have to see the operating situation as the actor sees it, perceive objects as the actor perceives them, ascertain their meaning in terms of the meaning that they have for the actor, and follow the actor's line of conduct as the actor organizes it—in short, one would have to take the role of the actor and see his world from his standpoint. This methodological approach stands in contrast to the so-called 'objective' approach so dominant today, namely, that of viewing the actor from the

perspective of an outside detached observer. The 'objective' approach holds the danger of the observer substituting his view of the field of action for the view held by the actor.

A switch to the pupils' perspective will prove to be very difficult since it is our natural tendency to look at problems of teaching and learning from the teacher's perspective and on the basis of teacher assumptions. Let me illustrate the difficulty with reference to an almost universal teacher practice, that of filling in pupils' reports. When the teacher writes, 'This pupil has shown very little interest this term', this seems a relatively uncomplicated matter. In effect the teacher is criticizing the pupil for showing little interest in the lessons as taught by the teacher. The fault is assigned to the pupil. If only the pupil would develop an interest in the subject and in the lessons, the pupil would make the desired progress. The teacher is perceiving and evaluating the problem on the basis of his own definition of the situation. But the alternative perspective is at least equally important, even though we do not normally look for it. The pupil might fill in his report with 'I have not made much progress this term because I am bored to death with the teacher's lessons in this subject. With different lessons or a different teacher next term I might make more progress.' Here the fault is assigned to the teacher. Neither perspective will in itself be sufficient to bring about a change in the pupil's interest. Although the pupil might say what he thinks to his parents, he cannot tell the teacher for to do so would be to cause offence and to displease the teacher. Thus there is no pressure on the teacher to take the pupil perspective.

In the above example the teacher may be unaware of the pupil perspective partly because the pupil cannot afford to make his perspective explicit and partly because the teacher would regard such a perspective as illegitimate. In other situations the teacher is more likely to be aware of the pupil perspective—though no less likely to regard it as illegitimate. Suppose the form 'clown' is giving one of his typical performances to the glee of his classmates. The teacher's interpretation and reaction will be related to such problems as: is this going to prevent work continuing? is he going to threaten the class discipline? is he going to make me look a fool? The pupils' interpretation and reaction is governed by such problems as: will he make us laugh? will he help us to escape any work? how can we help to prolong his clowning? The perspective of the pupil may or may not be evident to the teacher. If it is, then it will give more rather than less weight to the problems deriving from his own perspective, since his awareness of the pupil perspective will be interpreted in terms of the support it offers to the clown's behaviour in opposition to his own definition of the situation.

Now I suggested earlier that the pupils appear to accept the teacher's definition of the situation, at least for the most part. If we can indeed capture the pupils' perspective on life in the classroom more adequately, we may discover that their acceptance of the teacher's definition is only superficial. This is because in taking the 'outside' perspective of the scientific observer we are in fact taking a perspective which is much closer to the teacher's than to the pupils' —a very unhappy position for an allegedly neutral observer—and in so doing may be led into making quite fallacious inferences. To put it more explicitly, if the pupils do not appear to be in *open conflict* with the teacher about the definition of the situation, we may infer that the definitions of the teacher and the pupils are congruent. This need not necessarily be the case at all, for the pupils may accept the teacher's definition of the situation in the sense that they do not express any overt rejection of it yet at the same time hold a definition of the situation which is based on a quite different set of assumptions, perceptions and interpretations. Lack of conflict between definitions of the situation does not mean that they must be the same. That by and large teacher and pupil definitions of the situation interlock is not in doubt, but they need not be the same definition merely because they interlock and *look* congruent. We need to delve deeper.

My argument will be that teacher and pupil definitions of the situation appear to be more congruent than they are and that in reality they are more divergent than appearances would suggest. The evidence is based mainly on those writers and researchers who have taken a more phenomenological approach to classroom interaction. However, it must be admitted that at this stage our knowledge of the pupils' perspective is extremely primitive. There are very few works like *Letter to a Teacher* (1970) where the pupils make their own perspective disturbingly explicit. Here we are faced with evidence that does not stem from questionnaires designed and administered by teachers or researchers.

I am not, in the rest of this chapter, undertaking to make a phenomenological analysis of teacher-pupil interaction, much as such an analysis is needed. Rather, I have the more limited objective of emphasizing the need for such an analysis by demonstrating how little we have been concerned hitherto with the pupil perspective and by indicating the rich stores that await future investigations and analyses. Such an approach needs to take account of Schutz's (1954) view that

The observational field of the social scientist, however, namely the social reality, has a specific meaning and relevance structure for the human beings living, acting, and thinking there. By a series of common-sense constructs they have pre-selected and pre-interpreted the world which they experience as the reality of their daily lives. It is these thought objects of theirs which determine their behaviour by motivating it. The thought objects constructed by the social scientist, in order to grasp this social reality, have to be founded upon the thought objects constructed by the common-sense thinking of men, living their daily lives within their social world. Thus, the constructs of the social sciences are, so to speak, constructs of the second degree, namely constructs of the constructs made by the actors on the social scene, whose behaviour the social scientist has to observe and to explain in accordance with the procedural rules of his science.

and that

all scientific explanations of the social world *can*, and for certain purposes *must*, refer to the subjective meaning of the action of the human being from which the social reality originates.

Otherwise the social world of the social scientist will be pure fiction, having no relation to the social reality as experienced by men.

My present analysis of the pupil perspective hangs on one dominant concept, with several subordinate concepts. It is essential to note that these concepts are *not* the concepts and constructs that the pupils themselves (or their teachers) use to interpret and give meaning to their life in school. I shall refer very little to the dominant concepts used by children, so I cannot claim that my dominant concepts are drawn out of the pupils' constructs; indeed, their relationship to them may be very tenuous. Until we have developed constructs out of the pupils' constructs—a daunting but challenging research task— I offer the present analysis, partly in the hope that it will stimulate such research, partly in the hope that the dominant concepts used here will be close to second order constructs, and partly in the hope that it will encourage teachers and educationists to recognize and rectify our traditional neglect of the pupil perspective.

In reading the following analysis the reader may well profit if he asks himself these three questions as he goes along. What are the constructs of the pupils in the areas examined? What is their relationship to the constructs used by the writer? What is the relationship between the writer's constructs and those of other theoretical approaches (e.g. how does 'pleasing teacher' relate to 'reinforcement theory'?).

The pupil's definition of the situation: a different perspective

In maintaining an exchange theory and symbolic interactionist approach and simultaneously trying to capture the pupil's classroom perspective, then the pupil's principal task, and the method by which he can maximize his rewards and minimize his costs, is *to please teacher*. This recognizes the power differential between teacher and pupil and the contingent nature of classroom interaction, since it is from the teacher that the majority of classroom rewards and punishments flow. It is the teacher's rewards and punishments, mainly in the form of approval and disapproval, that form the pupil's principal rewards and costs. The centrality of this phenomenon from the pupil perspective has been noted by several writers. Miles (1964) suggests: 'The student's here-and-now task, as classroom learning goes forward, is, in effect, to please—or at least not to displease—the teacher.' The American anthropologist Jules Henry wrote an article in 1955 entitled 'Docility, or giving teacher what she wants,' in which he suggests that in school pupils must learn to find out what the teacher wants and then satisfy her. This is not always easy, since the teacher does not always make it clear what she wants. The pupils must hunt for signals from the teacher and then direct their behaviour accordingly. They must 'gropingly find a way to gratify the teacher' in order to 'bask in the sun of the teacher's acceptance'. Jerome Bruner (1966) has made the same point. 'Young children in school expend extraordinary time and effort figuring out what it is that the teacher wants.' The matter has been put most succinctly by Jackson (1968), when he points out that the pupil

> must learn how the reward system of the classroom operates and then use that knowledge to increase the flow of rewards to himself. A second job . . . consists in trying to publicize positive evaluations and conceal negative ones.

We are all familiar with the pupil reaction to having pleased teacher, with the pupil who joyfully and boastfully tells his neighbour, 'Look, sir has given me a silver star!' with the pupil who wallows in the admiration of his fellows when it is announced that he is top of the class. We are also familiar with the pupil reaction to having displeased teacher. A good example is available from Barker & Wright's (1951) study, *One Boy's Day*.

> [A second pupil] came back and asked him brightly and eagerly. 'What did you get in your spelling?' Raymond blushed and looked down at his desk. He fidgeted with his hands a moment before he answered. In a swift hoarse whisper he said crossly,

'None of your business.' He seemed quite embarrassed as he spoke. . . .

Mrs. Logan, who was recording the grades, said, 'Let's see the hands of those who didn't get 100.'

Raymond sat there for a moment. He then very slowly and reluctantly raised his hand.

The teacher called out the name of each child whose hand was raised, asking in each case for the number missed.

After hearing the first child, she said pleasantly, 'Raymond, you missed two didn't you?'

Raymond mumbled in embarrassment, 'Yes.' He looked very unhappy and blushed again.

He looked blankly at his desk for a moment. While Mrs. Logan went through the rest of the names, Raymond continued to appear crestfallen at his failure to get 100.

The concept of pleasing teacher can readily draw on the language of exchange theory since it assumes that a significant part of life in school is concerned with the ways in which the pupil maximizes profits and minimizes losses in his relationship with the teacher. In other words, he must develop techniques whereby he can maximize the possibilities of receiving a favourable evaluation and minimize the chances of being evaluated negatively. Thus the most obvious way in which the pupil can maximize the rewards he receives from the teacher is by learning what sort of behaviours in fact please the teacher, i.e. *by learning the system* then by behaving accordingly and making it evident to the teacher that he is indeed behaving in the approved manner i.e. *by playing the system.* For unless the teacher is aware that the pupil is conforming to the expected pattern, there can be no reward— except insofar as by conforming he cannot receive a punishment for not conforming. Similarly, he can succeed in avoiding negative evaluations if he learns what displeases the teacher and then either declines to indulge in such behaviour or takes steps to hide such behaviour from the teacher. There are, then, three laws to guide the pupil in learning the skill of pleasing teacher.

First law: *find out what pleases and displeases the teacher.*

Second law: *bring to the teacher's attention those things which please the teacher and conceal from him those behaviours which will displease him.*

Third law: *remember that it is a* competitive *situation. The pupil must try to please the teacher and avoid displeasing him more than other pupils.*

The importance of the third law reveals itself in the well-known fact that pupils disapprove of the person who too openly conforms to the second law. The pupil who persistently and openly brings his own merits and efforts to the teacher's attention is dubbed a 'show-off' or an 'apple-polisher' or other less polite tags. Similarly, the pupil who brings the hitherto concealed misbehaviours of other pupils to the teacher's attention ('Sir, look what Smith's doing!') is rejected as a 'tell-tale' and a traitor. The existence of the third law creates difficulties in abiding by the second law because informal norms develop among the pupils to inhibit pleasing teacher at the expense of other pupils. Only the pupil who has learnt the third law can develop the requisite subtle skills to fulfil the second law without endangering his relationship with his peers.

In effect the pupil has to be highly competitive not only because competition is inherent in the pleasing teacher process, but also because the teacher will actively stimulate competition towards good work and exemplary behaviour, towards the marks and prizes that are given for outstanding achievements in the disciplinary and instructional areas. Yet the pupil must not involve himself too openly in the competition; it must not take on 'cut-throat' characteristics. If the competition is espoused with too much enthusiasm, with too overt an individualism, it will meet with teacher disapproval for it threatens the teacher's desire to create a spirit of harmony and co-operation within the class. Because pleasing teacher is competitive, every time one pupil succeeds, by implication all the other pupils fail. In short, he must learn to compete but to avoid the teacher disapproval given to those who appear to be too little or too much involved in the competition. At the same time he must learn to accept and control his disappointment at his failures in the competition, avoiding giving the impression that his failure is of no consequence to him or is profoundly distressing to him. It requires elaborate learning and the acquisition of subtle self-presentation skills.

The teacher's desire simultaneously to create a spirit of competition and co-operation expresses itself in a variety of classroom practices. An obvious example is the classroom quiz, where the class is divided into two teams. From the teacher's point of view such a device increases motivation; the pupils always enjoy a quiz. It encourages competition between the teams but co-operation within the team. As Henry (1966) has noted, this situation has two consequences which are not always noticed. We have the tragedy of the left-overs, the pupils that each team picks last and often reluctantly; their self-esteem is hardly improved by the situation. Further, when a pupil fails to answer his question correctly, the failure is a double one since he lets down not only himself but also all the other members of his team.

When a teacher asks questions of his pupils, he is sometimes amazed by the quite absurd wild guesses made by the children in the answers they offer. He should not be surprised, for the child is giving evidence that he has learned what the teacher expects of the 'good' pupil. If his guess is right, he will be rewarded with praise. If he is wrong, he will at least have tried and the teacher is often reluctant to condemn the pupil who tries. But if he sits in silence he cannot but receive disapproval from the teacher, for it is only 'bad' pupils who neither know the answer nor exhibit the spirit of enthusiasm. The 'wild guess' is a very reasonable response of the pupil, since such behaviour avoids certain punishment and carries a potentiality for high reward.

Probably the most notorious strategy for pleasing teacher is that of cheating. In cheating, the pupil is trying to give the teacher the impression that he has performed the desired behaviour when he has not in a situation such as an examination. Teachers usually exhibit anger and moral indignation towards the pupil who cheats because he has used what, in the teacher's view, are quite illegitimate methods of meeting the requirements imposed by the teacher. But to the pupil who has not done his homework and who in consequence is certain to fail in the test, cheating is a means of gaining a possible reward and avoiding a certain punishment. The only risk in cheating is that of being found out.

All practising teachers soon become aware of these strategies used by pupils. They learn to spot the pupil who appears to be poring over his Latin grammar book but is actually reading his comic hidden under the desk. Nor are they always taken in by the professional sycophant or 'greaser' who so ostentatiously displays his conformity, who always says the right thing, and who so anxiously offers his help when it is not needed. Yet sometimes the strategies are more subtle and succeed. After all, what use is a strategy if it does not pay off? In one school I know the pupils would write the date in pencil in their mathematics exercise books. If the teacher did not mark the work in that particular lesson, then on the next occasion the pupil could rub out the old date, substitute the new one, and spend the rest of the lesson on matters more important than mathematics, secure in the certainty that if called to account by the teacher he could produce incontrovertible evidence of having worked that day. Sometimes even the common strategies pay off, not because the teacher is unfamiliar with them, but rather because the teacher is so engrossed in his own task that he forgets about the strategy. For example, the teacher wishes each pupil to answer one question from those set in the textbook. To make sure that no child is accidentally left out, he assigns the questions in a systematic way according to the pupil's seating position. However, this gives the pupil the opportunity to use the strategy of 'looking ahead'. He counts up the pupils who will have to respond before him, calculates which question—all being well—will be his, and then prepares his answer in advance. With such preparation his chances of giving a right answer are substantially improved. He succeeds in pleasing the teacher, but in so doing he defeats the teacher's wider purpose of giving him practice at a whole set of questions rather than at a single one. Ironically, the teacher, in approving the correct answer, reinforces the use of the strategy and sabotages his own goals.

One of the most insightful exponents of children's classroom strategies is the American John Holt (1964). From his own experience as a teacher he illustrates the techniques that children adopt to mask their ignorance. Let us take the case of one pupil. She knows the teacher's attention is divided between all the pupils. She knows the teacher tends to ask questions of those pupils who are not paying attention or who seem confused and uncertain. Therefore she feels fairly safe in putting up her hand, waving it excitedly as if she is bursting to give the answer, even when she does *not* know the answer. Yet she does this only when other hands are raised. To be the only person with a raised hand, or to be the only person without a raised hand, greatly increases the probability of being the target of the teacher's question. The best means of avoiding the teacher's attention is to follow the majority. One might add, as a gloss on Holt, that pupils quickly learn that teachers very rarely ask a pupil two or more questions in succession. When the teacher does so, especially when the first answer was correct, it is a cause for surprise because it is unusual. Sometimes it is a source of complaint—'But I've just done one, sir!' In being given the privilege or the chore of answering, pupils feel there should be fair shares for all. Thus after having answered one question, the pupil can afford to relax his attention.

A second strategy noted by Holt is that of 'hedging one's bet'. If a pupil who is unsure of the answer finds himself called upon by the teacher, he has a chance of gaining approval by giving an ambiguous answer which the teacher might take to be right. In a French lesson, ignorance of the gender of a word can at times be overcome by pronouncing a word which is somewhere between *le* and *la*. Similarly if the pupil is not sure whether a word is spelled with an *a* or an *o* he can write a letter which could be either of them. I remember as a child writing the word *receive* with two letters somewhere between *e* and *i* and with the dot exactly between the two. Other forms of this strategy are the mumbled answer and the addition of 'sort of' to an answer, which allows one to say, 'Yes that's what I meant', when the correct answer is supplied.

The strategies we have examined so far are examples of particular techniques for dealing with specific situations. If the pupil is to succeed in pleasing the teacher he must have at his disposal a whole range of self-presentation techniques which are readily available and adaptable to all eventualities. Some indication of the depth of the self-presentation skills acquired by pupils is given in this entertaining account by a pupil (Willans & Searle, 1958).

The only way with a maths master is to hav a very worried xpression. Stare at the book intently with a deep frown as if furious that you cannot see the answer. at the same time scratch the head with the end of the pen. After 5 minits it is not safe to do nothing any longer. Brush away all the objects which hav fallen out of the hair and put up hand.
  'Sir?' (*whisper*)
  'Please sir?' (*louder*)
  'Yes, molesworth?' sa maths master. (*Thinks: it is that uter worm agane*)
  'Sir i don't quite *see* this.'
nb it is essential to sa you don't quite '*see*' sum as this means you are only temporarily bafled by unruly equation and not that you don't kno the fanetest about any of it. (Dialog continue:)
  'What do you not see molesworth?' sa maths master (*Thinks: a worthy dolt who is making an honest efort*)
  'number six sir i can't make it out sir.'
  'What can you not make out molesworth?'
  'number six sir.'
  'it is all very simple molesworth if you had been paing atention to what i was saing at the beginning of the lesson. Go back to your desk and *think*.'
This gets a boy nowhere but it shows he is KEEN which is important with maths masters.
  Maths masters do not like neck of any kind and canot stand the casual approach.
  HOW NOT TO APPROACH A MATHS MASTER
  'Sir?'
  'Sir Sir please?'
  'Sir sir please sir?'
  'Sir sir please sir sir please?'
  'Yes molesworth?'
  'I simply haven't the fogiest about number six sir'
  'Indeed, molesworth?'
  'It's just a jumble of letters sir i mean i kno i couldn't care less whether i get it right or not but what sort of an ass sir can hav written this book.'
  (*Maths master give below of rage and tear across room with dividers. He hurl me three times round head and then out of window*)
Sometimes the pupil's strategies of trying to please the teacher, of

seeking to present a favourable image of himself, can take the form of a game among the pupils. When I was at school I and a friend of mine used to play a game called 'Constructive Questions' in music lessons. The object of the game was to ask questions which *sounded* impressive and which would be taken seriously by the teacher. The winner was the person who asked the teacher the most questions during the lesson before the teacher 'cottoned on' to the game. I can still remember the questions—about 'double stopping' and 'grace notes' and 'plagal cadences'—but I have no idea what the terms mean. It was simply good fun, a means of passing the time with a little skill and much amusement. And I did get 'very good work' written on my report instead of the 'satisfactory' that was traditional in Music. Actually my friend usually won, so you will not be surprised to learn that he is now a well known actor.

It is no exaggeration to suggest that the art of pleasing teacher forms an important part of schooling which the pupil must learn. It can be said, in Jackson's (1968) words, to constitute

> a hidden curriculum which each student (and teacher) must master if he is to make his way satisfactorily through the school. The demands created by these features of classroom life may be contrasted with the academic demands—the 'official' curriculum, so to speak,—to which educators have traditionally paid most attention. As might be expected, the two curriculums are related to each other in several important ways . . . Indeed, many of the rewards and punishments that sound as if they are being dispensed on the basis of academic success and failure are really more closely related to the mastery of the hidden curriculum.

This interpenetration has been neatly exemplified by Jules Henry (1966).

> A child writing the word 'August' on the board, for example, is not only learning the word 'August' but also how to hold the chalk without making it squeak, how to write clearly, how to keep going even though the class is tittering at his slowness, how to appraise the glances of the children in order to know whether he is doing it right or wrong, et cetera. If the spelling, arithmetic or music lesson were only what it appeared to be, the education of the American child would be much simpler: but it is all the things that the child learns *along with* his subject matter that really constitute the drag on the educational process as it applies to the curriculum . . . School metamorphoses the child, giving it the kind of Self the school can manage, and then proceeds to minister to the Self it has made. . . It is simply that

the child must react in terms of the institutional definitions or he fails. The first two years of school are spent not so much in learning the rudiments of the three R's as in learning definitions.

Pleasing teacher is a useful concept for bringing together both the 'formal' and the 'hidden' curricula and their collective impact on the pupil. It is a useful device for throwing light on the pupil role, not merely in its formal definition of what is expected of the pupil by the teacher, but in its wider sense of the total demands of the classroom situation with which the pupil must come to terms. Most of the previous work on the pupil role has been concerned with teacher expectations. In turning to the pupil perspective we must make a shift in emphasis towards the pervasive features of classroom life as they are experienced by the pupil. As Jackson has shown, the pupil has to cope with the problems of *delay* (e.g. waiting one's turn), *denial* (e.g. being refused permission), *interruption* (e.g. being called to the teacher's desk), and *social distraction* (e.g. not being free to talk to one's peers)—and these problems must be solved in such a way that teacher will be pleased.

The content of the pupil role can be outlined, as we did earlier, in very general terms such as obedience, industry and politeness etc. In reality each of these dimensions is highly complex. Disobedience, for instance, is by no means a simple, unitary dimension. There is *passive* disobedience, failing to do as the teacher instructs; *negative* disobedience, indulging in an activity of which the teacher disapproves in direct contradiction of the teacher's orders; and *positive disobedience*, doing something of which the teacher approves, but at a time or in a manner which is forbidden by the teacher. The pupil is under pressure to make such distinctions, for the teacher's disapproval tends to be greatest for negative disobedience and least for positive disobedience. Moreover, whilst there is a general pupil role common to most classrooms at any given age level within most schools, each teacher has his own unique set of expectations and demands, his own hierarchy of approved behaviours and of misbehaviours, and his own system of reward and punishment. By the time a pupil has reached the secondary stage of his education, he must, if he is to succeed in pleasing teacher, be able to learn the pupil role that is peculiar to each teacher as well as the general pupil role that is common to most teachers. To learn the pupil role and to succeed in pleasing teacher, the pupil needs the ability to recognize what is expected of him, a set of skills to meet these expectations, and a set of strategies that will allow him to depart from these expectations without incurring disapproval or to give the impression that he is meeting the expectations when he is unable or unwilling to do so.

As an illustration, one of the central elements in the pupil role

185

requires that he absorb himself in academic tasks. Once the pupil has recognized the centrality of task-absorption, he must bring to bear or acquire the art of concentration and perseverance. Task-absorption also requires the pupil to eschew various distractions, both auto-involvements (Goffman, 1963b) such as day-dreaming and gazing through the window, and the social distractions afforded by other pupils. Yet frequently these distractions are more attractive to the pupil than the academic tasks on hand, so he must learn how to indulge in them without the teacher being aware of it. He must acquire the skills of looking as if he is listening or reading or writing when he is not, of talking to or passing notes to his friends without drawing the teacher's attention. He must also be armed with appropriate escape routes in case his deception is unmasked—'Please miss, I was only asking him if I could borrow his rubber.'

We will not understand the reasons why children develop and use these self-limiting and self-defeating strategies unless we see life in the classroom from the pupil's perspective. As Holt puts it:

> Children see school almost entirely in terms of the day-to-day and hour-to-hour tasks that we impose on them. . . . For children, the central business of school is not learning, whatever this vague term means; it is getting these daily tasks done, or at least out of the way, with a minimum of effort and unpleasantness. Each task is an end in itself. The children don't care how they dispose of it. If they can get it out of the way by doing it, they will do it; if experience has taught them that this does not work very well, they will turn to other means, illegitimate means, that wholly defeat whatever purpose the task-giver had in mind.

We, the teachers, set them tasks and ask them questions and mark their exercise books to get them to learn and develop an interest in the subject. This is our objective and our definition of the situation. But it is not the pupils' objective; it is not their definition of the situation. Their problem is to 'make the best of a bad job'. To survive, to adapt to that from which they cannot escape; they develop strategies to keep us happy, to please us. For in pleasing the teacher the pupil protects himself and maintains his self-esteem. He keeps the stream of approval flowing towards him, and avoids the embarrassment, shame, disapproval, trouble and punishment which follow when he does or says the wrong thing.

The fundamental difference between the pupil perspective and the teacher perspective becomes clearer when we remember the teacher's view of the curriculum. The teacher thinks in terms of the course of study as a whole, extending over weeks, terms and years. He imagines a cumulative development of knowledge and understanding by the pupil, each lesson or step adding to the total building which he calls his subject syllabus. Every lesson is one more brick, cemented onto earlier bricks and serving as the solid foundation for the next brick. The pupil's perspective is often of a very different nature. As each child receives a brick, he simply pushes it to one side once it has been dealt with. There is never a foundation for the next brick and instead of the child ending up with a secure edifice he merely stands in the middle of a bomb site.

The pupil, especially if he is middle class in origin, is willing to accept these discrete 'bricks' dispensed by the teacher. He is essentially passive, respecting the teacher's authority and responsive to the teacher's rewards that are offered for so doing. His obedience, patience and deference inhibit rebellion against the disjointedness, lack of coherence and even meaninglessness that tend to characterize a surprising amount of what he does in school. Any signs of rebellion (e.g. careless work, lack of concentration, a failing interest) are soon quelled by an appeal to the future examinations, to which he is also highly responsive—and which can be used to persuade him to cement together some of the loose bricks that have collected in his mind. The lower working class child is less responsive to the reward system of the teacher, less responsive to the examinations, (if he is in a position to take them), and more ready to rebel against the meaninglessness of the loose bricks that make up the lessons.

All pupils at times indulge in those attractive alternatives to what the teacher wants. Because the teacher finds it difficult to take the perspective of the pupil (whatever his social class), much pupil behaviour is perceived by the teacher as 'irrational' or 'stupid'. Because the teacher cannot see its purpose, he wrongly infers that there is no purpose to it. The power and age differentials between teacher and pupil conspire against the dissolution of such misinterpretations. The pupil, being anxious to please and not to displease and being relatively inarticulate, is unlikely to volunteer information that will make the rationality of his conduct apparent to the teacher. Even when the pupil does attempt to do so, the teacher's inability to change perspective leads him to perceive the proffered explanation as a mere excuse. In addition, because the pupil is a child, the teacher can put down the child's 'irrational' behaviour to 'childishness' or 'immaturity'.

In Holt's view these strategies are dictated above all by *fear*—fear of not pleasing, of saying something foolish or stupid, of being laughed at, of being the object of sarcasm, disapproval and punishment. Most teachers react to such a claim by thinking of other teachers —it is never our own pupils who are afraid. But the point is that we as teachers do not recognize the fear in our own pupils. We can recognize only the grossest symptoms of fear, but not the fear which pervades the classroom. For the fear is controlled by the children. To superficial observation they do not *look* as if they are afraid. Are we, the teachers, really aware of this? John Holt, typically forcing us to face the brutal truth, claims that we like children who are a little afraid of us, for it is this which makes them docile and deferential. The ideal pupil is one who is sufficiently afraid of the teacher to do as he is bidden, but without making the teacher feel that it is fear which is the motivating force, for that would threaten the teacher's image of himself as kind and lovable.

Holt goes much further than this, suggesting that a teacher is like a man in a wood at night with a powerful torch. He turns on the light in order to see, but in the beam of the torch the creatures do not behave as they would in the dark. Teachers cannot learn very much about children in classrooms if they look at them only as part of the teaching process when the children are aware that they are being looked at. Teachers will discover how the pupils feel and think and 'tick' only when the observation takes place when the pupils are not aware that they are being observed. This is a hard saying. Teachers always like to feel that they know how the pupils feel, that they understand them, and that they are sensitive to the nuances of their classroom behaviours. For what teacher finds it easy to confess that he has but a limited understanding of his pupils and at the same time retain his self-esteem and sense of competence? But it seems to me that Holt's observation is correct—at least as far as the vast majority of teachers is concerned. We believe we understand our pupils because we want to, but in truth we greatly over-estimate the extent of our understanding.

In my discussion of the pleasing teacher phenomenon with various groups of teachers and students during the past year or two, some critical members of the audience have suggested that I am over-estimating the degree to which pupils have to acquire an appropriate front, their ability to deceive the teacher and the teacher's naïvety in being deceived. This view fails to recognize the implications of Holt's statement. The teacher is indeed very sensitive to certain forms of pupil concealment, namely those which from the teacher perspective have been defined as illegitimate. A good example is cheating. The teacher will be on the look out for cheating since he regards this as a moral offence and a threat to his instructional role. The pupil's lack of skill in this area of deception can soon become obvious as in the self-evident lie of the pupil who protests, 'I did the homework myself, honestly, sir.' The point is the teacher is insensitive to a much larger area of pupil deception because he is not motivated to recognize it as such. In their attempt to please teacher the pupils feign enthusiasm and interest, which the teacher does not recognize as such because he *wants* the pupils to be enthusiastic and interested. Were he to look behind the mask, the boredom he would see would threaten his self-esteem and sense of professional competence. The pupil does not need to be very skilled in his deception because in effect the teacher is in collusion with the deception. The pupil need only avoid showing his real feelings whilst he is being directly observed by the teacher.

The teacher himself teaches the pupil that he must hide his real feelings. The pupil learns that when he yawns with boredom he is rebuked by the teacher for being inattentive. I remember that in my first week of teaching I was very cross with a first-year pupil for handing in a very poor piece of work. At the end of my sermon the boy burst into tears. Inevitably my heart melted and I tried to comfort him, though telling him at the same time that he was too old to cry. Unintentionally I was teaching him to avoid revealing his true feelings of acute distress. He must, I was saying by implication, not distress me by showing that he was distressed, otherwise I would be unable to be the sort of teacher I wanted to be. It was easier to require him to change than to change myself. He did not cry again.

The teacher, in performing his teacher role, inevitably finds it difficult to perceive the situation from the pupils' perspective. Because the teacher is not fully aware of the effects of his own behaviour on the pupils, he persists in certain forms of conduct which, from his own point of view, seem likely to promote his instructional objectives, but which have the effect of driving the pupil into defensive strategies that actually inhibit learning. The teacher frequently fails to realize what the pupils are doing, and even when he does, he tends not to recognize the pupils' reactions as a response to his own behaviour.

The best exemplification of this, given by John Holt, is the teacher's addiction to the *right answer*. It will be recalled that earlier the teacher's instructional role was defined as the task of getting the pupils to learn and to *show evidence of that learning.* This rider is very important because much of the teacher's behaviour in his instructional role consists in looking for the evidence of the effect of his teaching on the child's learning. He must check that he is indeed promoting the pupil's learning. This is a surprisingly difficult task, for the teacher has to rely on what are essentially indirect indices of the learner's restructured mental faculties. Examination results can provide such a measure, but examinations take place somewhat infrequently. The teacher needs more regular and immediate feedback, evidence which is more readily available in the hour-by-hour activities of the classroom. Such evidence takes two basic forms— what the children write in their exercise books and what they say in response to his questions. At root the evidence on which the teacher relies is the child's ability to produce, in written or oral form, the right answer to the problems imposed by the teacher.

The teacher's reliance on right answers, and his desire to obtain plenty of them, indicates that from the pupil's point of view much of his behaviour is *answer-centred*. The teacher sets the problems and the task of the pupils is to find the right answer which will please the teacher. They know that the teacher knows the answer; it is there in the teacher's mind. Their job is to hunt around until it can be found. The focus is not so much on the problem itself as a problem, but on chasing the answer. The result is that schools encourage *producers*, the pupils who can get the 'right answers', and may thus be discouraging places for *thinkers*. The pupils who wrote *Letter to a Teacher* make the same point.

> During [oral] exams the whole class sinks either into laziness or terror. Even the boy being questioned wastes his time. He keeps taking cover, avoids what he understands least, keeps stressing the things he knows well. To make you happy we need know only how to sell our goods. And how to fill empty spaces with empty words. To repeat critical remarks read in [books by established critics], passing them off as our own and giving the impression that we have read the originals.

This means that the best strategy for the pupils is to find a technique which will successfully produce the right answers and please the teacher. Any technique which works, which produces right answers fast, will be a good one. Sometimes in order to find a successful technique the pupils must understand the problem. But at other times no real learning is necessary for the pupils can find a formula or a recipe which will produce the right answer but which does not require any understanding of the problem at all. In such a case the teacher infers that pupils are learning, for the answer was right. Approval is given and the use of the strategy is strengthened. Yet the learning is only *apparent* learning, for no understanding of the problem has taken place. The recipe is a substitute for real learning. All teachers have experience of this, though they may not recognize it as such. On Monday all pupils seem to be able to solve the problems. On Wednesday, when the same work is covered again, half the pupils get the answers wrong. The teacher tends to assume that the children are being lazy or forgetful or careless and he rebukes them accordingly. After all, this is the most obvious explanation. But what may often occur in such circumstances is that the children never really understood in the first place. They found a formula or recipe which worked on Monday, but which they have forgotten by Wednesday.

Perhaps more children than we imagine can 'get by' for a surprisingly large portion of their learning by means of such recipes. Often, of course, they are actually suggested by the teacher himself in order to help the learner. I did quite well in mathematics at school but my understanding of the subject is so small that I consider myself innumerate. I passed my examination in trigonometry largely on the basis of the formulae *Sweet Peas Hot, Cold Boiled Ham* and *Tea Pot Brown.* These were easy mnemonics, invented and drilled into the pupils by my teacher, from which one could derive *Sine* equals *Perpendicular* over *Hypotenuse* and so on. I did not need to understand about sines, cosines and tangents to solve the problems. In consequence I never really tried to understand them and never did understand them. Perhaps that is why, in spite of my apparent ability at the subject, I always disliked it. It is not much fun being able to get the right answer when one does not know why it is right. But there were two advantages. It did keep the teacher happy and me out of trouble; and it did get me through the examination. For many pupils pleasing teacher is not just an end in itself, a survival kit for passing through school unscathed. It is also a necessary part of passing examinations and obtaining some qualifications, either into a job or onto the next rung of the educational ladder. This forms one of the most important rewards which accrue to the pupil who tries to please the teacher. Without the incentive of the examinations the pressure to please teacher would be substantially diminished. As the pupil-authors of *Letter to a Teacher* write:

But your pupils' own goal is also a mystery. Maybe it is non-existent; maybe it is just shoddy. Day in and day out they study for marks, for reports and diplomas. Meanwhile they lose interest in all the fine things they are studying. Languages, sciences, history—everything becomes purely pass marks. Behind those sheets of paper there is only a desire for personal gain. The diploma means money. Nobody mentions this, but give the bag a good squeeze and that's what comes out. To be a happy pupil in your schools you have to be a social climber at the age of twelve.

The art of pleasing teacher as we have so far described it is essentially the orientation to life in school of the middle-class pupil. The values derived from home are consonant with this orientation and he possesses the requisite values, attitudes and social skills—all of which have in recent years become of interest to sociologists of education—that enable him to learn the art with relative ease and to be successful in pleasing teacher. Sometimes he learns that art so naturally and so thoroughly that he is entirely identified with the 'good pupil' role. In other cases he is aware that the good pupil image which he adopts towards the teachers is to some degree a misrepresentation of his true character, but he is able to play his part without effort and without a sense of hypocrisy at being 'two-faced'. One might assume that the 'successful' working-class pupil undergoes a similar socialization at school, whether or not he is supported in this endeavour by his

191

home life. Certainly large numbers of working-class pupils are handi-capped in this aspect of life in school as in others, so it is not sur-prising that many of them conclude their school life in playing the game in a very half-hearted manner. Who in this world ever retains enthusiasm for a game in which one is a consistent loser?

There are, however, alternatives to pleasing teacher with varying degrees of eagerness. The first is the delinquent or 'delinquescent' orientation. Here the rules of pleasing teacher are turned upside down and the principal objective becomes to *displease* the teacher. The underlying causes of this orientation are complex and I have tried to deal with some of them elsewhere (Hargreaves, 1967, 1971), but with respect to this argument it can be said to result from a persistent failure to please teacher successfully and a realization that for the pupil it is an unprofitable activity. The delinquent orientation is a very difficult game because the pupil cannot sustain a continuous and open rebellion against the teacher and the system, for to do so would be to provoke the teachers into calling into force the most severe penalties at their disposal, including corporal punishment. In coming to terms with the reality of his life in school, the delinquent has to play a more subtle game so that, in practice, the situation is not so much one of pitched battle but of intermittent skirmishes, in which the occasional flouting of the school rules is detected and punished. The delinquent spends much of his time in school in terms of a sullen compliance to the rules of the game, not because he accepts the rules but because he fears excessive sanctions. Such compliance is regarded as legitimate by other members of the delinquent group who do not expect their friends to become martyrs. But this com-pliance is *expedient compliance*. When punishment for non-conformity is both certain and severe, then the delinquent complies. Where the probability of severe punishment is relatively low, as in the classes of a teacher with 'weak' discipline, the delinquent can and does flout the rules in an overt expression of his rejection of the system. From time to time the delinquent will also indulge in disobedience when apprehension and punishment seem certain and this will command the admiration of his fellows and enhance his peer group status to a greater degree than will non-conformity in more 'safe' situations. So the delinquent does not seek to displease the teacher indiscrimi-nately. Rather he takes a calculated risk. With luck he may be able to express his resentment and rejection of the system with very little personal cost. The occasional severe punishment remains a small cost relative to the high rewards that have accrued to him, in the form of peer group status, in the interim.

When the delinquent complies with the rules he is still distinguish-able from those who are trying to please teacher, since he conforms only to the very minimal requirements of the teacher. This is because his object is to avoid punishment, not to obtain approval. Indeed, compliance is a very dangerous state precisely because if he conforms beyond the minimal requirements he is in danger of receiving some sort of reward from the teacher, who may be trying to encourage the boy. But for the delinquent to receive approval is to spoil his game and to undermine his peer group status. This arises only occasionally in practice, partly because the delinquents learn how to conform minimally without risk of receiving approval, and partly because if a boy mistakenly or accidentally oversteps the minimal requirement, his friends will warn him by 'kidding' him about working too hard or being too friendly with the teachers.

The second alternative to pleasing teacher, and the most difficult to elaborate in detail, is what might be called the indifferent orientation. In this case the pupil desires neither to please nor to displease the teacher. He is simply 'not bothered'. School is neither gladly accepted nor bitterly rejected. If he is rewarded by the teacher, it gives him little pleasure. If he is punished, it is regarded as an unfortunate and unfair intrusion. Essentially it is an orientation of withdrawal or of 'serving one's sentence'. At times he appears to be trying to please the teacher, but the resemblance to the normal game is superficial. He makes no attempt on his own initiative to please the teacher, but when challenged by the teacher he reacts with the expected and pleasing answer not in order to receive a reward but in order to fend off a challenge or a potential punishment. This attitude is betrayed in the 'I don't know' strategy that is typical of the indifferent orientation. When challenged by the teacher, the pupil seems either not to know the 'right' response or lack the motivation to hunt around for the pleasing answer. He responds with an 'I don't know' as a relatively safe way out until the 'right' answer becomes obvious from the teacher's prompting. A good example of this is given in Barry Hines' masterly novel, *A Kestrel for a Knave* (1968) in which Billy Casper comes very close to the indifferent orientation.

'You were asleep weren't you? . . . Well? Speak up, lad!'
'I don't know, Sir.'
'Well I know. You were fast asleep on your feet. Weren't you?'
'Yes, Sir.'
'Fast asleep during the Lord's Prayer! I'll thrash you, you irreverent scoundrel!'
He demonstrated the act twice down the side of the lectern.
'Were you tired, lad?'
'I don't know, Sir.'
'Don't know? You wouldn't be tired if you'd get to bed at night instead of roaming the streets at all hours up to mischief!'

'No, Sir,'
'Or sitting up 'til dawn watching some tripe on television!
Report to my room straight after assembly. You will be tired
when I've finished with you, lad.'

'Casper!'
Billy sat up and put his hands away.
'What, Sir?'
'What, Sir. You'd know if you'd been listening. Have you
been listening?'
'Yes, Sir.'
'Tell me what we've been talking about then.'
'Er . . . stories, Sir.'
'What kind of stories?'
'Er . . .'
'You don't know, do you?'
'No, Sir.'
'You haven't heard a word of what's been said, have you,
Casper?'
'Yes, Sir—some of it.'
'Some of it. I'll bet you have. Stand up, lad.'
Billy sighed and pushed the chair away with the backs of his
knees.
'Right, now you can do some work for a change. You're going
to tell us any story about yourself, the same as Anderson did.'
'I don't know any, Sir.'

One of the most tragic aspects of the indifferent orientation is that
the pupil reacts in a very defensive way against the teacher who, full
of good intentions, tries to offer help. For instance, the teacher may
try to find out what really interests the child by gentle questioning.
Typically the pupil responds with the other standard techniques of
'I'm not bothered' or 'I don't mind' whenever the teacher tries to
probe his interests or problems. This is easily interpreted by the
teacher as apathy. But it may also be that the pupil is reluctant to
commit himself or give his true view because he suspects (correctly),
on the basis of past experience, that the teacher is asking from an
ulterior motive, i.e. seeking a means of getting the pupil involved in
school work. If the pupil does commit himself then he may find that
he is 'conned' into something he would rather not do. It is simpler and
safer to say, 'I'm not really bothered.'

'Middle-class' pupils, namely those who share the teacher's basic
values, are clearly at an advantage with respect to the three laws of
pleasing teacher. 'Working-class' pupils, namely those who espouse
the teacher's middle-class values to a much smaller extent or who

194

have values which conflict with those of the teacher, are likely to be at a disadvantage with respect to many aspects of pleasing teacher. There is a great range of research evidence to show that the deficiencies of these pupils in linguistic and cognitive abilities and skills inhibit them from succeeding in the formal curriculum. Here I wish to show how they also lack some of the *social* skills that are essential to pleasing teacher.

Ability to meet expectations

		High		Low
		Meeting of expectations		
		Yes	No	
Recognition of expectations	High	1	2	3
	Low	4	5	6

Fig. 6.4

Theoretically the situation can take several forms, as a function of the degree to which a pupil recognizes what is expected of him, the pupil role, and of the degree to which he possesses the abilities and skills to meet these expectations (Fig. 6.4). The 'good' or 'middle-class' pupil typically falls into box 1, where he has learned the content of the pupil role, is able to meet these requirements, and does in fact meet them. In his earliest days at school such a pupil may at times fall into box 4, where he accidentally meets the teacher's expectations without realizing that such behaviour is normative. Or he may fall into box 5, where he fails to meet the expectations because he has not recognized them as yet; once they are learned then the pupil can meet them because he has the ability to do so, and is thus free to move into box 1. The 'working class' pupil is more likely to find himself in box 6, finding it more difficult to learn the content of the pupil role and lacking the abilities that are required to succeed in it. Here he remains until he learns the content of the pupil role through his experience and then he moves into box 3. He has learned the laws of pleasing teacher but still lacks the ability to succeed. He has learned the nature of the game but finds it difficult to prevent himself from losing. If he has the capacity and motivation to acquire the necessary skills, then he will be free to move into box 1 and become a 'good' pupil. But not all pupils make this move. Some remain in box 3, but adapt to the situation by giving up the attempt to please teacher and by setting up an alternative delinquent orientation.

Others may move into box 2 where they are fully able to please the teacher but intentionally decline to do so.

We can now illustrate the kinds of social skills which are lacking in pupils in boxes 3 and 6. When pupils misbehave and deviate from the teacher's expectations, they are not normally punished immediately. More typically, the teacher asks the offenders to give an explanation of their misconduct. They are, to use the term of Scott & Lyman (1968), asked to give an *account*—'a statement made by a social actor to explain unanticipated or untoward behaviour.' The success with which they can avoid negative evaluation or punishment depends upon their ability to provide an account which will be accepted by the teacher as satisfactory. Some 'working-class' pupils seem to be deficient with respect to the social skills required to give such an acceptable account. For instance, when a teacher asks them why they committed the misbehaviour, they not uncommonly respond with a silence or 'I don't know'. In the young child this complete failure to attempt to provide an account may be either a failure to recognize the misbehaviour as such or a lack of social skill in providing an account that will be acceptable to the teacher. In the older pupil it is more likely to be a lack of social skill. When the pupil offers no account of his misconduct, there is nothing to inhibit the teacher's disapproval, for the silence may be interpreted as a silence of guilt, the teacher's inference being that the pupil is unwilling or too ashamed rather than unable to provide an account. At other times these pupils may attempt to make an account, but do so ineptly, that is, they offer accounts which will be regarded as unacceptable and illegitimate by the teacher. Accounts such as 'I was only *borrowing* his football', or 'He deserved to be hit because he's soft', or 'Smith's always doing it but you never tell him off' are unlikely to be acceptable to the teacher. Suppose a pupil drops a bottle of chemicals and is called to account. The 'working-class' pupil's 'I couldn't help it, honest,' is much less acceptable than the 'middle-class' pupil's 'Sorry, sir, but my hands were a bit slippery with washing test-tubes and the bottle was much heavier than I thought.'

In giving accounts these pupils are also disadvantaged with respect to the *manner* in which the account is given as well as to its content. They will sometimes deny absolutely that they have committed the misbehaviour even when the teacher has seen them in the very act. They are also often weak in presenting a congruent front, as in the case of the pupil who cannot control his grin as he makes a penitent apology. Such pupils are also prone to inappropriate account styles. In Scott & Lyman's terms, they adopt intimate or casual styles in place of the formal style which is appropriate to teacher–pupil relationships. Their inability to switch to a different account style for the teacher and their persistence with the casual or intimate account styles used in the home or with the peer group has the effect of making many of their accounts quite unacceptable to the teachers. Thus their accounts are carefully scrutinized and checked by the teacher and are more likely to be regarded as 'mere excuses' rather than as 'valid reasons'. When the pupils give up the struggle to please the teacher and adopt the delinquent orientation, then there is often little pressure to direct the account at the teacher and it may haply be orientated to peers—'Well, it was a bit of a laugh, wasn't it?'—to the great annoyance of the teacher.

These three orientations, the 'normal', the 'delinquent' and the 'indifferent' have been outlined at a fairly crude level. We need to know very much more about the variant forms of coming to terms with pleasing teacher, about the range of strategies and skills involved, and about the different interpretive schemes employed by the pupils. (In Chapter 11 we shall try to distinguish two forms of pleasing teacher in terms of 'compliance' and 'identification'.) One of the main problems is that of showing how these pleasing teacher orientations develop and change during the career of pupils in the educational system, and in particular we need research on the ways in which and the means by which pupils acquire an orientation and its associated strategies during the early years of schooling.

Using the concept of pleasing teacher I have tried to outline one form of a more phenomenological approach to the study of teacher-pupil interaction which contrasts with the approaches of the interaction analysts and the role theorists. This approach, which we have barely begun to explore, has several important advantages. First, it opens research to the amateur, the practising teacher. Most teachers think of educational research as a very specialized concern of certain non-teachers that we call professional researchers. Research is not, in other words, something that the ordinary teacher can do. Yet John Holt shows that research is not the exclusive preserve of professional researchers. Indeed, he is the best evidence I know of to justify the assertion that professional researchers are too far from the realities of the classroom. John Holt writes as a teacher, without rarified techniques, without technical language, without elaborate conceptualization. Yet his work reveals an almost unique insight into the nature of teacher–pupil interaction and in particular into the nature of the pupil perspective. Second, this approach offers potentialities of real interdisciplinary research. The philosophers' work on phenomenology has influenced both sociology (e.g. Harold Garfinkel, Howard Becker, Aaron Cicourel, John Kitsuse, David Matza) and psychology (R. D. Laing, Carl Rogers), though at present most work has been done on deviant or abnormal behaviour. Third, this approach offers an interesting orientation for teaching education students about classroom interaction. We all agree on the importance of sensitivity in the art of being a teacher. This approach emphasizes the ability of the teacher to put himself in the pupils' shoes and to see classroom events from their perspective. It suggests, in other words, a justification for the training of teachers in empathy.

The pleasing teacher phenomenon throws considerable light on certain aspects of life in the classroom which presumably are neither desired nor intended by the teacher. But how can they be avoided? Whilst it may be true that the sensitive teacher is more aware of these effects than most, being aware of their existence does not in itself provide a means of obviating them. If we are to find a way in which teachers and pupils can escape the difficulties associated with pleasing teacher, we must first analyse the phenomenon in greater depth and get closer to its sources.

Pleasing teacher: further analysis and a way through

The strategies and effects associated with pleasing teacher do not spring directly or inevitably from the teacher's greater power in the classroom. Rather, they derive from two aspects of the teacher's role, two particular aspects of his greater power, namely his functions as *evaluator* and *motivator*. These represent two of the most central and pervasive elements of the teacher's role which we must now analyse further. The teacher must evaluate the pupil's learning and academic progress (instructional role): and he must evaluate the pupil's behaviour (disciplinary role). He must motivate the pupil to work hard and to be interested in what he is supposed to learn (instructional role); and he must motivate the pupil to behave well and to conform to the rules of conduct and procedure (disciplinary role). Now I want to suggest that the principal way the teacher evaluates and motivates is by *giving approval*. In giving approval the teacher reveals to the pupils the extent to which they are succeeding in pleasing teacher. In seeking to please the teacher, the pupils are competing for the teacher's approval. The teacher gives approval mainly when he evaluates and motivates the pupils. In evaluating the pupils' learning or behaviour, the teacher gives approval to those who live up to his expectations. In motivating the pupils to learn or behave appropriately, the teacher gives approval to the pupils as a reward for their conformity or as an incentive to further conformity.

The distinction between teacher as evaluator and teacher as motivator is an analytical one. In real life these two functions interpenetrate, just as do the instructional and disciplinary roles. This will become clear as we illustrate the problem with reference to the practice of giving *marks* or grades. Why do teachers give marks? There are several possibilities.

1   Marks are rewards for learning-achievements. In giving a high

mark the teacher wishes to reward the pupil who has done the required work correctly and thus, by inference, learned.

2 Marks are incentives to learn. They are intended to motivate the pupil to maintain his achievements or to increase his efforts.

3 Marks are 'feedback' for the pupil. In other words marks are intended to inform the pupil about the depth and accuracy of his learning.

4 Marks are 'feedback' for the teacher, giving him some indication of the degree to which he has succeeded in getting the pupils to learn successfully.

5 Marks are measures of progress. They allow the teacher to keep a record of the pupil's learning and achievements.

6 Marks can be used for making comparisons between pupils, in order to estimate relative achievements and relative rates of progress.

The same six functions of marks could be applied to 'House points' or 'merit marks' which are related to pupils' conduct within the disciplinary area. The argument which follows tends to apply to both forms, but for the sake of simplicity I have confined the argument to marks given for academic work, or for appropriate behaviour within the instructional area.

The central point of the argument is that the teacher, in giving marks, is dispensing approval—whether or not he intends to do so. Marks are merely a common example of how the teacher's behaviour is pervaded by his evaluative and motivational functions which involve the meting out of approval. Almost all teachers use marks as a measure of progress. A 'mark-book' is one of the teacher's basic possessions. However, almost all teachers wish to do more than keep a record, for typically they inform the pupils of the marks they have been awarded, which would not be necessary simply for keeping a record. If marks are given as rewards or incentives—two functions which are hard to disentangle—they are clearly an indication to the pupil of the teacher's approval. The higher the mark, the greater the teacher's approval. In fact, teachers frequently strengthen the approval implications of the mark by adding a comment with it, such as 'very good work' (evaluation) or 'keep it up' (motivation). Similar comments with approval loadings, both evaluative and motivational, accompany low marks—'You can do better than this.'

The teacher may or may not intend to imply approval or disapproval in his marks or comments. (If he has been taught in an educational psychology course to 'reinforce' correct responses by pupils it will probably be intentional.) Yet in both cases the teacher will be seeking to give the pupil 'feedback' on his behaviour, to inform him of the accuracy or quality or success of the work. But from the pupil's perspective the function of marks is not so much to give him feedback

199

on his learning abilities or achievements; rather the main function of marks is that they indicate the degree to which he is succeeding in his general objective of pleasing teacher and obtaining approval. When a teacher gives work back, the pupil is less concerned to discover whether or not he has mastered the concepts or principles necessary to a successful solution of the set task than to find out whether or not he has been awarded a high mark and a favourable comment. The pupil's first reaction on receiving corrected work, a reaction which betrays his basic goal, is to estimate the degree of approval implied in the teacher's mark and comment and then to compare the amount received with that bestowed on other pupils. Even if his mark is low in absolute terms, he will feel considerable satisfaction if he knows that he has received most of the available approval or least disapproval. The disjunction between the teacher's intentions and the pupil's response is clear when we remember that the pupil normally pays little attention to the detailed corrections and suggestions that the teacher has written for the pupil's edification. This 'feedback' on which the teacher has spent so much of his time is largely ignored by the pupil since for him it is the approval which matters much more than the advance he is making or could make in the understanding of what is being taught.

I want to suggest, then, that the majority of pupils become addicted to the teacher's approval during the process of formal schooling. When they learn, it tends to be as a means of obtaining approval rather than as an end in itself. Indeed, paradoxically, the pupils' desire to obtain approval may subvert the learning process. Approval-seeking may become a *substitute* for learning. Even in the sixth form or at the university, where the teacher puts such stress on the 'autonomy' and 'independence' of the learner, most students do not lose their need for the teacher's approval—partly because they have become 'hooked' on it, partly because the teacher's alleged concern with student autonomy and independence is a euphemistic fiction surrounding the fact that the student is under considerably less close supervision than in the past. There is no real attempt to create conditions in which the student can acquire and develop autonomy and independence; he is merely thrown back on his own devices. Many students are severely distressed by this condition. They are left to 'dry out' from their addiction without any help or support. They are not cured. Typically, they adapt to the situation, not by placing less importance on the teacher's approval, but by learning to receive less approval in absolute terms and to concentrate their need for approval into the available channels, such as the tutorial discussion group. Here the students, in anxious competition for the teacher's approval, become defensive about expressing their puzzlements and problems too openly. As in school, the desire to please teacher inhibits any genuine exchange of views and difficulties. The addiction remains—and the teacher wonders why his student discussion group is so unproductive.

If it is true that the pupils' concern to obtain approval and to avoid disapproval leads them to respond in ways which inhibit learning, then we must ask ourselves to what extent does learning require approval from the teacher? It is certainly true that enormous amounts of childhood learning are associated with the bestowal of approval by socializing agents, especially the parents. Approval in its various forms is without doubt one of the most common sources of reward and one of the most basic elements of childhood socialization. Yet even if approval is essential in this respect, it does not necessarily follow that it is an essential or desirable tool in the promotion of classroom learning. Human beings can learn without approval, and if our earlier analysis bears any truth, then learning based on teacher approval may be undesirable in the classroom.

Can approval in the classroom be abolished? If we could do so, would we achieve a more desirable and more significant sort of learning? I doubt very much whether we could abolish approval in the classroom. Before they come to school, the pupils have become accustomed to expect and desire approval from adults. Inevitably, this need will be transferred to the teacher. Furthermore, much normal interaction involves the giving of approval and disapproval; it is a universal aspect of human relationships. It would be an impossible task for the teacher to try to abandon it completely. The question is rather whether we can reduce and minimize approval giving in the classroom and what effect this might have on the learning process.

The effects of such a course of action are far reaching, since to minimize the teacher's approval giving requires us also to minimize the teacher's functions of evaluation and motivation, for it is principally these functions which stimulate him into approval giving. The basic problem with approval is that it is *personal*. Whilst the teacher may intend to approve or disapprove the pupil's efforts or achievements or interests, it is difficult for him to do so without at the same time approving or disapproving the pupil as a person. When I tell John that his work is well done, I am also implying that John is a good person. The approval spills over from the activities to which the approval is directed to John as a whole. Similarly, if I tell John that his work is bad, the implication is that John as a person is being judged unfavourably. Approval becomes, from childhood, associated with feelings of liking and love. It is for this reason that it is exceptionally difficult for the teacher to give 'pure feedback' to a pupil with reference to his learning—to tell him whether he is right or wrong, succeeding or failing, progressing or regressing, without

201

conveying his judgment of the person. In all these cases the teacher tends to reveal the extent to which he values the pupil as a person. It is only in an impersonal relationship that feedback in a pure form is possible—the only other case will be mentioned shortly. A teaching machine can give pure feedback because the machine does not form a personal relationship with the pupil. (Even in this case approval may not be entirely absent, since the pupil may approve or disapprove himself when the machine tells him that he is making the right or wrong response.) Similarly it is often easier to accept criticism (negatively toned feedback) from a stranger than from an intimate. In contemporary schools, the problem is particularly acute, since the teachers are trained to make 'good personal relationships' with their pupils. In so doing, they *increase* the degree to which the feedback they offer to their pupils is loaded with approval. The more personal the teacher–pupil relationship is, the more approval-loaded the feedback becomes. In other words the cold impersonal teacher, ironically, is more able to offer the pupil feedback without valuing the pupil as a person.

If we are to solve this problem we shall need to find a different definition of the situation in the classroom and a re-interpretation of teacher and pupil roles and relationships in which the functions of evaluation and motivation can be dissociated from approval-seeking and approval-giving. This is obviously an enormous task, but a solution is available in the work of Carl Rogers (1951, 1961, 1969). Rogers developed a form of psychotherapy which has become known as Non-directive or Client-centred therapy. This method rests on three main assumptions—the therapist's genuine respect for the client as a person; the belief that the client has within him a basic force working towards growth and the realization of his potentialities; the belief that this self-realization is promoted in non-threatening personal relationships. It is assumed that the person seeking the therapist's help, who is always referred to as a client not a patient, has within him the capacity to understand the factors causing him distress and pain which make him 'ill' or unable to function adequately, and that the client has the capacity to reorganize himself in such a way as to overcome his illness. It is assumed that these powers in the client will become effective if the therapist can create a relationship which is marked by respect, warmth and acceptance.

Such a view of therapy stands in marked contrast to most traditional psychiatric methods. The question is not the traditional one of: How can I treat, cure or change this person? The question becomes: How can I provide a relationship and atmosphere in which the person can come to redirect his own growth as a person? The therapist is no longer concerned with the diagnosis of the disorder, with evaluation of the client's personality, with the collection of a case-history, with the seeking out of causes, or with the probing of the unconscious. Instead of exploring and interpreting, the therapist tries to provide a warm, empathic and acceptant relationship in which the client can express, explore and clarify his own thoughts, feelings and behaviour.

The therapist thus approaches the client with genuine concern and care, with 'unconditional positive regard'. He tries to create a warm and acceptant relationship. He tries to develop the skill of *empathy*; to see the client as the client sees himself, to stand in the client's shoes and see the world from his perspective, with all the associated problems, confusions and fears. His aim is not to explain, interpret, and evaluate or judge, but to understand by empathy. Within such a secure relationship the hidden and suppressed thoughts and feelings of the client can come to light without the client fearing that he will be hurt or condemned. It is clear that this form of therapy has a Meadian basis. The client takes to himself the attitude that the therapist takes to him. Because the therapist accepts him, the client can accept himself. Once the client accepts his own thoughts and feelings, however terrible or unnatural they seem to be, he is then free to explore himself and to integrate the rejected or distorted aspects of his experience and re-orientate his life on more positive lines.

At this point we have to exercise some caution, for we have to remember that we cannot solve our problems by trying to turn the teacher into a therapist. Teaching is not the same as therapy. The class is there not to promote the mental health of its members, though this may be one of several objectives. The principal objective of the school class is a different sort of learning. Fortunately, Rogers himself has in recent years devoted a great deal of his writing more directly to educational problems. In the end there is no substitute for reading at firsthand the views of Carl Rogers himself on education. Perhaps the best introduction to his thinking is one of his earliest writings on education, a highly dramatic and stimulating paper entitled 'Personal Thoughts on Teaching and Learning' (Rogers, 1961). There are few articles which in so small a space—the article is but a few hundred words long—raise so many issues for teachers. Let us consider some of the central ideas of this paper, in which Rogers tries to tell his audience how he feels about teaching and learning.

Anything that can be taught to another is relatively inconsequential, and has little or no significant influence on behaviour . . .
    I have come to feel that the only learning which significantly influences behaviour is self-discovered, self-appropriated learning.
    Such self-discovered learning, truth that has been personally appropriated and assimilated in experience, cannot be directly

communicated to another . . . I realize that I have lost interest in
being a teacher.

When I try to teach . . . I am appalled by the results, which
seem little more than inconsequential, because sometimes the
teaching appears to succeed. When this happens I find that the
results are damaging. It seems to cause the individual to distrust
his own experience, and to stifle significant learning. Hence I
have come to feel that the outcomes of teaching are either
unimportant or hurtful. When I look back at the results of my
past teaching, the real results seem the same—either damage was
done, or nothing significant occurred. This is frankly troubling.
As a consequence I realize that I am only interested in being a
learner. . .

Having let his audience into the inside of his feelings in this way,
having shown us something of his experience of being teacher and
learner, Rogers then considers some of the principal implications of
his feelings. These are that he would do away with teaching: that
examinations, grades, credits and degrees would be abolished. In
this exposition of his feelings and their implications, Rogers recognizes that he is led in directions which are regarded as disturbing
and absurd. At the same time he warns us that what he says should
not be regarded as conclusions or guides to action for anyone else.
He ends by turning to his audience with the words:

I want to know primarily whether anything in my inward
experience and thinking as I have tried to describe it to you,
speaks to anything in your experience in the classroom as you
have lived it, and if so, what the meanings are that exist for
you in your experience.

This is very unusual in a paper on teaching and learning. He does not
do what we might expect of him. He does not tell us the educational
implications of his experience as a psychotherapist on which his fame
rested, nor the way such implications might relate to the writing of
illustrious educationists. Moreover he makes little attempt to
define precisely what he means by the terms teaching and learning,
even though we all know that these terms are vague and very difficult
to define. He simply tells us what he feels and asks if what he says
speaks to our experience.

For my part, these ideas speak to me both as a learner and as a
teacher. I recall many hours spent as a pupil in school or as a student
in a University during which I listened passively to lectures of various
sorts, some excruciatingly boring, others brimming with ideas and
ingeniously structured, but all conveying vast amounts of information
and ideas which I meticulously noted down for further reference.

Usually I left the lecture with a feeling of accomplishment, for I had in my hands the essence of the labours of my masters. When it came to writing an essay, however, I usually found the notes to be of little value, for they did not really answer the problems I was trying to solve. Too often I had to begin again and discover for myself. On re-reading the notes prior to the examination, it was a shock to find how little of the content I remembered or understood. The meaning and the structure of the lecture had somehow evaporated. In short, I felt I had really learnt something when, either because of my own interest or because of the demands of my teachers, I had to grapple with the problem for myself.

As a teacher I have had a similar experience. I have spent many hours preparing lessons and lectures. I have planned in detail the ground that must be covered, the content and structure of the lesson, and the nature of the fundamental concepts involved. But I have also had the disturbing experience of reading essays and examination papers of my own pupils and students in which I could hardly credit the misunderstandings, distortions, omissions, irrelevancies and sheer inventions based on the lessons and lectures I had so carefully prepared. I often blamed this on the inattentiveness and laziness of the pupils rather than on the possible defects of my own teaching, for how could I fail where I had tried so hard? I was encouraged in this view by the fact that some pupils wrote excellent essays and did seem to have profited from my lessons. It was not until later that I realized that good work was produced by the able and motivated pupils who were prepared to learn the material again by struggling through the problems in their own time. It took an even longer time for me to recognize that I could hardly hope to succeed with my own pupils and students by using methods which I as a pupil and student had found so singularly lacking. It is an interesting paradox that as soon as we become teachers we tend to become so involved in teacher-problems that we completely forget what we felt and thought as students.

In other words once I examine whether or not what Rogers says does speak to me, I find myself in essential agreement with him, exploring for myself those feelings about my teaching and about my learning which I would prefer not to recognize and accept, partly because they make me feel insecure and anxious, and partly because they are leading me to conclusions which seem ridiculous.

I take it that the essence of Rogers's paper is the feeling that what he does as teacher or learner fails to contribute to the realization of himself or actively inhibits it. More specifically, he feels that nothing of what he does as a teacher promotes his own self-realization or that of his pupils, and that only certain sorts of learning do contribute to self-realization. He is not concerned primarily with such matters as the effectiveness of his teaching, but rather with the more existential question of the effect of his teaching on himself as a person and on his pupils as persons. It is a question that we usually do not bother to ask. When we do ask it the results are rather frightening. But I believe that it is a question that we should ask ourselves and try to answer with honesty, regardless of the implications or apparent absurdity of them, for I suspect that very large numbers of teachers are in essential agreement with Rogers, sharing a similar experience and coming, reluctantly, to similar conclusions.

But before we turn to the question of how the therapeutic situation can be translated into the educational situation, we must consider what Rogers has to say about *learning*, which is one of the elements common to both therapy and education. Neither can proceed without learning.

Rogers believes that we cannot teach another person directly: we can only facilitate his learning. With a little thought most teachers will agree with this statement. What we do as teachers, whether we make the pupils work in groups or alone, whether we lecture or initiate a project, whether we praise or blame, is directed to the facilitation of learning. And the pupil will not learn until he involves himself in the activity of learning, a process that only the pupil himself can do. Rogers also believes that the only really important sort of learning is that which influences behaviour, learning which affects the learner's self. Rogers seems to be making a distinction between different sorts of learning, or perhaps between education and knowledge. As Peters (1967) points out, the acquisition of knowledge may be an important part of education but it is in itself not the same as education.

> The problem of education is to pass on this knowledge and this understanding in such a way that they develop a life of their own in the minds of others and transform how they see the world, and hence how they feel about it.

Learning is not merely the acquisition of knowledge, but a pervasive process of change which affects how a person sees himself, the world, and his relationship to the world. Rogers calls such learning *significant* learning: it is learning which is self-discovered, self-appropriated, learning which contributes to and enhances the structure of the self. It is self-discovered in the sense that it takes place in response to the learner's problems and purposes and in this way makes a difference to him as a person. It means grappling and struggling with a problem. Learning which is not significant tends to be the acquisition of knowledge which is simply stored—or more typically, knowledge which is inadequately grasped, then distorted and finally stored.

If my experience as a learner and teacher is not atypical, this raises some crucial issues. Why is it that, though I subscribe to the notion of significant learning, I seem so unable to promote it? Why do I put obstacles in the path of my own goal? Why as a teacher have I become so insensitive to my own past experience, *and* my present experience, as a learner? (How many lecturers feel that they learn much from listening to other people's lectures?) What has gone wrong? Let us consider the problem from the point of view of the learner. Carl Rogers may be right in suggesting that significant learning, in therapeutic or educational settings, is most effectively promoted in situations where the threat to the self of the learner is at a minimum. Such a situation is achieved, as in therapy, when the teacher shows an unconditional positive regard for the learner, when the climate is acceptant of the learner. It is in such a situation that the learner is most likely to drop his defensiveness and to open himself to his experience. He will feel free to express his uncertainties and difficulties and thus be able to come to terms with the problems and ideas at issue.

This long account of the work of Carl Rogers has provided us with a solution to the problem of how the teacher can divest himself of his evaluative and motivational roles. The answer is that the teacher must try to promote a relationship which is characterized by *acceptance*. Now acceptance differs very markedly from approval. Indeed, normally the two are mutually exclusive. Rogers found that to show the client (or learner) acceptance, he had actively to restrain his tendency to approve or disapprove the learner's feelings, ideas or behaviour. For in accepting another, one has to offer him a non-threatening atmosphere in which he is valued as a person. He must be given 'unconditional positive regard'. Approval, on the other hand, is highly threatening precisely because it is conditional. If you do *x* I will approve you; but if you do *y* I will disapprove you. Approval is something which is deserved or won; acceptance is something which is simply given. It makes no demands and sets no conditions from the giver on the person to whom it is given.

The answer then would seem to be that we should train teachers to learn the art of acceptance and in so doing to unlearn the natural tendency of a teacher to load all his reactions to pupils with approval and disapproval. Clearly this would be a more difficult process for some teachers than for others, but for the vast majority it would involve a radical change in their conception of the roles of teachers and pupils. Equally, it would involve a radical change in the role conception of those who train the teachers. This is so because if acceptance becomes the key problem of the teacher, many of his traditional assumptions and practices need to be revised.

At the heart of this revolution stands the transfer of evaluation and motivation from largely teacher-centred functions to largely pupil-centred functions. Let us take first the case of evaluation. In divesting himself of his approval-bound evaluative function, much of the evaluation of the pupil's efforts, abilities, and achievements must be *self-evaluation* by the pupil. It is he who must be his own judge—with the help of the teacher where this is necessary or sought by the pupil. Experiments along these lines have already begun (Rogers, 1969). Interestingly, this is as difficult for the pupil as for the teacher. The average pupil feels disturbed when he is asked to estimate the effort he has put into a piece of work, or its quality and worth. At the same time, pupils are usually in surprising agreement with the teacher in assessing the work. Once the pupil has been persuaded to evaluate himself, the teacher ceases to be the fount of all approval. Instead, he becomes the person with whom the learner can discuss his achievements and progress; his intellectual difficulties; his problems of distraction, concentration and lack of interest. They can be discussed more openly because the teacher accepts the pupil, who need no longer be afraid to discuss his work honestly because the relationship is no longer conditional. The strategies for obtaining approval or deflecting disapproval are now superfluous.

It should be clear that experiments in self-evaluation are not likely to succeed unless the teacher has created an acceptant relationship with the pupils. If the relationship is still approval orientated, then to initiate self-evaluation is to invite disaster because the pupil will persist in the old strategies. If an approval-orientated teacher asks the pupils to award themselves marks, then many of the pupils will give themselves unjustly inflated high marks in order to secure the teacher's approval. I have known of teachers in approval orientated classrooms who have tried to create a more 'mature' learning situation in which pupils mark their own work, for example by correcting their work from answers printed in the back of the text-book. The natural result is that some pupils alter their incorrect work but mark it as if it had been correct in the first place. With luck, they will convince the teacher that they have done the work correctly and merited his approval. Only when acceptance has replaced approval can one expect the pupils not to 'cheat' in such a situation. Only in a truly acceptant relationship is an attempt to give relatively pure feedback likely to succeed.

Let us now turn to the motivational function of the teacher. This consists essentially of creating and/or stimulating a desire in the pupils to make the efforts necessary for an understanding of the subject, to be persistent and interested in what the teacher requires them to learn. Yet we know that many of our pupils do not have a high level of motivation; so the teacher falls back on the incentives of examinations, competition and the various approval–disapproval techniques that have become 'second nature' to him. How, then, can the teacher transfer his motivation functions to the pupils?

208

Clearly the teacher must draw on the pupil's natural curiosity, interest and desire to learn. As teachers, we recognize the existence of such qualities in all the pupils. We are familiar with the backward pupil who seems to make no progress in his schoolwork, but who can easily remember the rank order of the First Division football clubs as it changes week by week. The difficulty is that what too many pupils seem to lack is not curiosity or interest *per se*, but a curiosity about and interest in the subject matter we want them to learn. Thus we are frequently faced with a situation in which we are trying to get the pupils to learn something that they do not really want to learn. In consequence, it is not surprising that relatively few of them ever really learn it, though many manage, under pressures of various sorts, to memorize some of the material on a temporary basis— enough, with luck, to pass an examination and to suggest that successful teaching and learning have taken place.

At times some teachers recognize, as did Carl Rogers, the futility of what they are doing. In John Updike's novel, *The Centaur* (1963), the hero George Caldwell, a teacher in an American high school, has such a moment of devastating honesty—devastating because he cannot see any alternative to the present position. A pupil has complained that he cannot see the point of memorizing lists of prehistoric animals. George concedes the point. The whole business depresses him just as much as it depresses the recalcitrant pupil. He can't justify what he is doing, either to himself or to the pupil. He's told to teach it and teach it he will. He knows how the pupil is suffering, but he has little choice in the matter. He has a family to support so he must keep down his job. And if that means he must strangle difficult pupils, he will do so.

D. H. Lawrence expresses much the same sentiments in his poem 'Last Lesson of the Afternoon'. He longs for the bell to ring for he is weary of trying to make pupils acquire knowledge that they do not wish to learn. There seems to be no point to the teaching and the learning; mutual harm appears to be the only result.

For my own part, I am not convinced that there is much to be gained by trying to instil knowledge into reluctant learners, with the vague justificatory hope that it will be 'good' for them and the consolation that they have passed their examinations. In truth it is not good for them, because even though we can pressure pupils to acquire skills and to memorize, it is doubtful if we can pressure them to *understand*, and also because in making them study against their wishes (a fact which pupils do not always make obvious to the teacher) we probably alienate them against the subject we are trying to teach and perhaps against schooling and study in general. The not uncommon tearing up of exercise books at the end of a course of study that will no longer be pursued has a more common counterpart in psychological abhorrence. In contemporary society, the rejection of learning by many pupils tends to be masked from teachers by the fact that they desire to learn not as an end in itself (so-called intrinsic motivation) but because they need to acquire a qualification to obtain a desired job (like approval-seeking, another form of extrinsic motivation). It is easy for us as teachers to accept this surface motivation of pupils as a genuine interest in learning. It protects us from knowing the truth which, if accepted, would plunge us into one of the most central problems of teaching: how to promote in the pupil a real desire to learn and understand.

If the answer to the motivational problem is that, as in the case of evaluation, it must be transformed into pupil *self-motivation*, how can this be done? It means that pupil learning must be largely *self-initiated* and *self-directed*, since we can expect the pupils' curiosity and interest to develop and to be sustained only in response to learning which they undertake and control. Such a view seems to be even more revolutionary than the notion of pupil self-evaluation, though the two seem to be inherently compatible. Again, it involves a radical reorganization of teacher and pupil roles. When the teacher attempts to make the change to self-directed learning there is considerable distress for both teacher and pupils, especially since some time has to be 'wasted' in which the pupils do nothing at all before their natural curiosity, interest and desire to learn lead them to begin to work and seek help from the teacher. Yet it can work and the results with 'difficult' classes, where the teacher is perhaps least prone to experiment, can greatly increase the productivity in work and produce a marked decline in disciplinary problems (Rogers, 1969, Chapter 1).

If it is self-initiated and self-directed learning which is significant learning, then one of the most important classroom transformations will be the change in stress from teacher-based problems to pupil-based problems. Assuming that the pupil is moving toward a realization of his potentialities under non-threatening conditions, the teacher's principal function is to facilitate learning by creating the right conditions. The teacher will become less concerned with such questions as: What information must I put over in this lesson? How can I ensure that the pupils are attending? How can I motivate them to work hard when they are not under my surveillance? How must I organize the material? How can I be sure to cover the syllabus in the time allotted to my course? These are questions deriving from an attitude of lack of trust towards the pupil, an attitude of believing that education is a process of *doing* something *to* people, even though it is for their ultimate benefit. It is a rejection of the notion that education is rather like pouring water into a bottle, if only the bottle can be held in the right position for long enough.

Once an attitude of trust towards the pupil is held, the major problems stem from the question: How can I create a non-threatening situation in which the learner's learning will be facilitated? The teacher's problems are related less to getting the pupils to do what he wants—'Teacher always knows best'—and more to adapting his behaviour to the needs of the pupils. The teacher must be prepared to adapt to the pupils rather than expecting the pupils to adapt to him. This involves a certain unpredictability in the course of study, which stands in striking contrast to the detailed planning which most teachers adopt. This lack of predictability combined with demands for role flexibility make the teaching situation an insecure one for the teacher accustomed to more traditional methods.

An important aspect of self-directed learning is that the pupil must begin to set his own goals in the process of learning. Each pupil must plan his own course—the ground he intends to cover, the problems he wants to try to solve, the avenues he wishes to explore. In this period of planning, the teacher can obviously be of great assistance in helping to make the plans more specific, in giving advice when the pupil runs into unforeseen difficulties, in providing resources and materials that will be of use to the pupil. Such a system of 'work contracts' is not new: the Dalton Plan is just one of several such ideas that have been devised in this century. Yet some system of self-directed learning is required if self-evaluation is to become meaningful, for we cannot expect the pupil to learn the art of evaluating himself and his learning if he is working on teacher-imposed tasks to which he may feel very little commitment.

Many of the basic problems implicit in such a programme have still to be elucidated. Hitherto most of the experiments have been conducted by individual teachers working in schools where there is little support from superiors or colleagues and where innovations are isolated in single classrooms. Yet the success of a programme may well depend in part upon a common school policy, to which all the teachers are committed, with the result that the impact on the pupils is maximized and they are not exposed to fundamentally opposed regimes in different classrooms. Yet it does not seem likely that a whole school will readily be prepared to adapt as a whole to the ethos of acceptance orientated learning. A universal commitment by a school staff involves not only a new conception of the teacher's class-room role, but also radical changes in the school's organization as well as in the structure of relationships among the teachers. Such changes are slow and difficult, especially in an age when teachers are still adjusting to the effects of an imposed Comprehensive re-organization.

It is significant that most of the present experiments are being conducted in primary schools, where the teacher, taking one class for the whole day, has considerable personal autonomy. In the secondary school a movement towards self-directed, self-motivated and self-evaluated learning would be much more difficult because of the interdependence of the various parts of the system. Each class of pupils is taught not by one teacher but by a whole range of specialist teachers. Every forty-five minutes or so most of the pupils and teachers change classes. The structure of secondary schools, so inimical to self-directed learning, is based on the assumption that it is a good thing for pupils to be given a different slice of the educational cake at regular intervals. It is argued that we cannot give them larger chunks of any one subject because they would be bored by their fare. The argument is sound; many pupils would be bored by a whole day of mathematics or geography because they did not want even forty-five minutes of it in the first place. Self-directed learning tries to deal with this more basic problem. The second argument in support of the traditional time-table is that by this means we ensure that pupils are given a *general* education, without excessive specialization. This is not a sound argument because one will have limited success in giving a general education to pupils who are very reluctant to be given it. For me the greatest tragedy of our present educational system is that though we manage to make most pupils learn to read, the majority read hardly any books for the rest of their lives. We give them the skill, but we also teach them that the skill is barely worth exercising, especially for pleasure.

Self-directed learning would let pupils learn what they want to learn for as long as they want to go on learning it. If a pupil wants to spend only half an hour on a subject and then turn to something else, he is free to do so. If he wants to spend six hours, or six days, or six weeks, or six months on a particular subject, then he is also free to do so. The assumption is that if the pupil wants to do this, then it is right for him at that moment and that all this time will be profitably spent in active learning. The pupil can and will turn to other fields when he is ready. If he is aiming towards a public examination, he will fully appreciate the danger of spending all his time on one subject. Ultimately the choice must be his. We cannot make all his choices for him and then wonder why he does not want to learn.

Thus the implication of self-directed learning is that the timetable of the secondary school would collapse. Teachers would simply be there, ready to teach and act as resources to those pupils who wanted to learn that subject. Almost all learning would be highly individualized, so the class unit would inevitably disappear. Similarly age-grading would also end, for within a classroom children of all ages who wanted to learn a subject would be present.

To the average teacher such a prospect is extremely terrifying. His basic assumptions about the structure of schools, of ordered classes of pupils of the same age coming to him at regular, specified intervals, have been eroded. Some of the fear would simply be fear of the unknown. In other cases teachers would fear that no pupils would voluntarily come to their classrooms to be taught. Yet it is just such teachers whose notions of teaching are most urgently in need of revision. Perhaps too for a few weeks most of the pupils would choose only one or two subjects, leaving many members of staff with no-one to teach. We would have to face the consequences of that, though for my own part I suspect that we would soon be restored to a fairly even spread. Yet these are all real fears, as real as are the risks that would have to be taken. But on the other side are all the gains from self-directed learning and a voluntary curriculum. Teaching could become so much more positive, since the teacher would be teaching pupils who really want to learn. As all teachers know, this is very hard work, but the effort is rewarding and worth-while. Too much of teachers' present energies are sapped by disciplinary or learning problems rooted in the pupils' lack of desire to learn what the teacher wants to (and feels he must) teach.

Another important area of concern is that the sociological implications of such modifications have not been adequately studied. In deprived urban areas the problems of a transformation to self-directed learning and self-evaluation are likely to be particularly acute. Partly this is because the ethos of such schools tends to be the most 'custodial' and the least 'progressive' and thus the change required of the teacher in role conception is much more radical than in the more middle class suburban school. A further reason is that the pupils of such schools are also least susceptible to change. Many of them are, relative to middle class pupils, apparently deficient in ability, motivation, interest and independence. Many of them have already been alienated from learning by their experience of school and the culture of the local community, and may regard such changes with suspicion or apathy. Even if they can be persuaded to embark on self-directed learning, their chosen fields of study may be very restricted and unimaginative, particularly in the early stages. These children frequently lack the wealth of interests and skills acquired by the middle-class child from home; they have a much smaller fund on which they can draw in a programme of self-directed learning. The reliance of such pupils on the teacher as a provider of resources and suggestions greatly increases.

The very small gains made when such pupils have been subjected to intensive compensatory education programmes do not inspire confidence. The spur to change may well rest on our knowledge that the quality of relationships and standards of learning in many of these schools in depressed areas cannot get very much worse. Almost *any* change is worth trying, and one which seeks to treat pupils as persons and to develop greater autonomy may well recommend itself to those teachers who have been unable to find a possible answer elsewhere.

The work of Carl Rogers is of fundamental significance. He provides us with an approach to teaching and learning which can help us to undermine the pleasing teacher phenomenon. In addition, his approach offers scope for a much more meaningful form of learning and, if teachers are willing to take the risks and have the courage to revise their role conceptions, a much more meaningful form of teaching. But Rogers also helps us to solve another problem with which we have been deeply concerned in this chapter, namely the inability or reluctance of the teacher to take the pupil perspective on classroom events. If teachers could develop the skill of empathy, which is obviously an important part of creating an acceptant classroom climate, then they would be more able and ready to capture the pupil perspective. A teacher with highly developed empathy skills would find it easy to do so and would in fact do so almost as a matter of course. As John Holt's writing makes clear, once the teacher does this a whole new range of teaching and learning problems and possibilities comes into being.

Holt ends his book with a plea for a more 'child-centred' approach to teaching. This is a very complex and diffuse notion as a recent work by Entwistle (1970a) makes clear. The child-centred approach has been a gospel preached by ardent innovators and their supporters (e.g. A. S. Neill) who are actual practitioners. Entwistle shows that at the academic level the central debate about child-centred education has been philosophical. Psychologists, social psychologists and sociologists have rarely been involved in the debate to any degree, though their theories and research have been used in polemical and philosophical arguments. This is perhaps because human scientists in education have rarely undertaken research which is of central relevance to the classroom problems of teaching and learning, and perhaps because human scientists have intentionally preferred the area which is open to 'scientific' analysis to the area where prescriptive debates arise. (The lack of overlap becomes clear when one remembers how many researches in education by human scientists leave the teacher with the feeling of 'so what?')

In a sense I have been trying to show in this book, and especially in this chapter, that many of the central areas of classroom life, which are thus central also to education as a whole, are important to human scientists as well as to those who wish to prescribe. The human scientist may not himself wish to prescribe, but this does not excuse him from ignoring those areas that are in the middle of prescriptive battles between various positions. On the contrary, it is often where the prescriptive battle rages most fiercely that the most central issues of the educative process are to be found. From a social psychological perspective it is clear that teachers and pupils in 'traditional' class-rooms make basic assumptions about the nature of teaching and learning and develop conceptions of roles and relationships that have very important consequences on what they actually do in the class-room. Some of these consequences are obvious to all; others, like the pleasing teacher phenomenon and its relation to the taken-for-granted evaluative and motivational roles of the teacher, are more covert. A (one cannot properly speak of *the*) child-centred approach makes a different set of assumptions and proposes different sets of roles and relationships. The present analysis has tried to show how some of the problems that permeate traditional classrooms may be solved to some degree by child-centred assumptions, roles and rela-tionships specified along Rogerian lines. The prescriptive elements inherent in this analysis are obvious. At the same time it is perfectly clear that were such an approach to be adopted on any scale, then we should have to begin our analysis afresh, by examining the con-sequences both overt and covert of such an orientation and by ex-ploring ways in which problematic, unintended or undesirable con-sequences could be mitigated or abolished by a further change of assumptions, roles and relationships.

It is easy to write books which offer analyses of teaching and learning situations and advocate changes but which are singularly lacking in detail about how these changes are to be effected in par-ticular classrooms. You may feel, quite understandably, that even though what has been said sounds splendid *in theory*, you have been left rather short on the *practical* detail. Some examples of what happens when such changes are executed are available in Rogers' book—but they are only examples. There can be no detailed prescriptions when radical changes in roles and relationships are required, because no blueprint, however detailed, can take account of the unique teacher, with a unique class of pupils, in a unique and dynamic situation. Teachers who wish to be told precisely how changes can be made are usually the teachers who do not really want to change and use the lack of detail as an excuse for not initiating change. If, however, the ideas suggested here 'speak' to you in the sense they help to clarify some of the problems you have experienced as a teacher, if they challenge or interest you, and if you *want* to change, then you will be prepared to begin the exciting process of experimentation by which change is introduced. Changes in schools depend much more on the enthusiastic, open-minded and sensitive experimentation by teachers in accord with their own needs, personalities and situation than on detailed prescriptions by either 'theorists' or other practi-tioners.

Instead of trying to make detailed suggestions about how these changes might be introduced, I want to summarize, from the work of

Carl Rogers, the basic assumptions and attitudes involved in a teacher –pupil relationship which is characterized by acceptance.

1 *The pupil wants to learn.* It is assumed that the pupil has a natural desire to learn, to discover, to progress. The teacher does not have to create or inject motivation into the pupil; his task is to release the motivation that is already there. (Traditional teacher attitude: whilst some pupils seem to be naturally motivated to learn, many pupils must be persuaded to learn with the pressure of my incentives and sanctions. Otherwise, they would waste time and do nothing. A few pupils become more motivated as they get older, which confirms me in my belief that making pupils learn pays motivational dividends.)

2 *The pupil learns most significantly when the content is seen by the pupil as relevant and instrumental to his own goals.* It is assumed that meaningful learning takes place not when the task is imposed by the teacher for the pupil's benefit but when the task is important to the pupil as a growing person. Thus significant learning must be self-initiated and self-directed. (Traditional teacher attitude: pupils do not know enough about my subject to choose what they should learn. Since I am more expert than they, it is my job to give the course a structure and to control what they learn and when and how it is to be learned. I try to make the subject as relevant as I can to the pupils, e.g. history with a local bias, mathematics in relation to family economics or cricket scores and so on.)

3 *The facilitation of learning rests on the nature of the teacher–pupil relationship.* It is assumed that learning is facilitated by a non-threatening, acceptant relationship. The learner under external pressure, threat and evaluation is inhibited in his learning. The acceptant relationship between teacher and pupil seems to be marked by four characteristics.

(a) *The teacher values the pupil.* The teacher cares for and respects the unique individuality of each pupil. (Traditional teacher attitude: frankly, there isn't much time to devote to individual children. Some of the pupils are lazy, badly behaved and generally 'difficult'. My job is to make them into something better, though I must admit my success is very limited. Certainly I don't and can't *like* them. I think all this talk about respecting pupils sentimental mush. In the end, we all have to conform, don't we?)

(b) *The teacher trusts the pupil.* The teacher believes that the pupil has the desire and potentiality to learn and grow; he cannot nor does he want to compel the pupil to learn. He sees himself as a resource to be used by the pupil. The most important evaluation is pupil self-evaluation. (Traditional teacher attitude: there are some pupils you can trust—but not the majority. You have to keep up a constant pressure on most of them, or standards of work and behaviour soon decline. It's my job to judge their progress; they don't know enough to do it themselves. On the few occasions when I have let them mark their own work a lot of them cheated.)

(c) *The teacher empathizes with the pupil.* The teacher seeks to understand the pupil, to develop a sensitive awareness of the depth and meaning of the pupil's feelings and thoughts. He seeks to view the world and the process of learning through the pupil's eyes. In consequence, he is good at *listening* to the pupil. (Traditional teacher attitude: I think I understand my pupils pretty well, but there are a few who are beyond my comprehension. There are times when I can't understand their attitudes at all. I do my best to explain things as clearly as I can, but often they simply don't see things the way I do. It may just be a difference between the generations, but I think the 'permissive society' has a lot to do with it.)

(d) *The teacher is himself.* He tries to be genuine and honest, a *real* person, rather than a front, or a façade or a mere role-performer. (Traditional teacher attitude: I can be myself with some of the better classes and in some of the extra-curricular activities, but one can't afford to be oneself with many pupils. A teacher has to be thick-skinned. Once they see a chink in the armour, they take advantage of you and you're on the slippery slope then.)

This notion of 'being one's true self' does, of course, raise a number of difficulties about the meaning of such a term. It is extremely difficult to define this meaning in relation to other concepts of the self. Whilst I recognize fully these difficulties, I also recognize that the expression does have a very real meaning in ordinary behaviour. Although I may find it very difficult to tell you precisely what I mean when I say that on a certain occasion I felt I was able to be my real self, you will probably understand what I felt, even though neither of us can give a detailed explanation of the term. So rather than trying to clarify the concept, something I cannot do, I would prefer to take account of a practical objection which is often felt by the teacher who wants to be genuine in the classroom. A teacher often feels that being genuine will inevitably conflict with being acceptant, for at times he may feel *un*acceptant. To be genuine requires him to express his feelings of unacceptance which then conflicts with the desire to maintain an acceptant relationship. This difficulty is probably more apparent than real. If the teacher feels angry and in his desire to be genuine expresses this anger, this does not automatically undermine the acceptant relationship, for if the teacher has created a general and underlying acceptant relationship, then the anger when expressed is not by any means as threatening as it would be in a relationship that is generally unacceptant. The pupils can 'take it' because anger is highly threatening only when there are no acceptant roots to the relationship to which the fundamental security of pupils is tied. (It is for this reason that the happily married man can cope more easily with anger from his wife than from his boss.) In the classroom situation the open expression of unacceptant feelings may, paradoxically, strengthen rather than weaken the general sense of being accepted by the teacher and may help the pupil to recognize that he in his turn must learn to accept the teacher.

4 *The teacher is open to experience.* The teacher is willing to experiment and change in response to the changing needs or situations and relationships. At the same time he is sensitive to the effects of his innovations and responds with flexibility rather than adhering rigidly to a predetermined plan. (Traditional teacher attitude: there are too many 'gimmicks' and 'bandwagons' in education at the present time. My methods are well tried and tested and I stick to them. Consistency is not a bad policy.)

In current educational writings there is something of a gap between those who try to *analyse* relationships in the classroom and those who *prescribe.* Analysis is usually seen to be the task of academics (psychologists, sociologists and philosophers) and prescription a task for practitioners (R. F. Mackenzie, A. S. Neill). Carl Rogers seems to me to be an almost unique and certainly the most important contemporary contributor to both areas. Teachers in Colleges and Departments of Education pay a great deal of lip service to the notion of 'good personal relationships' between teacher and pupils, but very rarely do they face up to the very real problems of how these relationships can be generated and sustained in the ordinary classroom. I have tried to suggest that the reason why so much of our preaching sounds vague and woolly is that we imagine that good relationships can be created in an approval based classroom where the pervasive climate is directly incompatible with such relationships. Often we try to maintain different—and irreconcilable—types of relationships in different aspects of classroom life and in consequence good relationships are stifled. After a successful school holiday I used to ask myself: why can't my relationships with the pupils always be like this? I did not realize that in school I tended to abandon the more Rogerian attitude I took to the children when away from school. I have known teachers who have maintained very good relationships with children in school and I suspect that these teachers have moved from an approval to an acceptance-orientated climate, probably without being able to explain how they achieved this because the change had been intuitively directed. A very few teachers have seen that the logic of acceptant relationships requires a movement toward pupil self-evaluation and self directed learning. Perhaps a more adequate analysis of classroom relationships will permit many more teachers to undertake the adventurous experiments intuitively pioneered by the few.

A radical solution

Today education is popular. Never before have its consumers (and their parents) been so likely to ask for more, from the nursery stage to undergraduate and postgraduate levels. In recent years increasingly large numbers of pupils stay on at school or enrol in other educational institutions beyond the statutory leaving age. The Department of Education statistics show a marked rise during the last twenty years in the numbers of pupils who remain at school on a full-time basis. Similar trends are to be found in the growth of the demand for full-time and part-time academic study in other educational institutions.

In response to this demand attempts have been made to offer additional or alternative means of acquiring education to those who for various reasons have fallen outside the traditional school-followed-by-higher education pattern. The Open University is an obvious example. But the public valuation of education is probably best expressed in the decision to raise the minimum statutory leaving age from fifteen to sixteen years. It is true that the raising of the school leaving age (RSLA) has been postponed, but this has been for economic reasons, not because it was thought undesirable by politicians, educationists or parents.

Yet the original announcement that the school leaving age was to be raised was not greeted with unalloyed enthusiasm by all teachers. The *Schools Council Enquiry 1* (Morton-Williams & Finch, 1968) has shown that no more than a quarter of the teachers perceive RSLA as offering major opportunities only, and only one in eight teachers believed that virtually all pupils would benefit from a suitable longer school course. Much opposition was expressed by teachers in secondary modern and comprehensive schools in the poorer districts of large towns and cities. They were already fighting a daily, and often losing, battle to preserve some semblance of order and learning in their classes of bored and sometimes recalcitrant teenage boys and girls, whose principal desire was to leave school at the earliest possible moment. These teachers felt alarm and dismay at the prospect of having to teach such pupils for a further year, during which the pupils would become increasingly bored and hostile—and in the case of boys both taller and stronger. Raising the school leaving age would, it was felt, simply transform reluctant learners into open rebels. A few teachers pinned their hopes on the development of new courses for 'Newsom' pupils and viewed the postponement as a heaven-sent developmental period. Yet in spite of the time, energy and money devoted to this cause, it is doubtful whether we have achieved enough to justify the early hopes. The postponement has, for our problem schools, been little more than a temporary respite from the evil day of RSLA.

Moreover the 'curriculum development' that has taken place often appears to represent an attempt to make the syllabus more attractive and interesting. This is, of course, a good thing; there is no reason to keep a syllabus dull and boring simply for the sake of so doing. The point is that it is frequently the same old traditional material which is being clothed in exciting forms. It is an attempt to motivate the pupils to learn the traditional content of the curriculum. It becomes a gigantic confidence trick by which we hope to cajole pupils into doing what we the teachers want them to do, namely learn the things we want them to learn and the things we have decided are good for them. I suspect that many children will soon see through the 'con' and realize that the sugar is coating the same bitter pills. Even where genuinely new material based on what we regard as the pupil's interests has been introduced into the curriculum, problems will still arise. Pupils are likely to recognize very quickly that it is the rejected failing pupils who are not bright enough to take public examinations who are given the New Exciting Course about Life. If they do, it would not be surprising if they are soon clamouring for the Dull and Irrelevant Course of the brighter children who have status in the school and end up with a qualification that will help them into a good job.

It is difficult to know how many teachers share this view, but it is recognized within the teaching profession that hand in hand with the growth of voluntary staying on at school there has been a concurrent growth in the number of pupils for whom school is an unpleasant experience, which alienates them both from the school itself and from other means of formal learning. Already for such pupils, mainly in the lower streams of our down-town schools, the last two years in school involve little in the way of intellectual growth or achievement. For these pupils, and often for their teachers, school is justly regarded as a 'waste of time'. If the effect of compelling them to stay on at school for a further year is to alienate them still further from taking any other type of formal education during the rest of their lives, then our action seems to be little short of criminal.

If the challenge of these pupils cannot be met simply by the creation of new curricula, we must seek solutions elsewhere. As always when we try to come to grips with an educational problem, we are rarely prepared to question our fundamental assumptions or to consider possibilities which would require radical changes within the educational system. (Teacher training is another excellent contemporary example.) Only solutions which are relatively simple, i.e. which will not disturb other elements in the system, are given serious consideration. The advocates of RSLA commendably wish to formalize the value placed on educational activities; to raise educational standards; to bring Britain into line with other 'advanced' Western countries with respect to length of schooling. What they advocate is relatively simple and it will reach the statute books. On the other hand we have a body of teachers who oppose RSLA because they know that they cannot compel unwilling teenagers to be educated. Being in school is not synonymous with being educated. The problem is a real one.

I want to suggest that increasing the length of *compulsory* schooling is not necessarily a good thing at all nor the mark of an advanced society. I want to suggest that it is an increase in *voluntary* schooling that is most likely to offer a solution to alienated pupils. But in creating a system of voluntary schooling a radical reorganization of the educational system will be required, and its effects will be felt in other areas of our social life and structure. At the same time, voluntary schooling may offer solutions to a whole range of educational problems, which, like the case of the alienated minority, are difficult to treat in isolation.

Let us suppose that we make all schooling for children of every age entirely voluntary. One's first reaction is to regard such a suggestion as extremely absurd and retrogressive. This response springs, I think, from our fear that if children are not compelled to attend school they will not go of their own volition. This may well be the case. If so, it is clear that children do not share our view of the intrinsic goodness of education. It seems that it is we adults and parents who believe that it is good for them, compel them to go to school, and hope that some day they will recognize, and be grateful for, the wisdom of our ways.

When compulsory schooling is ended millions of pupils, then, will rush joyfully out of school. One last glorious end-of-term and end-of-school celebration. But for how long would such massive absenteeism last? A week? A month? A year? For ever? I suspect that we are likely to over-estimate the period for which children will decline to go to school. After all, we know that very many children have become bored by the end of the long summer holiday—in spite of the good weather. Very few of them, of course, admit it. Yet many do look forward to the beginning of school in September. I feel confident, then, that if we introduced voluntary schooling most, if not all, children would be ready to queue at the school gates for readmission in the September or October following the summer holiday. As far as the majority is concerned, attracting them back into school is the least of our problems.

But what of the unknown numbers, probably a minority, who do not return to school and give no indication of a desire to do so in the foreseeable future? We know, in general, which children these will be: the alienated pupils we have just discussed; the children who, when they were forced to go to school, disliked and rejected what we imposed for their own good, namely urban working-class children.

What would we do with them? Certainly we should need to provide additional social workers to prevent mischief and delinquency, to provide alternative facilities such as play centres where they can be supervised whilst Mum is out at work and other children are in school. But as long as a midday meal is provided for *all* children, I cannot see that the situation will be more problematic than the present school holiday periods.

A more serious difficulty is that of integrating the child into school once he makes the decision to return. It would be impractical to try to run a school where children turned up for lessons on an intermittent basis. Numbers in school would be unpredictable; no continuous learning could be guaranteed; and children would be free to withdraw at whim. Teachers would find themselves in an impossible situation, unable to prepare work for such a variable clientele and unable to keep track or records of individual progress. One condition would need to be imposed: the child must make a work contract with the school. When the child first returns to school, he would be obliged to make an agreement, either to come to school regularly for an agreed period, or to complete a given work schedule planned with the help of a teacher. No doubt such work contracts would be broken in many cases. When two or three work contracts had been broken, the school would then penalize the offender by barring him from school for a fixed short period. This would pressure the child to increase his commitment and probably enhance the general attractiveness of the school.

A few working-class parents may well discourage their children from returning to school, especially where there are younger siblings to be cared for whilst Mum is at work or where some illegal employment can be found for the child. These are not insuperable difficulties and I shall suggest one means of minimizing this latter danger shortly. For the moment we must remember that under the present system it is already difficult to keep a child of such parents in school on a regular basis. Local Education Authorities are reluctant to prosecute and rely rather on the persuasive powers of Education Welfare Officers. Under a system of voluntary schooling these and other social workers would have a vitally important role to play. Strange as it sounds, a more subtle problem than the parents who try to prevent their child from attending school might be the parents who compel, under various threats, their child to attend school. A system of voluntary schooling could be subverted if most of the pupils who do attend do so because their parents will allow them to follow no other course of action. Parental encouragement, on the other hand, would be as useful under a voluntary system as under the present system, but it is extremely difficult to distinguish the various sorts of pressures that parents can exert.

Implicit in what has been said so far is one of the most fundamental objections to voluntary schooling, namely its effect in exacerbating what Robert Merton has called the St Matthew effect: 'To him that hath shall be given'. It is the working-class child who is most likely to withdraw from school for long periods of time. The middle-class children would make most use of school and thus increase their present domination of all, but especially the higher, levels of learning. The current struggle to maximize the equality of opportunity for working-class children—the comprehensive school, de-streaming, compensatory education programmes, Educational Priority Areas—would apparently be undermined. Moreover, it might be argued, even those working class children who 'opted in' might be predisposed to choose narrow vocational courses in preparation for a career in skilled or semi-skilled manual occupations, for a system of voluntary schooling would seem to require to some degree a voluntary curriculum. The effect might be to reduce the number of highly gifted working-class pupils who under the present system are inducted into higher education and thus into the professions.

This objection can be met only in the radical changes in the system that are essential concomitants of a scheme of voluntary schooling. The first concerns the age at which children can legally be employed in a full-time job. This would be lowered to, say, twelve years. Again this looks, superficially, like a very reactionary proposal, especially at a time when we are celebrating the centenary of the 1870 Education Act. It may be the reverse. In 1870 the struggle was essentially about the *right* of certain sections of the population to education. In 1970 the right of everyone to education is universally acknowledged. Today the question is not about rights as such but about how people can be persuaded to exercise their rights more fully and about how the system must be shaped to create an equal opportunity structure for all. Compulsory schooling, and especially an extension of it, may paradoxically be inimical to that goal.

The minimum age of employment would be reduced to meet the needs of adolescents in the twelve to sixteen age group who absolutely decline to go to school. Under compulsory schooling, these pupils unwillingly conform to the minimum requirements of the system in order to survive, but they learn little. Under a voluntary system, they withdraw from school and presumably 'hang about' until they are old enough to take a job. Neither is very satisfactory. So perhaps the most reasonable alternative is to open to them the experience of work, with all its satisfactions and dissatisfactions. Certainly it seems preferable to a vacuum of inactivity either in school or on the streets. However, a condition of employment, which would have to be rigorously enforced, would be that if such a child whilst working develops a wish for either full-time or part-time education, whether

223

or not it is directly relevant to his job, then he has a right to it. Part-time education could continue on its present day-release pattern, or full-time education in school could be resumed. In both cases the pupil would make the usual work contract with the school.

Working-class pupils, the ones most attracted to early employment, might be reluctant to return to school because of the loss of income from their jobs. The short-term financial gain of a job would be most attractive to such children—as it is under the present system to the fifteen-year-olds—and in some cases parents might actively support this view. This could be solved by paying all children, whether in school or at work, a fixed token wage between the ages of twelve and sixteen years. This is obviously one of the most difficult of the present proposals, but it could be argued that the disparity in income between middle-class and working-class occupations is likely to make some such incentive system a necessity sooner or later.

The major change inherent in the notion of voluntary schooling and the reduction of the minimum employment age is that the school and the education system of which the school is a part would have to become very much more flexible than it is at present. Suppose a child, who has withdrawn from school for a long period and perhaps forsaken the job to which he was initially attracted, decided to return to school. How would he be able to resume his education? In the first place we should need a body of advisers or counsellors to whom the child, and his parents, could turn for information and guidance. It would have to be determined what the child wanted to learn and where within the system such learning could be pursued. Within one geographical area the school system would have to offer children of all ages courses in the subjects they wished to learn. Presumably different schools would offer different specialist subjects. What the schools offer, the content and length of particular courses, would be known to the counsellors. Courses would recruit their members simply on the basis of those who wished to take what was offered. Whatever the age of the applicants, they would be admitted if they possessed the necessary qualifications. Schools could no longer be age-graded as at present, for within a few years pupils of very different ages would be enrolling for the more basic courses.

Schools would have to operate on a 'credit' system. When a child successfully completes a given course, he would be given a credit which certified his level of attainment. He could then withdraw from school, use his credit to gain entry to a more advanced course, or turn to a different subject. No major changes in the national examination system would be required immediately, though perhaps examinations could be held more frequently than at present.

The major gains of such a flexible system of voluntary schooling are obvious. We could abandon the pretence that if you compel children to attend school they are automatically being educated. We could stop forcing certain subjects on children who do not wish to learn that particular subject at that particular time, but who are otherwise anxious to learn. We could give up for good the monster of a compulsory 'general' education. Schools would no longer need to make pupils change subjects and create a Victoria-station-in-the-rush-hour scene every thirty-five minutes, which springs from a determination to give all the pupils a little taste of everything on a regular basis. Pupils would not find themselves permanently trapped in a particular sort of school or course from which there is no escape until it is too late, and then be condemned to a wistful recollection of the chances they missed or the wrong choices they made. We would no longer have to be concerned to provide 'second chances', for the system would be sufficiently flexible to offer an endless series of further chances and opportunities.

One of the most outstanding gains would be the commitment of the learners. There would be a reduced sense of failure for those who need to return to basic courses at a later stage than normal. Pupils of different ages could and would soon learn to work together on a co-operative basis along the lines of 'family grouping' developed in the primary schools. Because the pupils have voluntarily entered the course, because they actively want to learn what is being offered, the motivation to learn will be high. The satisfaction of both pupils and teacher will inevitably soar. A. S. Neill, that passionate proponent of the voluntary, once pointed out that there is little point in forcing a child to read if he does not desire the skill; but when he does want it, he will probably acquire the skill in a very short time and with a minimum of help from the teacher. For this reason, one would expect the general level of attainment to rise.

One important potential consequence of a system of voluntary schooling might be that schools could and would become more thoroughly and consistently differentiated. At present there is relatively little differentiation between state schools in terms of the social and educational philosophies and principles that determine the organization, curriculum, teaching methods and social relationships in the school. It is true that some schools are more 'progressive' than others, that some schools 'stream' whilst others do not. Yet such labels usually tell us more about the views of the headteacher than the school as such. The headteacher may be progressive and try to encourage his staff to adopt such an approach, but he cannot easily get rid of those members of staff, often a majority, who do not agree with him and it takes a long time for him to replace such teachers with new appointees whose thinking is more in line with his own. A headteacher can decide whether his school will be streamed or not, but he cannot compel teachers to operate the imposed system in the way he would wish. Schools within the state system work on the basis of compromise and however much we try to emphasize their differences, there is little to choose between them. The early comprehensive schools that were purpose built were able to select staff committed to certain educational principles. Most current comprehensive schools contain many teachers still committed to the tripartite system. As a result parental choice is, at least for the majority who cannot afford the greater variation within the private school sector, virtually non-existent. It is hardly suprising that parents differentiate schools on other criteria, a 'good' school being one with middle-class pupils, many facilities and a good public examination record.

If pupils were free to opt in or out of schools, then some schools and some teachers would find that they had no pupils. The effect would be salutary. Whilst the teaching profession is reluctant to concede the point, it has to be admitted that some schools and some teachers do not deserve to have any pupils. Under a voluntary system, the pupils would, like their counterparts in adult education, vote with their feet. I suspect that the long-term effect, once teachers had changed their role conception or left the profession would be that schools would be more free to develop particular educational approaches, from the extreme 'progressive' to the extreme 'traditional'. Teachers could operate their schools from consensus, rather than compromise. Teachers, as well as pupils and their parents, would have some real choice of school. If teachers could join a school with whose goals and philosophy they were in sympathy, they would find it easier to co-operate with their colleagues, become more committed to their job, and gain greater satisfaction from the exercise of their profession. We would find, in other words, that the teacher's relationship with his pupils would become a contractual relationship with a client, as Musgrove & Taylor (1969) have advocated. In my view only under a system of voluntary schooling can the pupil be said to have the status of client and only then are we likely to break free from what Musgrove and Taylor have called the 'thoroughly unadventurous despotism, shaping nothing more outrageous than a standardized utility product'.

There seems to be little doubt that the automated society will demand more knowledge and skills and a generally higher level of education from its workers. As some occupations decline or disappear and new ones emerge or expand, this occupational fluidity creates a need for elaborate retraining procedures. This in itself suggests a need for a much more flexible system of education which is closely integrated with those educational institutions orientated towards the young. As automation proceeds and technological development accelerates, the working week grows shorter and leisure time increases proportionately. The relationship between education, work and leisure in the future has been badly neglected, with a few notable exceptions such as Entwistle (1970b). Educationists have approached education for leisure half-heartedly and without much success. Perhaps we should now be concerned not with education for leisure but leisure for education. We are moving into an age when people of all ages will have more time to devote to formal education, either as part of occupational retraining or as a means of furthering their interests. But men and women will use their leisure for more education only if they have come to realize that formal education is not simply a compulsory and rather unexciting part of childhood and adolescence, not simply the acquisition of undesired knowledge, but rather an enjoyable means of developing one's interests and attaining a wide variety of personal goals, including occupational ones. A highly flexible system of voluntary schooling based on self-directed learning seems more appropriate than our present educational structure based on different assumptions. Voluntary schooling may, in other words, help to fulfil important sociological and economic functions as well as social psychological ones.

The changes implicit in the concept of voluntary schooling and self-directed learning are complex and far reaching. They cannot be treated fully or systematically in this book. What does matter is that we should be debating the issues involved, not only within narrow educational circles. As the reception of the Black Papers has shown, it is the relatively parochial and minor issues like the grammar school versus comprehensive school that consume much of our present energies. There are much more fundamental and urgent issues in education than this.*

Recommended reading

E. J. AMIDON & J. B. HOUGH (eds), *Interaction Analysis: Theory, Research and Application*, Addison-Wesley, 1967.

J. HENRY, *Culture Against Man*, Tavistock, 1966.

J. HOLT, *How Children Fail*, Pitman, 1964 (Penguin, 1969).

P. W. JACKSON, *Life in Classrooms*, Holt, Rinehart & Winston, 1968.

C. R. ROGERS, *Freedom to Learn*, Merrill, 1969.

W. WALLER, *The Sociology of Teaching*, Wiley, 1932.

School of Barbiana, *Letter to a Teacher*, Penguin, 1970.

* See Addendum on p. 421.

7 Discipline

When it is said that a teacher has a 'discipline problem' the meaning usually intended is that the teacher has in some way failed to master the task of creating or maintaining order in his classroom. Yet every teacher, whether he is a 'success' or a 'failure'—in his own eyes or in the eyes of his colleagues or pupils—has a discipline problem in the sense that he must come to terms with the requirements of his disciplinarian role. We suggested that discipline comprises one of the teacher's two basic sub-roles for this very reason: no teacher can escape it. Most teachers, when asked to explain how one achieves or maintains discipline, find it difficult to give a very meaningful account. One is offered either trite and superficial accounts of incidents in the teacher's own experience ('Well, I always . . .' or 'I remember once when . . .') or vague generalizations that duck the question altogether ('It's all a matter of personality') and give no enlightenment to the seeker after knowledge. Tutors in Colleges and Departments of Education who are deeply involved in trying to help their students to cope with disciplinary difficulties with generous advice and criticism, know only too well how complex a phenomenon discipline is. For it seems to be of limited help to the students to tell them what or what not to do. Disciplinary 'tips for teachers' simply do not meet the needs of the novice. Discipline is a problem because we understand so little of the processes by which it is established or maintained, whether successfully or unsuccessfully. Although the way in which we evaluate the quality of a particular teacher's discipline depends very much on our own values and educational philosophy, we have little difficulty in recognizing 'good' and 'bad' forms of it. The difficulty is in analysing its composition. Here I shall be treating discipline basically from the point of view of the beginning teacher, though I hope nevertheless that the analysis will be of interest to the more experienced.

Discipline is of interest to a social psychologist because it is an *interpersonal* concept. Discipline refers to a set of rules or norms, specifying acceptable forms of classroom conduct, which is either imposed by teachers on pupils or agreed between them. It is a form of social control. Many of these rules are designed to regulate encounters between persons, between the teacher and the pupils and between pupil and pupil. So what we call discipline forms a major part of the 'definition of the situation' in the classroom. For this reason we cannot say very much about classrooms without encroaching into the territory of discipline, which is part of essence of life in the classroom for both teacher and pupil. Even to attempt to discuss discipline as distinct from the other basic feature of classroom behaviour that we call instruction is a dangerous pursuit. Discipline and instruction are the warp and woof of the fabric of classroom interaction; to pull out one for isolated inspection is to risk turning the whole into a meaningless bundle of threads. Yet that is a risk we intend to take.

We could approach an analysis of discipline in several ways. We might, for example, observe a number of different classrooms and then examine the variations in discipline that we see. We could arm ourselves with a set of questions which would reveal some of the more important differences. We would probably want to include questions such as:

What behaviours are subject to rules?
What sort of rules are they? (e.g. do's or don'ts)
How many rules are there?
Who makes the rules?
Which rules are part of a written code?
How explicit are the rules?
Are the rules accepted and conformed to?
Which rules are broken? How often? For what reason?
Are the rules enforced? By whom? In what ways? On what occasions?
What are the penalties for rule-breaking?
Who determines the penalties?

This might make a very interesting research, though so far as I know it has not been undertaken. Certainly it would tell us much about the operation and variability of classroom discipline. Unfortunately we do not have such information. Even if we did, I suspect that the outcome would be like the work of the interaction analysts we considered in the last chapter, namely a static and desiccated set of numbers in categories that loses the flavour of discipline in real classrooms.

A preferable approach would be to treat the establishment and maintenance of discipline as a *dynamic* interpersonal process. Because in many respects schools and classrooms are like society in miniature,

229

such an approach raises the many problems that are involved with any variety of deviant behaviour—crime, delinquency, mental illness, homosexuality, drug addiction and so on. In the interactionist perspective such deviance is seen as both the product and process of social interaction, the outcome of a complex set of transactions between a person who behaves in a given way and a person or group who responds in a particular way (see Rubington & Weinberg, 1968). Additional questions with respect to classroom discipline suggest themselves.

On what occasions and in what ways are the rules made?
How are the rules defined and formulated?
What constitutes a misbehaviour?
Is there a hierarchy of misbehaviours?
Is there consensus about the defined rules and misbehaviours?
How are misbehaviours detected?
What are the social reactions (by teachers, pupils, perpetrator) to the misbehaviour and the perpetrator?
How are the misbehaviours dealt with? For what reasons?
With what effects?

No systematic consideration of all these problems can be given in this short chapter, where the analysis is intended to be of 'practical' value to the teacher as well as 'theoretical'. After reading the rest of the chapter the interested reader might return to reconsider these questions and examine the ways in which a study of classroom discipline from this perspective might enlarge our understanding of interpersonal relations in school.

Let us begin with a concrete example of a disciplinary situation, taken from Waugh (1928).

The masters went upstairs.
'That's your little mob in there,' said Grimes; 'you let them out at eleven.'
'But what am I to teach them?' said Paul in a sudden panic.
'Oh, I shouldn't try to *teach* them anything, not just yet, anyway. Just keep them quiet.'
'Now that's a thing I've never learned to do,' sighed Mr. Prendergast.
Paul watched him amble into his classroom at the end of the passage, where a burst of applause greeted his arrival.
Dumb with terror he went into his own classroom.
Ten boys sat before him, their hands folded, their eyes bright with expectation.
'Good morning, sir,' said the one nearest him.
'Good morning,' said Paul.

'Good morning, sir,' said the next.

'Good morning,' said Paul.

'Good morning, sir,' said the next.

'Oh, shut up,' said Paul.

At this the boy took out a handkerchief and began to cry quietly.

'Oh, sir,' came a chorus of reproach, 'you've hurt his feelings. He's very sensitive; it's his Welsh blood, you know; it makes people very emotional. Say "Good morning" to him, sir, or he won't be happy all day. After all, it is a good morning, isn't it, sir?'

'Silence!' shouted Paul above the uproar, and for a few moments things were quieter.

'Please, sir,' said a small voice—Paul turned and saw a grave-looking youth holding up his hand—'please, sir, perhaps he's been smoking cigars and doesn't feel well.'

'Silence!' said Paul again.

The ten boys stopped talking and sat perfectly still staring at him. He felt himself getting hot and red under their scrutiny.

'I suppose the first thing I ought to do is to get your names clear. What is your name?' he asked, turning to the first boy.

'Tangent, sir,'

'And yours?'

'Tangent, sir,' said the next boy. Paul's heart sank.

'But you can't both be called Tangent.'

'No, sir. *I'm* Tangent. He's just trying to be funny.'

'I like that. *Me* trying to be funny! Please, sir, I'm Tangent, sir; really I am.'

'If it comes to that,' said Clutterbuck from the back of the room, 'there is only one Tangent here, and that is me. Anyone else can jolly well go to blazes.'

Paul felt desperate.

'Well, is there anyone who isn't Tangent?'

Four or five voices instantly arose.

'I'm not, sir; I'm not Tangent. I wouldn't be called Tangent, not on the end of a barge pole.'

In a few seconds the room had become divided into two parties; those who were Tangent and those who were not. Blows were already being exchanged, when the door opened and Grimes came in. There was a slight hush.

'I thought you might want this,' he said, handing Paul a walking stick. 'And if you take my advice, you'll set them something to do.'

231

> He went out; and Paul firmly grasping the walking-stick, faced his form.

This entertaining, fictional account of a situation in a small private preparatory school many years ago captures in a masterly way the basic ingredients of the problem. We first notice that Paul, who is about to teach his first lesson, is nervous and apprehensive ('dumb with terror'). He does not know where to begin. He is the victim of the situation, unable to manipulate it to his own ends. He does not take the initiative in defining the situation, even though the pupils will expect and probably accept it if he does so. By his failure to take the initiative, the definition is given into the pupils' hands and they quickly turn it into a direction which is profitable to them, fun rather than work, but which is extremely costly to Paul as the situation becomes progressively worse.

The pupils gain control in defining the situation almost at once, before Paul has the chance to realize what is happening. Even as he enters the classroom, the pupils await him eagerly, 'their eyes bright with expectation'. Innocently he responds in a friendly way to the pupils' opening move of a 'Good morning' greeting, only to find himself trapped in a disconcerting chain reaction. A few minutes later the boys use the same technique successfully in response to Paul's attempt to discover their names. In not knowing how to define the situation, Paul hesitated, and was lost.

It is for this reason that most experienced teachers insist that the teacher must, if he is to survive, define the situation in his own terms at once. Basically this initial definition is not so much a statement of the rules that will govern behaviour in that class, but rather a clear indication that the teacher is completely in charge and not to be treated lightly. If the teacher can establish himself as the imperturbable lord and master of the classroom, who is not to be disobeyed or fooled, then all will be well. Once the teacher has thus exerted his authority, he can at a later stage create the kind of relationship with the pupils that his personal philosophy of education requires.

> You can't ever let them get the upper hand on you or you're finished. So I start out tough. The first day I got a new class in I let them know who's boss. You've got to start out tough, then you can ease up as you go along. If you start out easy-going, when you try to get tough they'll just look at you and laugh. (Becker, 1952)

> If you started with a mailed fist, you could later open that fist to reveal a velvet palm. If you let them step all over you at the beginning, there was no gaining control later. (Hunter, 1955)

Either you murder them, or they'll murder you. Either you
win or they win. And I'll tell you, mate, I'm the one that's
going to win. (A teacher's comment to the author)

This is a lesson learned the hard way by countless generations of
student teachers who, believing that the pupils ought to be treated
with respect as mature persons, try to create a definition of the
situation that is congruent with their beliefs. Almost always the result
is disastrous. The pupils do not respond in the expected way. Soon
the teacher finds himself only nominally in charge of a collection of
noisy, disobedient, rude and irresponsible children, who are quite
unwilling either to listen to the teacher or to work. It is a fascinating
fact that pupils, even those in 'good' middle-class schools, respond so
often in this way towards a teacher who is sincerely seeking to make
life at school more pleasant for them. This is why so many student
teachers fall so regularly into the trap. It seems incredible that the
pupils should reject what is so clearly intended for their benefit. As
we discussed in the last chapter pupils like a teacher who is friendly,
patient and humorous, who takes an interest in the pupils as individuals and who makes his lessons interesting. Why, then, should the
pupils react so badly to a teacher who tries to live up to these ideals?
Clearly the pupils' perspective or definition of the situation is not
congruent with that of the teacher, whereas the student teacher
assumes that it is. The teacher, finding his 'reasonable' assumption
undermined, has to find some explanation of the pupils' behaviour.
He can do this by switching to a common teacher assumption that
'kids are little devils' or by taking a view of human nature which
suggests that children are essentially anti-social and need to be
restrained and directed into acceptable social behaviour. From a
social psychological view the problem is one of explaining the disjunction in teacher and pupil perspectives.

Perhaps part of the answer is that children do not like school very
much, at least in the sense that they can think of preferable activities.
This is not to say that they do not wish to learn, but rather that they
do not wish to learn the subjects imposed by the teacher in the ways
and at the time and place that the teacher demands. As has already
been pointed out, unless we re-organize our schools very radically,
the pupils are unlikely to find life at school rewarding for most of
the time. Since the majority of the teachers manage to compel their
pupils by various means, to do things they do not particularly like,
they are in a sense constrained by the system to turn the lessons of a
teacher who is unwilling or unable to achieve a dominant position
over them into periods of rest and relaxation, or into 'fun'—anything
that is not work. If this is the case, the teacher who is not prepared to
be 'tough' is rejected not so much because the children do not want

what he offers, but because his refusal to dominate serves as an emergency exit out of the prevailing boredom of school through which the pupils joyfully stampede. Such an interpretation might be confirmed by the fact that pupils do not usually give the teacher very much time to explain what he is offering them. As in the case of Paul above, the pupils are so eager to find a way out that they are unprepared, when they see one, to wait and see what things would be like if they stayed behind. And once a stampede has begun, it is difficult to stand still.

Other interpretations are possible. It may be that the pupils simply distrust the teacher who does not define the situation in terms of his own dominance. After all, since most of the other teachers dominate in the classroom successfully, it is only a matter of time before this teacher follows the customary pattern and becomes like the rest of them. In the meantime, the most profitable course for the pupils is to make hay while the sun shines—for it will be rain and storm very soon.

Whatever the causes, the phenomenon itself remains. If the teacher does not establish his own dominance, the children are likely to turn the classroom into a circus without a ringmaster and the teacher will become rapidly exhausted and demoralized. Many of the authorities on discipline interpret this situation as an attempt by the pupils to find the limits of the situation. In 'seeing how far they can go' the pupils are exploring what the teacher will allow and what he will forbid. I am sure there is much truth in this view, for it is natural enough that pupils should try to discover a new teacher's code of acceptable behaviour, which will be unique to that teacher. But I am equally sure that this phenomenon is more complex than this view allows. My reason is that even the non-dominating teacher who makes his rules and expectations quite explicit in the very early stages seems to suffer the same unhappy fate as the vague non-dominating teacher. Every teacher, in short, is pressured by the pupils to exert his own dominance, sooner or later.

The pupils do this in a variety of ways. The whole class may appear to challenge the teacher with an uncanny, almost telepathic ability to co-operate against him. Alternatively, just one or two pupils may play the active role before an encouraging and appreciative audience. Sometimes a student or beginning teacher, finding the pupils polite and obedient, believes that all is well, that he is in control, and that the warnings of other teachers were but complaints of sadistic cynics, or even that he is such a good teacher that he has managed to avoid the trials and tribulations which the rest of us have to endure. Usually in such a case he has simply fallen for what I call the Disciplinary Illusion. The pupils do not always issue an immediate challenge to the teacher. Instead they play the game of watching and waiting.

While the teacher imagines that he is firmly in control, the pupils are silently sizing him up. The first few minor infractions to test the situation are casually dismissed by the confident teacher, and then the infractions begin to assume major proportions. Soon the apparent discipline has been transformed into an obvious and uncontrollable indiscipline.

If the teacher feels that he must establish his dominance in the early stages of his first encounter with the students, he is faced with a problem of self-presentation. He must express himself in such a way that the pupils are correctly impressed that he is in charge and that he is a person not to be trifled with. At the first meeting with their new teacher the pupils are seeking to discover, either by a calculating silence or by tentative, probing assaults on the teacher's dominance, precisely what sort of person the teacher is and what sort of disciplinary regime is to be created. They are highly sensitive to the teacher's behaviour, to the expressions he gives and gives off. The 'first impressions' made by the teacher are thus especially important in establishing the definition of the situation. The teacher's image and the tone of all future interactions emerge within the first few minutes of the encounter. It is often thought by student teachers that the teacher's physical size is of great importance in this respect. This is without foundation. As an experienced headmaster puts it (Boas, 1963):

> For over thirty years I have interviewed applicants for staff vacancies and at the end I was no more capable of judging from mere appearances and demeanour whether an applicant was likely to be able to keep order than when I started. A six-foot Oxford 'blue' would appear, seeming to have all the manly virtues and social graces, and the next day he would be found impotent in the classroom with a company of small boys dancing ring-a-ring a roses triumphantly around him.

The pupils are far too skilled in the character analysis of teachers to be misled by mere physical appearances. They can detect those clues to the teacher's real character, intentions and abilities with the cunning and patience of a wild animal stalking its prey. The advantage of the tall well-built teacher, like the disadvantage of small stature, lasts for a very short time, since the impression conveyed by the body is soon discredited (or confirmed) by other aspects of the teacher's self-presentation. What counts is not the single isolated 'bit' of the teacher's image, but the total, overall picture of himself constructed by the teacher in the fine detail of his behaviour. He must present himself as a unified whole to which all the bits, each congruent with the rest, contribute. Any bit which is discrepant with the whole will be identified by the pupils as a vulnerable chink in the armour and become the focal point of future attacks. Let us now consider the case of a teacher who, in contrast to the irresolute approach of Paul given above, is determined to impose his own 'tough' definition of the situation and tries to present himself accordingly. We join the teacher, Rick, as he receives his class on the first day of term (Hunter, 1955):

he picked up his roll book and his briefcase, walked quickly to the steps, and mounted them with his shoulders back and his head high. He paused dramatically for a moment, and then began calling the roll in his best Sir Laurence Olivier voice.

'Abrahams,' and he saw movement out there in the seats, but he did not pause to focus the movement.

'Arretti,' and another blur of movement.

'Bonneli,' and 'Casey,' and 'Diaz,' and 'Di Zeffolo,' and 'Donato,' and 'Dover,' and 'Estes,' and on, and on, until he flipped over the last card in the book. There had not been a murmur while he spoke, and he was satisfied that he had been accorded the respect due to an English high-school teacher. He slapped the roll book shut, and walked down the steps and then into the centre isle, conscious of the curious eyes of the kids upon him.

When he reached his official class the same curiosity was reflected in their eyes.

'Follow me,' he said unsmiling. 'No talking on the way up.'

That, he figured, was the correct approach. Let them know who's boss right from the start, just the way Small had advised.

'Hey, teach,' one of the boys said, 'what did Mr. Halloran say your name was?'

Rick turned his head sharply. The boy who'd spoken was blond, and there was a vacuous smile on his face, and the smile did not quite reach his eyes.

'I said no talking, and I meant it,' Rick snapped.

The boy was silent for a second, and then Rick heard him say, 'Dig this cat. He's playin' it hard.'

He chose to ignore the comment. He walked along ahead of the class, feeling excitement grow within him now, feeling the same excitement he'd felt when he got the job, only greater now, stronger, like the times at school when he'd waited in the wings for his cue. Like that, only without the curious butterflies in his stomach, and without the uncon-

scious dread that he would forget all his lines the moment
he stepped out on to the stage in front of all those people.
He felt in complete control of the situation, and yet there
was this raging excitement within him, as if there was something he had to do and he simply could not wait to get it
done . . .

He led the class to the stairwell, and aside from a few
whisperers here and there, they were very orderly, and he
felt that everything was going well. . . .

When they reached the door to Room 206 he inserted the
key expertly, twisted it, removed it, and then pushed the
door back.

'Sit anywhere,' he said brusquely. 'We'll arrange seating
later.'

The boys filed in, still curious, still wondering what sort
of a duck this new bird with the Butch haircut was. They
seated themselves quickly and quietly, and Rick thought,
This is going even better than I expected.

He walked rapidly to his desk, pulled out his chair, but
did not sit. He looked out over the faces in the seats before
him, and then sniffed the air authoritatively, like a blood-
hound after a quarry. He cocked one eyebrow and glanced
at the windows.

'What's your name?' he asked.

The boy looked frightened, as if he had been accused of
something he hadn't done. 'Me?'

'Yes, what's your name?'

'Dover, I didn't do nothing, teach. Jeez . . .'

'Open some of the windows in here, Dover. It's a little
stuffy.'

Dover smiled, his lips pulling back over bright white
teeth. He got up from his seat and crossed behind Rick's desk,
and Rick congratulated himself on having handled that
perfectly. He had not simply given an order which would
have resulted in a mad scramble to the windows. He had
first chosen one of the boys, and then given the order. All
according to the book. All fine and dandy. Damn, if things
weren't going fine.

He turned and walked to the blackboard, located a piece
of chalk on the runner, and wrote his name in big letters
on the black surface.

MR. DADIER.

'That's my name,' he said. 'In case you missed it in the
auditorium.' He paused. 'Mr. Dad-ee-yay,' he pronounced
clearly.

237

Although Rick's troubles with this class are by no means over, his initial encounter with his pupils can be regarded as a success, at least insofar as he attempts to assert his own dominance. How has he done it? In the first place he shows considerable foresight in *planning ahead* against possible disruption which might undermine his control or upset his performance. He selects a single boy to open the windows; he prohibits talking before they set off from the hall to the classroom; he has a store of pencils readily available. Second, *he refuses to allow the pupils any opportunity of taking the initiative in defining the situation* from him. He silences at once the boy who asks for his name on the corridor, and who in so doing both makes the first assault on the rules ('No talking on the way up'), and seeks to tempt the teacher into the mine-field of an informal interaction. Third, in the very early stages of the interaction *he outlines a general code of expected behaviour*. He forbids shouting out without permission and fixes the modes of address. Fourth, he indicates that *he is not to be treated lightly*. He threatens a punishment at the first sign of insolence. Finally, and perhaps most important of all, *he tries to present himself generally in ways which are consonant with his total intentions*. He makes no attempt to give clues to a liberal, child-centred philosophy. He shows no sign of wishing to court popularity. He is cold and unsmiling; he speaks 'brusquely', 'coldly', 'gruffly'. His expressions make the right impression on the pupils—'Dig this cat. He's playin' it hard.'

There is a further contrast in the behaviour of Rick and Paul. Rick is very conscious that he is 'playing a part', that his behaviour is somewhat artificial and contrived. These feelings are so strong that he recalls actual performances in plays during his own schooldays. Paul, on the other hand, is 'being himself'; his behaviour is not staged. This raises the perennial problem of student teachers: is it right for the teacher to 'act a part'? Most beginning teachers rightly want to endow their relationships with their pupils with honesty and integrity, which seems inconsistent with pretending to be something one is not. To 'act a part' is associated with feelings of intense strain and guilt. I do not think the student teacher should worry about this problem. The contrived performance is a means to an end, namely the establishment of his dominance. It is essentially a *temporary* measure. For reasons we shall examine shortly, he will soon be able to drop his mask. The only danger is that of permanently 'playing a part' where the contrived performance becomes an end in itself. A few teachers do become life-long classroom actors, dreary cardboard caricatures who become incapable of any adequate and satisfying relationships with their pupils. But they are exceptions who are trapped by the opening move of the game; the majority advance to a much deeper personal involvement in the intricacies of educating children.

One of the essential features of 'playing a part' is an ability to make a simultaneous check on the effect of the act on the pupils. A contrived performance is extremely dangerous if the actor is so bound up in his art of presenting himself that he takes no account of the audience. It is essential that only a part of the self is involved in the act, whilst the more central self evaluates and guides the performer. This is fully exemplified in Rick's account. At times he stands back from his performance to see how well he is doing, to congratulate himself on his success, or warn himself that he is losing ground.

The second danger of 'playing a part' is that of failing to make the performance *convincing*. Unless the teacher is consistent in his performance, unless he attends to the minor details with the same care and attention that he lavishes on his general approach, then he will be 'seen through' by the pupils. A discredited act is worse than no act at all. It is for this reason that if the teacher is to play a part it must be of a simple unelaborate kind. If the act is too ambitiously conceived then he is almost certain to be unable to sustain it, especially if as is likely he is nervous and apprehensive about meeting the pupils. If the ambitious act breaks down, the teacher will look as foolish as the amateur actor who fluffs his lines or trips on the staircase and causes the set to collapse. Drama is soon turned into farce.

In advising student teachers to take the initiative in defining the situation in terms of their own dominance and to be prepared, if necessary, to 'play a part' to this end, I have been accused of being reactionary, Victorian and inconsistent with my own educational philosophy. Such accusations stem from a misunderstanding of my position. By 'tough' I do *not* mean to suggest a ruthless and punitive autocracy. A 'tough' approach is characterized by a maintenance of the teacher's dominance by firmness, clarity and consistency. In other words, the teacher must remain in charge, be firm and consistent in his treatment of the pupils and be prepared to punish pupils who call his bluff. This is *not* achieved, I might add, if the teacher's first action is to bring out the three largest pupils and punish them quite arbitrarily to prove that 'there will be no nonsense in my classroom'. In being 'tough' the teacher can, and should, be humane and reasonable.

My advice is also based on the fact that most state secondary schools, in which most of my students will teach, contain teachers whose attitudes are conservative and traditional rather than liberal and progressive. Their approach to the pupils will not be very 'tender-minded' or 'child-centred'. Any teacher who does not behave in accordance with the 'party line' is likely to be regarded as 'soft' and incompetent by both staff and pupils. Yet if the teacher is to survive in his chosen profession he must have a minimum of respect and support from his colleagues as well as from his pupils. Both pupils and staff are unlikely to respond favourably to his new ideas simply because they are so different to what is customary in that school and because they will not have been tried there. If the student or beginning teacher proclaims to a wary and cynical staff room his child-centred theories (which in *their* eyes are irrelevant and impractical theories invented by people quite out of touch with the real world of teaching) and yet in the classroom cannot maintain even basic control, then his disciplinary failure will be seen as the perfect refutation of the theories. On the other hand, if the new teacher starts out 'tough' he has more chance of achieving adequate discipline in the classroom and of earning the esteem of the more experienced teachers.

Starting out with a 'tough' approach does not mean that the teacher must persist in maintaining such a regime for the rest of his career. Many teachers do so, for they know that the 'tough' approach works and gains the respect of colleagues and pupils. In consequence they become afraid to try any other approach, lose their ideals, and join those teachers who actively oppose any new policies. In truth, the 'tough' approach should be regarded not as a terminus, but as a point of departure. Once the pupils know that the teacher is in charge, that trying to 'play him up' is a profitless activity, the teacher can begin to put into practice his own ideas and ideals about how pupils should behave in the classroom. The situation is, in other words, open to redefinition. The pupils will probably respond with pleasure and gratitude, regarding the teacher's innovations as new and exciting privileges rather than as opportunities for having fun or for taking a rise out of the teacher. Such changes must, however, be introduced very slowly. A sudden change in regime is likely to have disastrous effects on the pupils, who will respond to the new uncertainties with misbehaviour and abuse of unfamiliar freedoms. The definition of the situation must be altered very gradually to allow the pupils to adapt to the new activities, new roles, new relationships and new codes of behaviour. In the Lewin, Lippitt & White study described in the last chapter, when the boys changed over from an authoritarian to a democratic leadership style, there was a very marked increase in aggression—the so-called 'blow-off' effect. It is a good lesson for teachers. If the redefinition of the situation is introduced very slowly, it has a high chance of success with the pupils. Equally important, it is likely to be accepted by the teacher's colleagues, for once he has earned their respect by producing a very competent performance in *their* terms, they are much more likely to respect new forms, and perhaps even learn from and imitate them. Radical changes are desperately needed in so many schools. In my view this is one of the best ways of bringing about the reforms. But I must repeat the importance of timing. Many students who accept the necessity of an initial 'tough' approach are so keen on redefining the situation that they instigate the redefinition too quickly and long before it is due. In the term (or less) of a teaching practice there is barely time to establish the initial definition of the situation. Student teachers, in their impatience with the old ways and in their enthusiasm to try out the new, easily persuade themselves that the ground is adequately prepared for change. In this, I must confess, they are often encouraged by their tutors who share their disrespect for the conventional and surpass them in their eagerness to experiment with new ideas and teaching methods. It never ceases to surprise me that so many students do indeed succeed in assessing the situation in the classroom so well. Perhaps it is that they make a compromise between the conservatism of the serving teachers and the radicalism of the tutors, and the right course may well be somewhere between the two.

One of the most important reasons why the definition of the situation must be initially made in terms of the teacher's dominance is that the detailed definition takes some time to establish. It cannot all be done at the first meeting. In accomplishing his own dominance immediately, the teacher is ensuring that he has the right to make the rules that form the code of acceptable behaviour and also the power to enforce them. If the teacher fails to establish his dominance this process will be extremely difficult.

In the very first encounter the teacher will try to lay down some of the most basic rules of the code, covering such areas as modes of address, movement about the room, distribution of equipment and so on. It is obviously best if these basic rules are few in number. They should also be fairly comprehensive, though not so general as to offer little guidance in specific situations. (Making the first rule 'Love thy neighbour as thyself' will not help very much.) During the following weeks the teacher will have to ensure that these rules—and the reasons for the rules—are fully understood, learned and conformed to. Both teacher and pupils may find this difficult, for many of the rules may be specific, at least in their detailed application, to one particular teacher. No two teachers ever have precisely the same code of acceptable conduct, though there will probably be much in common. New teachers are frequently besieged by cries of 'But sir, Miss Smith always said . . .'

The rules which the teacher enjoins at the first meeting cannot be more than basic generalized rules. In the ensuing weeks these rules will be clarified with respect to particular events. The pupils will learn what behaviours are covered by the rules and in what ways. Further, the basic rules will not cover many eventualities. New rules will be promulgated and brought into force as relevant to the requirements of novel situations as they arise. Often the rules will be made in response to a pupil's misbehaviour. In such cases the rule arises to prevent the continuance and recurrence of something of which the teacher does not approve. In creating rules in response to pupil misbehaviours the teacher must beware of manufacturing a code that consists of a list of *don'ts.*

The definition of the situation takes some weeks to establish because the rules must be often created and always clarified in relation to concrete incidents where the rules are applicable; because the pupils need time to learn the rules and how they apply in given situations; and because the teacher must be able to demonstrate his power to enforce the rules and gain conformity to them. The definition of the situation is, in short, a progressive and cumulative process. It is built up, day by day, incident by incident, into a consistent whole.

Once the rules are mastered by the pupils, that is learned, understood and conformed to (at least by most of the pupils for most of the time), the teacher's relationship with his pupils can enter a new phase. The rules become part of the taken-for-granted aspect of life in the classroom. As the teacher has asserted his dominance and generated a set of clear-cut rules and penalties, there are unlikely to be any serious attempts to undermine the rules or to play up the teacher. So although the teacher will have to assert his own dominance, reiterate the rules and impose punishments from time to time, generally speaking his dominance can begin to fade into the background. The classroom situation and relationships are open to redefinition in accordance with the teacher's personal philosophy of education.

An important part of this change concerns the teacher's attempt to *legitimate* his authority. Hitherto the teacher's ability to establish and maintain discipline and order in the classroom has rested on his power to dominate and to impose sanctions. The teacher relied on his *formal* authority as the possessor of adult status, legal rights and so on. Now the teacher is in a position to change this to a more *personal* authority which is based on the teacher's personal qualities and the special relationship he develops with the pupils.

Whilst the teacher is concerned with his own dominance, the pupils comply to the teacher's demands and rules without in any way being necessarily committed to them. The pupils tend to be concerned with the effects of their behaviour on the teacher and comply largely to avoid punishments of various kinds. As the teacher moves to the exercise of personal authority, his formal power becomes increasingly less prominent. The sources of his new power are based on the pupils' desire to establish and maintain a good relationship with him, and on the liking, respect and admiration that the pupils have for him. The pupils are now less concerned with simply avoiding punishments and more concerned with maintaining their self-image as good pupils, worthy of the teacher's approval, and with abiding by values which are congruent with those of the teachers. They behave well, not because they are pressured to do so, but because they want to. On the surface the order and discipline of the classroom may appear to be unchanged; in fact, the sources of the teacher's power and the pupils' motives for conforming have been radically altered. We have moved from an obedience based on fear and the desire to survive unscathed to a relationship of deep trust and mutual concern. It is at this point that discipline has in a meaningful sense become self-discipline. It is only now that the teacher, faced with misbehaving pupils can say, 'I'm disappointed in you; you've all let me and yourselves down', and find his pupils truly penitent and eager to make amends.

This new definition of the situation has important repercussions on the profitability of the interaction for both teacher and pupils. The initial dominance of the teacher is very costly to the pupils; life is not very satisfying under such a regime. It is equally costly to the teacher —though much less costly than is the total loss of control caused by a failure to establish dominance in the first place—because it is often exhausting and alien to the teacher's philosophy to exert dominance. The new definition of the situation, in contrast, is highly profitable to both. The pupils find themselves in a situation where their needs are more fully met under a teacher whose behaviour falls much nearer to their ideal of the good teacher. The teacher himself no longer feels the strain of applying his formal authority unremittingly and can begin the rewarding task, with all its ups and downs, of developing a warm, friendly atmosphere in which relationships can prosper and the pupils' potentialities can grow and mature.

Redefining the situation in this way is, as all experienced successful teachers know, much more difficult and more an uneven process than it sounds. If it were easy, so many teachers would not become trapped for life in the stage of initial dominance or, even worse, of attempted dominance. The truth of the matter is that the seeds of the redefinition are sown in the initial stage of teacher dominance. It is absurd to expect a harvest if these seeds are not sown at the right time and in the right way. The way in which the redefinition is prepared for consists in the form of the dominance exerted by the teacher and the techniques he uses to attain it. Advice to be 'tough' at the outset is dangerous precisely because such advice rarely specifies in what ways the teacher should be 'tough'. Unless we can formulate further detail, the advice to be 'tough' will produce quite as many disasters as successes.

Educationists not infrequently tell their students that there is no discipline problem if the teacher's lessons interest the children. In my view this is pernicious nonsense. It is true, of course, that when the teacher's lessons are exciting to the children there are less likely to be infractions of the classroom code. Boredom is one of the major sources of minor infringements of the rules and fascinating lessons reduce boredom to a minimum. But boredom is not the only cause of

243

rebellion and the ability to plan fascinating lessons is not the only skill required of the effective teacher. The 'discipline problem' is, as I have tried to show, a much more fundamental and pervasive issue than many educationists allow. When they argue that interesting lessons eliminate the discipline problem they ignore the wide variety of interpersonal skills that are the mark of the 'good' teacher. It is these interpersonal skills that we must now consider.

When a pupil breaks one of the rules, the teacher who is aware of this deviation is called upon to respond. Typically a wide range of actions is at the teacher's disposal, but only one or two of the possible responses are actually enacted. In selecting a particular response the teacher usually seeks to stop the deviation from the rules and inhibit its recurrence and its 'spread' to other pupils. Essentially, then, the teacher is attempting to *control* the pupil's behaviour by means of certain *techniques*. Following Redl (1951, 1952), we shall regard these control techniques as a kit of tools. Like tools they can be used properly or improperly. One can successfully use a chisel as a screwdriver, but in so doing one damages the chisel. The screwdriver might be used as a chisel without harming the screwdriver, but the wood will not escape serious damage. The same is true of control techniques. If they are used improperly, there will be deleterious effects on both the technique and on the pupil.

The 'good' teacher, presumably, is one who achieves morally commendable disciplinary objectives by the proper use of morally acceptable techniques. This is clearly a highly complex process by which the teacher solves the problems of what technique to use on what occasion in what way and for what reason. Thus there are four questions to be asked—which? when? how? why? *Which?* requires the teacher to select from his repertoire a technique that is appropriate to the particular misbehaviour committed, to the context in which the misbehaviour arises, and to the particular needs of the offending pupil. A technique cannot be regarded as appropriate or successful unless it has the desired effect. This requires that the teacher know how the pupil is likely to react to the technique in this particular situation, which calls for considerable imagination, insight and understanding on the part of the teacher. *When?* raises the matter of timing. Is this the right moment to intervene? Should action be postponed till later? The *how?* question is concerned with the manner in which the teacher applies the technique. For example, if the teacher decides to administer a rebuke, what tone of voice should the teacher use? Should the pupil be rebuked publicly or in private? *Why?* includes two questions. The first concerns the teacher's reason for wishing to apply the technique, his objectives in seeking to take action. Is it because he wants to help the pupil or is he motivated by anger or resentment or impatience? The second concerns the pupil's motives.

Why did the pupil misbehave in the first place? Did his misbehaviour spring from ignorance or misunderstanding of the rules? from boredom? from frustration? from an attempt to annoy the teacher? from an attempt to gain popularity with his friends?

These questions are essentially *short-term* ones. They are concerned with the immediate situation, that is, with the issues involved in the teacher's selection of a control technique to deal with a classroom misbehaviour by a pupil. However, the appropriateness of a control technique is not determined by these short-term problems alone. It is equally important to consider a number of *long-term* objectives which are less closely related to the immediate situation. Neither the misbehaviour nor the control technique that is adopted by the teacher can be regarded as isolated incidents. Every action by the teacher must be seen in the light of the contribution it makes to the general, on-going relationship of teacher and pupil. So if a technique is to be appropriate it should make a positive contribution to that relationship or at least not damage it in any substantial way. This is *not* to say that the pupil must always be pleased with the control technique adopted by the teacher. Frequently the pupil will be displeased with and resentful of the teacher's action. Rather I am suggesting that the teacher must beware of using techniques which damage his relationship with the pupil in a more than momentary sense. This raises the problem of how a teacher calculates such effects of his actions, a problem to which we shall turn shortly. Secondly, it is important to consider the effect of a control technique on the pupil's relationship with his peers. It should not, for example, cause the pupil to feel degraded as a person or to lose the respect of his friends if this is avoidable. A third long-term objective concerns the pupil's attitudes —to school, to learning, to the curriculum. A control technique cannot be said to be satisfactory if it promotes in the pupil negative attitudes of dislike and avoidance to these.

In attempting to analyse the disciplinary skills of teachers we have soon found ourselves compelled to make value judgments—to distinguish the 'good' from the 'bad' methods of achieving and maintaining order and discipline in the classroom. Yet the lack of a generally agreed definition of what comprises 'good' discipline is well known. To some teachers, 'good' discipline is defined in almost military terms, with domination by the teacher, unquestioning obedience from the pupils, and silent classrooms. Such a view is anathema to teachers of a progressive or child-centred approach. What is 'good' in one school of thought is very often the 'bad' of a different educational philosophy. The skills required to promote discipline according to those different views are probably not the same. The 'good' teacher that I am concerned with might be considered to be of a broadly child-centred persuasion—though in practice such a teacher might be very reluctant to describe himself in partisan terms. My analysis is based on the behaviour of many teachers whose classes I have observed and who have in common only the fact that they are teachers. Some I would regard as 'good' and some as 'bad' in terms of discipline. All of them would claim to be concerned with the pupil's welfare and his developing autonomy. In the end, then, what I mean by the 'good' teacher with respect to discipline is a teacher who possesses and uses the following skills.

1. *The teacher possesses an extensive repertoire of control techniques that are readily available for use and that are effective.*

If the teacher is to be able to apply to a misbehaviour a control technique that is appropriate to the misbehaviour, the particular individual and to the context in which the misbehaviour occurs, then he must have at his disposal a wide range of readily available techniques. It is important that these techniques are not just 'theoretically' available. A knowledge of what *might* be done can be gained from books or from observation of other teachers, but such knowledge is useless unless it can be called upon in practice and will actually stop the misbehaviour.

The repertoires of many experienced teachers are remarkably limited. In such cases the majority of available techniques tend to be in the punishment area—rebukes, lectures, tirades, sarcasm, ridicule, detention, removal of privileges, extra work, corporal punishment. The popularity of such techniques derives, I suspect, from the fact that many teachers, especially early in their career, see a misbehaviour as highly threatening to their control and mastery of the situation and thus as a personal affront. Almost instinctively, therefore, counter-attack seems to be the best form of defence. Such a reaction on the part of the teacher is encouraged by advice to be 'tough' in the early stages, since this advice is almost invariably interpreted as a need to be punishing. Further, since these punishing counter-attacks are not uncommonly effective, that is they do stop the misbehaviour and inhibit further immediate outbreaks, they can easily become adopted by the insecure teacher as the natural and basic method of maintaining discipline. In this way they quickly become over-valued by the teacher, who precludes himself from experimenting with alternative methods.

From my own observations I conclude that there is a Punishment Illusion, which is akin to the Disciplinary Illusion discussed earlier. Punishing techniques *appear* to be effective because they often stop the misbehaviour. The teacher's attack, verbal or physical, on the offending pupil has a 'stunning' effect, which, like the sharp blow of a boxer, disables temporarily and hinders immediate retaliation. But rarely (perhaps never) does it get to the root of the misbehaviour and in consequence new forms of misbehaviour arise or the pupil smoulders in resentment and humiliation. Like the boxer, he recovers and returns to the fight wiser for the experience or retires from the ring angry at the inequality of the contest.

It is for this reason that more 'positive' techniques are an essential part of the teacher's repertoire. By 'positive' I mean techniques which aim to cure as well as to stop a misbehaviour, for cure is the only adequate means of preventing a recurrence. Without doubt the greatest exponent of these positive control techniques is Fritz Redl. His provocative, stimulating and highly practical discussions of disciplinary problems are unrivalled. Although my thinking has been considerably influenced by Redl (1952, 1966), and my debts to him are evident throughout this book, I wish to refer to his work directly at this point by recounting some of the control techniques he has identified, named and fully described. It is a sample to whet the appetite.

Signal interference. Many misbehaviours can be inhibited in their early stages if the teacher can bring an incipient act to the offender's attention and indicate that it is not acceptable. The most common methods of signal interference are catching the pupil's eye and clearing one's throat. Another form of the technique is for the teacher to prowl in the area of an offence until the pupil 'gets the message'. Of course, if this technique is to be effective the teacher must anticipate the misbehaviour or recognize the early stages. It is not a technique for inhibiting a serious misbehaviour that is in full flight.

Interest boosting. Since many misbehaviours arise from the pupil's boredom or frustration they can be eliminated by an effort by the teacher to rekindle the pupil's interest in the work. This is often much more effective than a reproof ('Pay attention', 'Get on with your work', 'Stop fidgeting', etc.) which often merely suggest to the child that his inattention is obvious and must be concealed.

Restructuring the situation. Here the teacher introduces a 'twist' or sudden change of direction, content or method into his teaching in order to counteract more large-scale boredom or inattention. Commonly this takes a form of a change of activity, though the aim of the lesson remains unchanged; a lecture becomes a question-and-answer session or a quiz; a discussion is ended and the pupils are given an opportunity to write; unusual material is suddenly introduced.

Humorous decontamination. A slight lapse or even a threatening situation can often be controlled by an unexpected humorous reaction from the teacher. If the teacher makes a joke to break the tension it is important that it is done in a benign way, otherwise the humour becomes punishing (e.g. ridicule). At the same time, it must contain some hint of disapproval or the pupil may feel that his behaviour is not being taken seriously or even that it is being encouraged.

Non-punitive exile. Some misbehaviours can be overcome by sending the child out of the classroom. It is not a punitive technique ('Get outside and don't come back in here until you can behave') but rather an attempt to allow the child a strategic retreat in order to 'get over it'. It is particularly effective in cases of the 'giggles' where attempts to stifle the offence can actually increase it.

2. *The teacher employs control techniques that meet both short-term and long-term objectives.*

I have tried to suggest that it is not enough in selecting his control technique, for the teacher to find one that simply stops the misbehaviour. In most cases the misbehaviour can be stopped by a whole range of techniques and if the intention is merely to stop it then the 'functional equivalence' makes the choice of a particular technique quite arbitrary and subject to the whim of the moment. Almost any method, particularly if it is punishing, will do the trick. But if the technique is to be appropriate, it must be selected as the most fitting teacher response to the particular offence, the needs of the offender, the factors of the immediate situation, the pupil's relationship with the teacher and with his peers, and the child's attitudes to learning and to the curriculum.

Many subtle skills are involved in learning to choose appropriate control techniques. It usually takes some time for the beginning teacher to acquire the capacity to know what misbehaviour committed by which pupil in which situation is best dealt with by a stern rebuke or by laughing it off—or by any other technique. Much trial and error seems to be needed. Moreover, every particular technique can be applied in different ways. For example, let us suppose that the teacher decides that a stern rebuke is the best response to a pupil's infraction of the rules. The question then arises: how stern should the rebuke be? If the rebuke is insufficiently stern it will not be taken seriously by the offending pupil and the misbehaviour will not be stopped. If the rebuke is too stern then the teacher has gone beyond what is necessary. Again much trial and error learning is needed to acquire the skill of applying 'just enough' pressure to stop a misbehaviour. Beginning teachers tend to swing somewhat erratically between under- and over-reaction—typically over-reacting when they find that a previous under-reaction has not been effective. In my own observations I have seen student teachers abandon a technique as 'useless', that is, it did not achieve the desired end of stopping the misbehaviour, and then switch to another technique (usually more punishing) when in fact the first technique was indeed an appropriate one, but failed because its application was marred by under-reaction. It is much more common, of course, for a teacher to conclude that a technique is useless as a result of an inappropriate application. The beginning teacher must beware of prematurely rejecting a technique before he has given himself the opportunity to assess its value by using it in an effective and appropriate way. It requires courage and persistence to continue to experiment with a technique which does not prove to be instantly successful until its usefulness can be fully assessed.

One of the most basic skills involved in choosing an appropriate technique is related to the teacher's *perception* of the offender. How the teacher reacts is in part a function of what motives and intentions the teacher infers and attributes to the pupil. Much less frequently than we imagine are we in possession of sufficient information to be certain of the causes of the child's behaviour because there are usually so few cues available to guide our inferences. Typically we fill in the gaps in our knowledge of the child's motives and intentions from our own assumptions of what might seem to be the *likely* or the most *reasonable* causes for the offence.

If one observes a 'disciplinary incident' in a classroom teacher–pupil interaction, one can observe simply the pupil's behaviour and the teacher's reaction to it. Teachers react in different ways to the same pupil behaviour because of factors which cannot be directly observed, namely the teacher's perception of the behaviour and, to use Schutz's term, the 'interpretive schemes' he brings to bear in assigning meaning to the behaviour. Now I want to suggest that all teachers share a common orientation to the pupil's behaviour and this orientation involves three areas that are regarded as constituting the problem. First, is the pupil behaviour to be defined as a misbehaviour? Second, if it is so defined, how serious is the misbehaviour? Third, what are the sources of the misbehaviour? The differences between teachers in their reactions to a common pupil behaviour spring from the different ways in which teachers answer these questions. The first two of these three questions are answered in relation to the *consequences* of the pupil's behaviour as perceived and interpreted by the teacher. There are three basic consequences:

(*a*) How threatening is the pupil's behaviour to the teacher? This concerns the degree to which the teacher interprets the behaviour as a challenge to his authority or as an affront to his person.

(*b*) What is the effect of the pupil's behaviour on other pupils? For example, will it distract other pupils from their work, or will it spread by contagion?

(*c*) What is the effect of the pupil's behaviour on the pupil himself? For example, is it inimical to the child's learning and academic progress?

These three areas or consequences are, of course, only analytically distinct. Whilst some pupil behaviours may have one main consequence, most are likely to have all three. For example, a child who is insolent threatens the teacher's authority, may influence other pupils to be insolent, and is less likely to accept the academic tasks assigned to him. Teachers differ in the weight they give to each of the three consequences. For example, it might be said that the 'child-centred' teacher will put more weight on consequences (b) and (c) than on (a). However, the general proposition is that the more negative the consequences of a pupil's behaviour in the eyes of the teacher, the more likely the teacher is to rate that behaviour as a misbehaviour and the more 'serious' he will rate that misbehaviour.

Considerable interpretation on the part of the teacher is thus needed before pupil behaviour can be defined as a 'serious misbehaviour'. Knowing that a teacher has so defined pupil behaviour tells us very little about how the teacher will react to it. We know only that there is a high probability that the teacher will react with some form of control technique. We must consider in addition the interpretation given by the teacher to the *sources* of the pupil's behaviour. There are two main sources:

(d) The pupil's intentions in committing the behaviour. This concerns what Schutz (1932) terms the 'in-order-to motive'.

(e) The pupil's motives in committing the behaviour. This concerns what Schutz terms the 'genuine because-motive'.

I shall use the term intentions to refer to the ground of the pupil's behaviour as felt subjectively by the pupil, and the term motive to refer to the ground of the pupil's behaviour as supposed by the teacher which has no relation to the pupil's intentions. If the teacher claims, 'The pupil did it just to annoy me,' he is referring to his perception of the pupil's intentions. If the teacher claims, 'The pupil did it because he comes from a deprived home,' then he is obviously not referring to his perception of the pupil's intentions but to the 'causes' of the pupil's behaviour. It is the second which will be defined as motives. The question of motives is important for the teacher because in interpreting the pupil's motives he is making an assessment of the pupil's responsibility for his actions. If the teacher says that the pupil committed the misbehaviour because of his bad home background, he is saying that the pupil was not in a position to control his behaviour and cannot be held to be fully responsible for it, whether or not the behaviour was intended. The teacher's interpretation of the pupil's behaviour is thus partly based on the ascribed intentions and motives. Whilst insolence is generally defined as a misbehaviour, if the teacher believes that the pupil did not intend to offend and that he had little control over his behaviour, then the teacher's reaction may not be the same as in the situation when the teacher believes that the insolence was intentional and that the pupil must be held to be responsible for it. In interpreting the pupil's behaviour the teacher may take into account either the pupil's intentions or his motives, or both.

My general argument is that when the pupil's behaviour is seen by the teacher to have many consequences (as defined earlier), and these consequences are perceived as 'serious', then the teacher will pay relatively little attention to the sources of the pupil's behaviour before making his reaction. This is because when the teacher perceives that the behaviour has many serious consequences, he feels under pressure to take immediate action in order to control the misbehaviour. A second condition under which little attention will be paid to sources is when the sources are not perceived as ambiguous or unclear, that is, when they are seen to be obvious or self-evident. (This second condition clearly calls for an investigation of the conditions under which the teacher perceives sources as self-evident.) Conversely, when the teacher perceives that the consequences of the pupil's behaviour are few, and they are not regarded as 'serious', then the teacher is in a position to pay more attention to sources, largely because he is not under the same pressure to make an instant reaction. He can take sources into account before he makes his reaction. If the sources are seen as unclear, then the teacher can afford to make a further investigation of sources. The problem arises when the consequences are perceived as many and serious and when the sources are seen as unclear. Here the teacher tends to react without taking the sources into account, because he gives a greater priority to controlling consequences than to being fair or just, which requires him to consider sources. Thus his reaction is determined by his interpretation of consequences irrespective of the sources. Having recognized the lack of clarity in the sources, the teacher may in the interests of justice seek to investigate the sources at a later time after the consequences have been dealt with. In practice, of course, he is often unable to undertake this source follow-up because of other pressing matters which require his attention.

Let us now illustrate this process in further detail. Sometimes I have set problems of discipline to student teachers, who were asked to state how they would respond to the following situations.

Case I

The children are working quietly on some work you have set. You have explained very carefully what you wish them to do and how it should be done. You have asked for questions and ironed out one or two difficulties. On your final check every pupil seems to be satisfied. After a few moments you notice one boy sitting and staring dumbly before him. When you ask what is the matter, he tells you that he doesn't understand what to do.

Most of the student teachers do not appear to see the consequences of the pupil's behaviour as very serious. Further, they recognize a lack of clarity with respect to the sources of the pupil's behaviour. Several possibilities exist. It may be that the pupil was wilfully inattentive; it may be that the pupil was too shy to inform the teacher of his difficulties earlier; it may be that he lacked the ability to understand the instructions; it may be that the teacher was at fault in not giving a clear enough explanation. Thus almost three-quarters of the student teachers reacted by giving a further explanation and/or by investigating the sources of the child's difficulties. If they do not investigate the sources, they give the pupil the benefit of the doubt. (In the following quotations R indicates the student teacher's reaction and E indicates what the teacher said in describing what led him to react in the way he did.)

R Explain it again after asking what he didn't understand.
E Probably shy about asking in front of the class.
R Explain it again individually.
E He might genuinely not have understood.
R Go over and explain again quietly.
E Because I may not have explained well enough.
R Explain it to him individually.
E Give child benefit of the doubt—assume he was not able to understand and too afraid to show ignorance before the whole class.

The problem for the teacher is with respect to sources and highly tentative inferences are made, as exemplified by the frequent use of 'may', 'might', 'perhaps', 'maybe' and similar qualifications in the teacher's explanations of his reaction.

A minority (about 20 per cent) of the student teachers paid little attention to the problem of sources, possibly because they defined the consequences as more serious than did the majority, or felt that they could make a highly probable inference about sources. Thus the minority tended to rebuke or threaten the pupil.

R Tell him severely to think about it again.
E He was probably lazy or awkward.
R Tell him, 'You'd better listen next time.'
E Care already taken to ensure that everyone understood.
R 'Haven't I just told you, you numbskull?' Then explain again.
E It would exasperate me, but I couldn't deprive him of another explanation.
R 'You haven't been listening, have you?'

E He couldn't have been paying attention to me.

A few students followed a compromise solution.

R Explain again. Would show exasperation.

E He may genuinely not have understood, but in case he's playing up must make him see that it won't be tolerated again.

The second disciplinary situation was as follows.

Case II

The pupils are handing in their homework. Everyone hands in the work on time except Barry, who is now failing to produce his homework for the third consecutive time. On previous occasions the reasons he supplied seemed very vague. When you ask him why he has not done the work this time, he says, 'I couldn't do it 'cos I forgot to take the right books home with me.'

In this case the majority of the student teachers perceived the consequences of the pupil's behaviour as very serious. Thirty per cent punished him and 20 per cent settled for a rebuke or a threat. The explanations offered by the students emphasize the focus that is placed on the consequences rather than the sources. If inferences are made with respect to sources they are quite unambiguous, as indicated by the frequent use of the term 'obviously'.

R Not accept the excuse. Let him do work after school.
E Quite obviously a slacker.
R Give him a detention.
E A clear case of consistent laziness.
R 'You didn't forget your head this morning!'
E Child is just being tiresome on purpose.
R Make him do it during break.
E In order to discourage would-be followers from that path.
R Anger. Make him do during day or next evening.
E It seems he's pulling a fast one to avoid work.
R Tell him to produce it next day, or a detention.
E He is a troublesome lazy child who must be taught more responsibility towards work and realize that I will not accept excuses.
R Give him a good telling off.
E To make him realize that he must do his homework for his own benefit.

In Case II a minority of student teachers were less involved with the consequences and gave some attention to sources. Whilst a very small number gave exclusive attention to sources, most of this minority tried to take consequences and sources into account in making their reaction.

R Try to find out why he didn't do the homework.

E There must be a special reason for it if he's the only one.

R In class remind him to remember next time. Later, talk with him and investigate the real reason.

E Because he's not likely just to keep on forgetting—could be cussedness, family problems, personality.

R Tell him to do the work at break and see me.

E He cannot be allowed to skip the task—but there may be genuine extenuating circumstances which have not been brought out.

R Get him to do it at lunchtime or after school.

E This is marking time before finding out if there is some under-lying difficulty at home.

It is worth noting that this minority, in making their consideration of sources, tend to be less punitive in the reactions they make. In allow-ing for the possibility of a justifiable motive, they are led to temper the reaction that is made when consequences alone are considered.

The student teachers were also asked if there was any information that they felt to be relevant to the situation. Almost without exception the minority of teachers who had raised the matter of sources in their reactions and/or explanations sought information on the home background (e.g. facilities for homework at home) or the ability of the pupil. In the case of the majority who had taken no account of sources in their reactions, there was a division of view. Some, apparently taking for granted that the pupil had committed the mis-behaviour intentionally and that he was fully responsible for his actions, either sought no further information or suggested such matters as

The other boys' reactions.

Does the same happen with other teachers?

What do other teachers do about it?

What is Barry's general attitude to work?

In other words, the further information that is sought is much more relevant to how the teacher should react to the consequences of the pupil's behaviour than to the problem of sources. Others, however, specified home background and ability factors. In this case it is not clear whether it was the question which provoked these student teachers into a consideration of sources or whether they had been considering sources but were unable to take account of sources in their reactions to the pupil. The main point remains. The majority perceived the behaviour of the pupil in Case II as very serious and directed their reactions to deal with these consequences, either genuinely inferring that the pupil intended to misbehave and did so with full responsibility, or not making such an inference but acting towards the pupil as if they had indeed made such an inference. The process of control technique selection, then, is highly complex, entailing a consideration of the past (the sources of the pupil's misbehaviour), the present (the immediate consequences of the pupil's behaviour), and the future (the long-term consequences of the pupil's behaviour). Once the situation has been diagnosed with respect to the sources and consequences of the misbehaviour, certain possible control techniques can be eliminated as inappropriate. The selection of the actual control technique is then influenced by the teacher's estimation of the short-term and long-term effects of the remaining possible control techniques, as defined earlier in the chapter.

3. *The teacher gets feedback on his own behaviour and learns from experience.*

The ability to profit from experience is implicit in the previous two points. If the teacher is able to apply a wide range of techniques that achieve both short and long-term objectives, then he must be able to judge the effects of the control technique once it has been applied, and estimate the extent to which it is appropriate and successful. There is obviously little point in applying a technique and then assuming that it is effective. If the technique is not effective and the misbehaviour does not stop, then the teacher will have to strengthen his reaction or switch to another technique. If he has made a mistake by selecting an inappropriate technique or applying the right technique wrongly, then he must be able to recognize the error and learn to avoid making the same mistake again. A surprising number of student teachers persist in endless repetition of the same mistake because they fail to recognize it as such. Sometimes the source of the difficulty is a failure in *seeing it through*. Because he feels insecure, afraid and self-conscious when faced with a misbehaviour, the inexperienced teacher tends to hurry on to other matters without checking that the technique has been effective. Such a teacher must resist the temptation to undue haste and take his time in seeing it through. Suppose a teacher calls for silence. If he then moves on before complete silence is attained, further disruptions are inevitable. To see it through, he must wait until complete silence reigns, then pause for a few seconds, and then move on. Seeing it through is no mean skill, for there is no point in seeing through a technique that is doomed to failure from the beginning. The skill consists in recognizing at an early stage that the technique is going to work and then seeing it through.

Whilst it is relatively easy to assess the short term effectiveness of a technique it is much more difficult to calculate the effectiveness or appropriateness of a technique in the long term. This is because long-term effects are inevitably dissociated in the teacher's mind from particular immediate situations. John may seem not to mind very much whenever the teacher says, 'Shut up—don't be such a chatterbox', and it may inhibit John from further chatter, but may not his general apathy in lessons be the cumulative result of the teacher's having said this twice a day for three years without having ever got to the root of his talkativeness?

A whole set of skills is involved in getting feedback on one's own behaviour as a teacher. Perhaps the most important is a sensitivity to knowing what is going on 'out there' in the classroom. This awareness of what is going on is called *with-it-ness* by Jacob Kounin (1966). It means that the teacher must have the proverbial 'eyes in the back of his head'. The teacher must not be so absorbed in his own behaviour or so intent on executing his own pre-determined lesson plans that he pays insufficient attention to how the pupils are reacting or behaving. In my view with-it-ness in its most subtle form consists of three elements: (i) close observation of the pupils; (ii) the pupils are not always aware that they are being observed, that is, the teacher is in fact observing them when he appears not to be; (iii) the teacher gives the children evidence that they have been observed. For example, if the teacher notices a misbehaviour, he does not always have to look at the child and tell him to stop. He can look in the *opposite* direction and then tell the pupil to put his comic away. When the offending pupil looks up he will see the teacher looking elsewhere and it will appear that the teacher can indeed see through the back of his head— and he will be duly impressed.

The new teacher may find it very hard to notice what is going on 'out there' because he is so pre-occupied with other things. As he gains more experience, what initially passed unnoticed soon becomes very obvious. This change in the teacher's sensitivity to pupil behaviour is noted by Blishen (1955) in describing the more developed awareness of a teacher after several months' classroom experience.

Williams was sniggering behind his hand. It was like an
elephant trying to hide behind a rosebush. I felt sorry again.
Concealment was impossible to them. Everything they did
was huge and instantly perceptible. It would go on like that
for a long time, until they learned adult techniques of
camouflage. They couldn't even whisper. Jenkins was trying
it now. His voice came out like a half-stifled klaxon. And
Jimmy Green was pinching his twin brother's bottom in what
was intended to be secrecy. The writhings of one, the mali-
cious plunges of the other's arm, could have been seen
clearly a hundred yards away.

An associated skill is the ability to recognize and distinguish a major from a minor deviance. The teacher who is so involved in explaining to Tom that he must not chew gum in lessons that he does not see or react to the fact that in the other corner Jim is busy tattooing with his pen the neck of an unwilling Mary, cannot be said to be with-it. In other words the teacher must never be involved in one activity to the exclusion of all others. He must be able to pay attention to one behaviour whilst continuing to monitor other activities. For example, when the teacher is marking work at his desk, he looks up regularly to check that all is well with the other pupils, that no one is seeking his attention or that no misbehaviour is brewing at the back. Occasionally new or student teachers discover, to their horror, that pupils have been taking photographs during the lesson or that an 'L' sign has been pinned on his back. Such ruses succeed because the inexperienced teacher has difficulty in *not* getting totally involved in one activity and in remembering to scan what is going on elsewhere. The experienced teacher keeps half an eye on activities and pupils that are not the immediate object of his attention. He can sense when mischief is in the making. Kounin calls this ability to pay attention simultaneously to two or more aspects of classroom activity *overlappingness*. The lack of this ability is obvious in the example of Blishen (1955).

I couldn't write on the blackboard without turning my back on them, nor without putting my whole energy into the business of being legible on a large scale. They knew at once that when I turned to the board I was effectively absenting myself from their midst. It was then that the trouble began.
'This sum . . .' I would say, and absent myself. There would be a buzz. Then I would hear the tail-end of an improper joke. There would be a shriek, the crash of a desk. But I would be lost in the complexities of handling chalk. 'Now, shut up!', I would murmur, but absent-mindedly, with evident incapacity to pay much attention to what was happening. When I did turn round, the class would no longer be in an arithmetical frame of mind.

As with many social skills, the ability to get feedback is most easily demonstrated negatively, that is by illustrating a lack of it. Another brilliant example is provided in A. J. Wentworth, B.A., the ingenious invention of Ellis (1964).

'This morning,' I remarked, taking up my *Hall and Knight*, 'we will do problems,' and I told them at once that if there was any more of that groaning they would do nothing but problems for the next month. It is my experience, as an assistant master of some years' standing, that if groaning is

257

not checked immediately it may swell to enormous proportions. I make it my business to stamp on it.

Mason, a fair haired boy with glasses, remarked when the
groaning had died down that it would not be possible to
do problems for the next month, and on being asked why
not, replied that there were only three weeks more of term.
This was true and I decided to make no reply. He then asked
if he could have a mark for that. I said, 'No, Mason, you
may not,' and taking up my book and a piece of chalk, read
out, 'I am just half as old as my father and in twenty years
I shall be five years older than he was twenty years ago.
How old am I?' Atkins promptly replied, 'Forty-two.' I en-
quired of him how, unless he was gifted with supernatural
powers, he imagined he could produce the answer without
troubling to do any working-out. He said, 'I saw it in the
Schools Yearbook.' This stupid reply caused a great deal of
laughter, which I suppressed.

I should have spoken sharply to Atkins, but at this
moment I noticed that his neighbour Sapoulos, the Greek
boy, appeared to be eating toffee, a practice which is forbidden at Burgrove during school hours. I ordered him to
stand up. 'Sapoulos', I said, 'you are perhaps not quite used
yet to our English ways, and I shall not punish you this time
for your disobedience, but please understand that I will not
have eating in my class. You did not come here to eat but
to learn. If you try hard and pay attention, I do not altogether despair of teaching you something, but if you do not
wish to learn I cannot help you. You might as well go back
to your own country.' Mason, without being given permission to speak, cried excitedly, 'He can't, sir. Didn't you
know? His father was chased out of Greece in a revolution
or something. A big man with a black beard chased him
for . . .'

'That will do, Mason,' I said. 'Who threw that?'

I am not, I hope a martinet, but I will not tolerate the
throwing of paper darts or other missiles in my algebra set.
Some of the boys make small pellets out of their blotting-
paper and flick them with their garters. This sort of thing
has to be put down with a firm hand or work becomes impossible. I accordingly warned the boy responsible that
another offence would mean an imposition. He had the impertinence to ask what sort of an imposition. I said it would
be a pretty stiff imposition, and if he wished to know more
exact details he had only to throw another dart to find out.
He thereupon threw another dart. I confess that at this point

I lost patience and threatened to keep the whole set in during the afternoon if I had any more trouble.

Poor Wentworth simply cannot see that his expression is having quite a different impression on the pupils to the one he intends. If he only would (or could) recognize that, for example, his threats are not taken seriously by the pupils, then the uselessness of the technique would be exposed and he himself would be made aware of the need to try alternative approaches. Like many teachers who never master the discipline problem, Wentworth is his own worst enemy.

4. *The teacher is able to make fast decisions.*

Earlier I tried to suggest the host of long-term and short-term objectives that must be satisfied if the control technique applied to a misbehaving pupil is to be regarded as appropriate. I do not imagine that anyone will dissent from these suggestions. The trouble is that when the teacher is faced with a disciplinary infringement in the classroom he must make an almost instant decision about how to react. Occasionally he has the opportunity to consider the problem, as when he can observe a minor misbehaviour in its early stages or when he asks a misbehaving pupil to come and see him at the end of the lesson. In such cases he can give some thought to the matter before taking action. But typically he must react at once. Classrooms are not like law courts. The teacher cannot hold a trial and then postpone judgment and sentence for a day or two. Even where the teacher postpones his final action until a later stage, the decision to postpone has to be taken at once, and must be a positive act rather than a lack of action based on indecisiveness. Moreover, the teacher has to decide whether he knows enough about the circumstances of the misbehaviour to act at once or to hold some sort of investigation into the precise details of the pupil's conduct and his motives for behaving so.

How can the teacher consider all the long-term and short-term objectives before he applies a control technique or holds a brief investigation? How can all this be accomplished in a few seconds? The answer is that the teacher does not consider all these matters— at least not consciously. The skill of good discipline is in some respects like the skill of riding a bicycle. When one rides a bicycle one does not have to think about where one's feet are, about how to turn the handlebars or about the delicate feat of maintaining one's balance. To the rider these skills are automatic. He simply *knows* how to ride the bicycle. Much of this knowledge is tacit (Polanyi, 1966). The rider, when asked for an explanation, does not find it easy to explain how he managed to control his feet, his balance, the handlebars, the brakes, etc., when he takes action to avoid a sudden obstacle. He can no more explain the details of his behaviour than can a pianist or a juggler.

The same is true of discipline. Ask an experienced teacher how he manages to make an appropriate reaction to a disciplinary infringement and he will probably tell you that it somehow seemed to be the right thing to do. The beginning teacher marvels, just as small children learning how to ride a bicycle admire the seemingly incredible skills of the experienced riders. In both cases a great deal of practice is required before the skill is achieved, before much of the necessary knowledge has become tacit and masked from our conscious thoughts. In both cases, some people seem to have more natural ability which helps them to learn faster than others. A few never learn. For them twenty years' experience is one year's experience repeated twenty times.

5. *The teacher is able to take prophylactic action.*

In other words, prevention is better than cure. Few misbehaviours arise in the classroom of a teacher who has mastered his disciplinary role because he takes action that prevents misbehaviours arising in the first place. Sometimes it *looks* as if there are no disciplinary problems. In reality, they have been forestalled. Many teachers seek to inhibit misbehaviours by establishing some basic rules at the first meeting with the pupils. Rules which are general enough to cover a variety of circumstances and specific enough to provide adequate guidance in a particular situation can prevent many of the misbehaviours which arise from ignorance about how to proceed rather than from a desire to flout the teacher's authority. A simple example of the value of planning ahead was given in Rick's story earlier. He asked one specific boy to open the window and avoided the chaos which might have resulted from a general request for fresh air. Similarly if the pupils are dismissed from a classroom row by row then a noisy stampede (and possible accidents or fighting) is easily prevented. Such examples may seem very mundane and obvious, but most beginning teachers take some time before they can put the obvious into practice. I have seen many student teachers distracted by incidents of the following type. The teacher tells the pupils to turn to page 24 of the textbook. His next words are drowned in a flurry of banging desk lids, raised hands, protests and mutterings—all because he had not checked that all the pupils possessed a copy of the book and they had it on the desks before them.

Yet it is by no means easy to take the necessary prophylactic action, as an example from Blishen (1969) shows. The teacher has taken his pupils on a visit to the Tate Gallery.

at last we reached the gallery, and made our way up the steps; somehow like the great scene in *Battleship Potemkin* back to front. There was a moment when I expected to see the attendants come out of the revolving doors in a body, brandishing bayonets. I drew the boys together for a moment with a quick flapping of my arms, and tried to give them a simple sharp summary of the proper etiquette for visiting a national art collection. 'If you feel strongly about any-thing, talk about it very quietly with a friend,' I said. 'Don't run. And don't touch anything.' I felt sure I'd missed several vital points, but by now we were on our way in. I *had* for-gotten about the revolving doors, of course, and I had to stand by while they discovered one of the oldest of comic routines.

One aspect of this skill concerns the ability to recognize a mis-behaviour in its early stages when it can be 'nipped in the bud'. Enormous problems can obviously accrue to the teacher who cannot recognize a misbehaviour until it is in full flower. The teacher who is aware of the early stages of boredom and frustration over work—and these are sources of much misbehaviour—can take remedial action before the misbehaviours actually arise.

6. *The teacher cares for and trusts his pupils.*

All the previous five factors we have discussed are skills rather than qualities. They can all be learned—though I am not sure whether they can be taught, as so many student teachers wish. They could have been described in quite different ways. To say that *sensitivity* and *flexibility* are the marks of the teacher with good discipline would simply be another way of saying the same thing. For example, Flanders (1964), describing the 'good' teacher (in general terms, not with special reference to discipline) writes:

First, the teacher was capable of providing a range of roles, spontaneously, that varied from fairly active, dominative supervision, on the one hand, to reflective, discriminating support, on the other hand. The teacher was able not only to achieve compliance but to support and encourage student initiative. Second, the teacher was able to control his own spontaneous behavior so that he could assume one role or another at will. Third, he had sufficient understanding of principles of teacher influence to make possible a logical bridge between his diagnosis of the present situation and the various actions he could take. Fourth, he was a sensi-tive, objective observer who could make valid diagnoses of current conditions. All of these skills, which seemed to characterise the most successful teachers, were superimposed upon a firm grasp of the subject matter being taught.

Now, with respect to discipline, I want to suggest that all these skills which the teacher acquires through experience will make him a successful teacher—at least successful in the sense that he will feel that he does not think that he has a 'discipline problem' with the pupils. In such a case the teacher's beliefs about himself will be confirmed by his colleagues, who may even give him that mark of respect which is implicit in the comment 'his discipline is good'. But this is not the ultimate success, any more than the pupils' passing the examinations is the mark of the ultimate success of the teacher in his instructional role. It is one element of being a successful teacher, one which is not to be despised, but it is by no means the ultimate criterion. The important and central criteria of our success as teachers are much less tangible, much less open to inspection by ourselves or by our colleagues.

The skill of teaching in the instructional area consists in facilitating the learning of the pupil. We cannot learn for the pupil; we can only help him to learn. It is very difficult to assess the extent to which our behaviour as teachers facilitates the pupils' learning. Because the evidence is not readily available we can never be really sure whether or not we are succeeding. Doctors can cure their patients or kill them; lawyers can win their clients' cases or lose them. We teachers do not find it so easy. Our problem is much more like the problem of the psychiatrist. Did the patient get well because of the therapy? or would he have recovered just as well without the treatment? Is he better in spite of the treatment? Or, even more disturbing, would he have got well faster without our interference? It is this insecurity which leads us teachers to rely overmuch on the tangible aspects of our work, such as marks and examination results. It is this insecurity which leads us to congratulate ourselves when the pupils pass the examinations, but to blame the pupils (or their homes) when they fail. It is this insecurity which has led educational researchers to evaluate teaching by a consideration of attainment—because attainment is most easily measured.

We can trust the pupil or we can distrust him. If we distrust him we are taking the view that the job of a teacher is to manipulate the situation and the pupil so that he does the things that the teacher wants him to do. Essentially, Tommy is an object which can change and develop only to the extent to which things are done to it. If we trust Tommy we act on the assumption that within Tommy is the capacity to learn and the capacity to want to learn. Tommy will change not so much because the teacher does something to him, but rather because the teacher creates situations which stimulate Tommy into drawing on his potentialities, exercising them and going through the painful process of coming to terms with himself.

This trust on the part of the teacher is extremely difficult to define and analyse. It has a mysterious, even a religious quality about it. But it can be learned. Because you do not trust or care for your pupils today does not mean that you will not do so tomorrow. But it cannot be faked. You cannot pretend to trust. If you do not trust and care for your pupils they will know it.

Caring for and trust in the child are not skills as such, but rather assumptions, attitudes or approaches taken by the teacher towards the pupil. Yet several social skills may be involved, especially in the way in which the teacher is able to *convey* his trust to the pupil and in the way in which the teacher is able to *sustain and develop* his assumptions. There is little point in trusting the pupil if the pupil does not realize that he is trusted or if the teacher's trust is short-lived. We saw in an earlier chapter that the student teacher tends to lose many of his ideals in teaching practice and during the early years of his life in the profession. One of the reasons for this may be that the teacher lacks the skills to put his attitude of caring and trust into effect. As a result he comes to change his assumptions because caring and trust yield little fruit. If the student teacher does care and trust then we must, in teacher training, help him to know how his assumptions can be used to good effect within his classroom teaching. We need, in other words, to define the skills associated with caring and trust and then train teachers in such skills.

We do not at present know very much about these associated skills. But it does not seem unreasonable to suggest that empathy is likely to be one of them. If the teacher can take the perspective of the pupil, if he can understand what it is like to be the pupil, then it is much easier to care. The teacher's generalized care and trust can be given a more specific focus, thus suggesting particular lines of conduct through which the teacher can express his care and trust. It is the teacher who cannot understand the pupil and who finds it impossible to take the pupil's perspective, who is most likely to condemn the child with an uncaring attitude. To train teachers in empathy would be to help them to develop their caring attitudes. In addition, for those teachers who seem to have little care and trust, a training in empathy would help them to acquire a sense of care and trust, for it is very difficult not to care about a pupil that one understands, however unattractive or offensive his attitudes or behaviour, however much one does not share his assumptions. As the French proverb has it, *tout comprendre, c'est tout pardonner.*

Clearly, too, skills in empathy are an obvious help in understanding the pupil's attitudes and motives, placing the teacher in a markedly advantageous position for selecting an appropriate control technique that will help the child to grow, rather than merely putting a stop to the misbehaviour, which is the teacher's most pressing need. The teacher is always in danger of taking disciplinary action towards the pupil to meet his own objectives, without taking account of the consequences of his action on the pupil as a person. Caring, with the associated skills of empathy, whether acquired through training or by intuition, helps the teacher to remember that what he sees as his disciplinary objectives must not be met at the expense of the pupil. There is a constant conflict, because of the nature of schools and the disciplinary expectations of colleagues, between the teacher's disciplinary objectives and the pupil's best interests. It is only the uncaring teacher, devoid of empathy, who thinks that his own disciplinary objectives are always in the interests of the pupil. It is only the uncaring teacher who believes that self-discipline by the pupil is compliance to the teacher's disciplinary objectives.

Let us try to clarify the concept of caring and trusting in relation to the problem of punishment. This is another concept which it is extremely difficult to define. I shall take it that any unpleasant experience imposed intentionally by a teacher on a pupil is a punishment. When we, as teachers, punish a pupil we like to think that we are doing it for the good of the child. But, if we are honest, we know that we often punish the pupil as a self-defence—because what the pupil has said or done is threatening to us. It is motivated by a spirit of angry revenge. It is as if we are saying, 'I'll get my own back and more. I'll make sure you don't do that again.' Sometimes we regret such punishments later, when our anger has cooled, when our resentment has evaporated or when our sense of dignity has been restored. We recognize that we punished in haste and without rational consideration. On other occasions we punish because we are frustrated by being unable to think of a non-punishing control technique that might work. We feel that we are at our wits' end, and the punishment we mete out is motivated by desperation. It is as if we are saying, 'You've got to learn and I can't think of any other way of doing it.' Some teachers wish to retain corporal punishment in schools, not because they like it, but because it can be a 'last resort'. They are afraid of meeting situations which they feel they cannot handle. Such types of punishment are common in schools because we have learned that, when we feel under attack or helpless, punishments are effective in the sense that they subdue and silence the pupil and give us a curious sense of satisfaction that we have won the day.

Most teachers at some stage in their careers are, I think, led to punish pupils on such grounds. But the teacher who cares about and trusts his pupils recognizes that he cannot justify such punishment and seeks to find alternative means of dealing with such situations.

He knows that Fritz Redl is right when he suggests three conditions which must be fulfilled before a punishment can be justified. First, the child must know that he was wrong to behave in the way he did and understand the reasons why his behaviour was at fault. Otherwise the child, though technically guilty, is really innocent. Second, the pupil should see the punishment as fair. If we punish pupils who are innocent or pupils who feel they are being treated unfairly, then the punishment cannot possibly help the child. He will react with anger, resentment and hostility. It will lead the child merely to conceal future misbehaviours or refrain from future misbehaviour simply in order to avoid the punishment. In neither case has the punishment helped the child to recognize the wrongness of the behaviour or the reasons why he did it. It has not led the child to a self-examination through which he can overcome the desire to do it again or take measures to exert some control over his desires. I once heard a teacher say to a child, 'Either you apologize to Miss Smith or you go into detention. You can take your choice.' This teacher is really saying, 'Miss Smith wants an apology. You must go and make an insincere apology to her to make her happy or I shall make life unpleasant for you by putting you into detention.' I do not see how Miss Smith or the child could possibly benefit from such a course of action.

Redl's third condition is that the punisher should care for the child. In a sense the first two conditions are implicit in this third one, for if the teacher cares, then he would not seek to impose punishments which are unjust or which cannot help the pupil. If the teacher cares, the justification of the punishment is that it provides an unpleasant experience which leads the child to work through and try to cope with the sources of the misbehaviour. It is the un-caring teacher who justifies a punishment simply on the grounds that it reduces the probability of a recurrence or has a deterrent effect.

Finally, let me say that caring for the child does not mean being sloppy, sentimental or mawkish. On the contrary, teachers with sentimental attitudes tend to exploit the pupils to satisfy their own emotional needs. Caring for children means believing in them and trusting them, from the beginning. It is this fundamental trust which allows the new teacher to move from a 'tough' initial definition of the situation to later re-definition such as that advocated by Carl Rogers, as discussed in the last chapter. The difficulty is that all teachers justify what they do on the grounds that it is for the ultimate benefit of the child—'This will hurt me more than it hurts you, Jones.' Even Mr Squeers believed that he had his pupils' interests at heart. We are all such experts in convincing ourselves of our altruism when in reality we are meeting our own needs. If the teacher is to act towards the pupils for their ultimate benefit, he cannot succeed without a ruthless self-honesty about his motives and intentions, both for the present and for the future. And that, as we well know, is the hardest thing in the world.

Recommended reading

R. FARLEY, *Secondary Modern Discipline*, Black, 1960.

F. REDL, *When We Deal with Children*, Free Press, 1966.

L. STENHOUSE, (ed.): *Discipline in Schools: A Symposium*, Pergamon, 1967.

W. WALLER, *The Sociology of Teaching*, 1932 (Science Editions, Wiley, 1965).

8 Friends

One irritating feature of most social psychology textbooks is that whilst their authors inform the reader quite fully about the problems that social psychologists have investigated, they often fail to point out those social psychological problems which have not been adequately or systematically treated at all. It may seem perverse to expect a writer to raise problems about which he can say little, yet it is often precisely these questions which are the most important and interesting to the layman. The social psychologist ought to reveal his awareness of these larger issues and explain the reasons why they have not been investigated. Most textbooks give the impression that the writer is unaware of such problems and/or that he cannot give an adequate explanation for their not having been investigated.

The study of friendship is a case in point. Social psychologists often speak of 'interpersonal attraction' rather than 'friendship'. This is not merely another use of jargon, but an accurate description of the central theoretical and empirical interests of social psychologists. In fact they have been relatively uninterested in different types of friendship and have fought shy of analysing the wide spectrum of love and liking relationships. Nor have social psychologists taken much interest in the origins, course and development of friendships. What crystallizes or breaks a developing friendship? What are the strains and conflicts which arise between friends and how are these resolved? How does a person's conception of friendship change with increasing age? Do males have a different conception of friendship than females? Do children, adolescents and adults have different conceptions of friendship? Do cross-sex friendships differ from same-sex friendships? Questions such as these have hardly been examined at all, even though they are among the most fundamental issues in the field.

Nor have social psychologists examined or tested the wisdom of writers, including poets and novelists, on the nature of friendship and love, that has accumulated over the ages. Such writers have been endlessly fascinated by the subtleties and infinite variety of liking relationships which form such a central part of the human condition. Take, for instance, the interesting assertions of La Bruyère, the French observer of human character:

Time, which strengthens friendship, weakens love.
Love and friendship are mutually exclusive.
The love that grows up little by little and by degrees is too like friendship to be a violent passion.
Antipathy may give place to love more readily, it would seem, than to friendship.
In friendship we notice only those faults that can injure our friends. In love, we notice only those faults in the loved one from which we ourselves suffer.

Rather than examining such insights or claims social psychologists have devoted their energies to two questions which are perhaps more elementary and certainly more subject to experimental control. First, under what conditions do persons come to have feelings of liking towards another? Second, what are the characteristics of friends as compared with persons who are not friends? Although both questions have been very narrowly interpreted and fairly crudely investigated, they do have some interesting implications for the study of children in school, especially since in recent years the relationships among the pupils, both formal and informal, have become of great interest to educators.

Feelings of attraction (and dislike) towards others are developed in relation to persons with whom we have some close personal experience. Whilst we may well have similar feelings towards persons we have not met, such as film stars, television and radio personalities and politicians, most of our likes and dislikes are directed towards people with whom we have some regular personal contact. Indeed, it seems to be true that it is difficult to maintain neutral feelings towards people with whom we interact on a regular basis. To like and dislike seems an inherent part of the human tendency towards the categorization and evaluation of others. One of the earliest propositions about liking and interaction is that of Homans (1951), who suggests that the more frequently two people interact, the more they will like one another. This is a very simple proposition which can be supported by a wide range of evidence.

In 1946 three American social psychologists, Leon Festinger, Stanley Schachter & Kurt Back (1950), in what has now become a classic study, investigated the friendship choices of married, war veteran students at the Massachusetts Institute of Technology. All were engineering students housed in a somewhat isolated community, creating greater constraints on the formation of friendships than would normally be the case. We shall be concerned with the part of the community called Westgate West, where the students and their wives lived in seventeen two-storey navy barracks, each containing ten apartments. The main finding was that the physical structure of the living accommodation makes a striking impact on the friendship structure of the students. Sixty-five per cent of all friendship choices are directed to persons in the same building, and of these choices over two thirds are given to persons living on the same floor of the building. If we consider the families living on the same floor, 60 per cent of their friendship choices are given to their immediate neighbours. This research is a clear demonstration of the *propinquity* or proximity factor in interpersonal attraction. The nearer two people are located in space, the greater the likelihood that they will come to like one another. Presumably this is because the nearer together two people are, the greater the *opportunity* for interaction and thus for developing feelings of liking. According to Homans, the more two people interact, the more they will like one another. The evidence seems to support Homans's proposition.

Further supportive evidence can be provided from educational contexts. Byrne & Buehler (1955) considered the impact of propinquity-opportunity on a group of students meeting three times a week for a psychology class. The students were seated in alphabetical order so that the seating pattern should be unaffected by previous acquaintanceship. At the first meeting 8 per cent of the students were acquainted with other class members, but only 3 per cent with those in adjacent seats. Twelve weeks later, 21 per cent were acquainted with members of the class as a whole, but 74 per cent were acquainted with neighbours. In a later study, Byrne (1961a) showed that it is not only acquaintanceship which increases between seat neighbours but also the intensity of the relationship. Neighbours were found to develop feelings of liking towards one another to a greater extent than non-neighbours. In another classroom study the French social psychologist Maisonneuve (1952) showed that pupils who occupy seats in corners or at the ends of rows tend to have fewer friends than pupils with seats in more central positions with greater potentiality for contact with neighbours. (When the pupils are seated alphabetically this finding cannot be explained in terms of the most introverted pupils choosing the less central seating positions.)

Thus far the evidence is in support of Homans's proposition, though the situation is perhaps better accounted for by three propositions rather than one. First, *the nearer two people are located in space, the greater their opportunity for interaction.* This proposition is, with obvious exceptions, generally self-evident. Under most conditions people cannot interact frequently if there is little physical proximity between them. The second step is to state that *the more opportunity people have for interaction the more they will in fact interact.* This is a more controversial proposition, for it is true that in some cases we do not make use of the available opportunities. One may, for example, regard one's neighbours in terms which will inhibit interaction—as foreigners, as socially inferior, as generally undesirable persons. We shall turn to this question later. If this proposition is allowed to have some general validity, then it leads to the third step, namely Homans's proposition that *the more two people interact, the more they will like one another.*

Certainly Homans's proposition seems to be generally supported by the evidence, though as in the case of the first two propositions there are also conditions in which the proposition does not hold. As Homans himself points out, it is not true of *all* interacting relationships. Where the two people are in an authority relationship—the boss and the subordinate—frequent interaction is by no means always accompanied by increased feelings of liking. Moreover, the evidence offers only *general* support for the proposition. In the Westgate West research not all neighbours liked one another. In Byrne's study a minority of students decreased their liking of students in neighbouring seats over time and increased their liking for non-neighbours. Further interaction with a person sometimes increases one's feeling of dislike. Because of this it might be more correct to suggest that the more two people interact the more intense their feelings of dislike or liking will become, and the more difficult it will be to remain neutral in one's feeling towards the other.

The weakness of this revised proposition is that it gives us no indication of the conditions that will give rise to a particular set of feelings. The strength of Homans's proposition is that it is more often true than false. (If we make an intensive study of the exceptions, we might be able to refine the proposition to take account of such exceptions.) In part this may be because we tend to like people more than we dislike them. Dislike may be a less common human phenomenon than liking. However, even where we come to dislike another, the proposition still contains an important truth, for the more we interact with a person to whom we have taken an initial dislike, the more we find that we discover new and likeable aspects of that person, leading us to revise our first impression. If, on the other hand, our first impression of dislike is confirmed on further acquaintance, we cease to interact with this person if possible. Strength is thus added to Homans's proposition by the fact that *the more two people dislike one another, the more they will avoid interaction with one another.* At the same time we can reverse the first proposition, as did Homans, and conclude that *the more two people like one another, the more they will interact.*

Having recognized the importance of propinquity or opportunity structure on the development of feelings of liking between two people, we must now consider why we do not frequently interact with all the persons with whom we have the opportunity to do so, and thus never develop a strong liking for them. Obviously we do not interact with some people simply because they fail to show any inclination to interact with us. The desire to interact must normally be to a fair degree reciprocal. Yet clearly from all the available people with whom we have high opportunities to interact we *select* some and reject others. It is no answer to say that we select those we like, for it is the liking which we are trying to account for.

An important part of the explanation concerns the *similarities* between people. When social psychologists investigated what distinguishes friends from non-friends (persons randomly paired), it was found that friends are more alike in certain respects than non-friends. This was shown to be true of values (e.g. Precker, 1952; Richardson, 1940) and attitudes (e.g. Winslow, 1937; Byrne, 1961b). In other words when two people share similar orientations to such matters as religion, politics, race, sex or any matter that is of importance to both, then they are more likely to be attracted to one another than if they had very divergent orientations. Many studies have researched into the significance of similarity of personality characteristics, but in this case the results have been inconclusive. We do not yet know in what situations similar personalities are mutually attracted and when 'opposites attract'.

We can now use the proposition that *the more similar two people are in attitudes and values, the more they will like one another* to account for why a person selects only some from all the people with whom he has high opportunity for interaction. A person evaluates others he meets according to the degree to which they are similar to him in important values and attitudes. Where there is similarity, he will tend to like and increase his interaction with them. Because the opportunity structure conditions potential interaction rates, a person is more likely to discover such similarities in those with whom he is brought into frequent contact, namely his 'neighbours'. At the same time, people who are in such close contact tend to be very similar in any case. Thus the people with whom we work tend to have similar jobs, a similar social background, similar education and training, similar styles of life. The same is true of residential location—we speak of a 'stockbroker belt', 'Jewish quarter', and 'solidly Conservative ward'. In a few cases occupation and residential location may be very closely related, as in the case of a group of houses occupied by policemen and their families, nurses' homes, students' halls of residence and so on. Thus the two factors of propinquity and similarity tend to operate simultaneously and in practice are quite difficult to separate.

271

From a symbolic interactionist position it is to be expected that people should choose their friends from among persons who are similar to them in many respects. Two people who share similar values, attitudes and interests are likely to offer support for one another's conceptions of social reality. Their sense of identity, the rightness of their values and attitudes, the significance and meaning of their activities, are all mutually confirmed and validated through extensive interaction. It would be somewhat surprising if they did not tend to like one another. We shall return to this point later in the chapter. Our present task is to examine the ways in which propinquity and similarity factors tend to operate hand in hand with respect to schoolchildren.

Let us consider the friendship patterns of the fourth-year pupils in a British secondary modern school (Hargreaves, 1967).

TABLE 8.1   *Friendship patterns in a secondary modern school*

(Figures are percentages)
Recipient

	Form	4A	4B	4C	4D	Others
Donor	4A	78	12	2	0	9
	4B	11	77	6	1	5
	4C	2	7	70	17	3
	4D	1	1	23	60	15

It can be seen from Table 8.1 that the majority of pupils choose their friends from the same year group. This is hardly surprising since schools are elaborately age-graded. It can also be seen that the majority of pupils choose their friends from members of the same stream or class. This presumably can be explained in terms of the opportunity structure. Each form is taught as a unit and remains together as an administrative and teaching unit for most of the school day. Since there is much greater opportunity for interaction with members of one's own stream than with members of other streams, it is not surprising to find that friendships are so heavily stream-based. However, this study also shows that within each stream there is considerable similarity in attitudes and values. Upper stream, particularly A stream, boys are more middle-class in their values and more favourable in their attitudes to school and to the teachers than are lower stream boys. So it could be argued that boys are making friendship choices according to the similarity factor. Obviously both similarity and propinquity factors are operating. The lower stream members have very similar values, yet there is a higher proportion of choices within each of these two lower streams than between streams. If similarity alone were at work, we should expect no greater tendency to in-stream than between-stream choice in the case of these two classes. If propinquity alone were at work, then there should be no greater between-stream choices for the C and the D streams than for the A and the D streams. Within each stream all the members had roughly equivalent opportunities for interaction, since the pupils were given freedom to choose their seating position. The structure of cliques within each stream shows that friendships developed between boys who were very similar in attitudes and values. When propinquity is held constant, it is the similarity factor which can be used as a basis for predicting which pupils will be attracted to one another.

The complex inter-relationship between propinquity and similarity factors is exemplified when we try to assess the impact of social class factors on friendship choices. In the United States some researchers (e.g. Bonney, 1946; Neugarten, 1956) have shown that among some, but not all, groups of school children there is a tendency to choose as friends members of a similar social class. In Britain the findings are very mixed. The studies of Oppenheim (1955) and Ford (1969) indicate that it is in the grammar schools where there is least evidence of a social bias in friendship choices. In Margaret New's (1967) study of several types of secondary school whilst there is little evidence of the selection of friends along social class lines, pupils did tend to choose friends with a similar score on a social class values test. This may explain why grammar school pupils fail to show a social class bias in friendship choices, since it can be argued that for a working class pupil to pass the 11 plus examination he needs to have values that are essentially middle class. In this case the majority of grammar school pupils, whatever their social class origins, would have similar class values and in this respect be relatively homogeneous. Since social class values would then not differentiate among grammar school pupils, they would have to select their friends according to other criteria of similarity. It is interesting to note that in the comprehensive school studied by Ford the greatest degree of in-group choice by middle-class children is in the lowest streams. Middle-class children in the higher streams are more likely to choose working-class pupils as friends than is the case in the lower streams, for the working class children assigned to the higher streams are precisely those children who under the tripartite system would have been awarded a grammar school place, i.e. those with middle-class values. Lower streams will consist of the less able middle-class children and working-class pupils with a markedly lower commitment to middle-class values. Since the lower stream pupils are heterogeneous with regard to social class values, we would not be surprised to find—as is the case—that here the middle-class children tend to choose one another rather than

273

working-class pupils. It is not known whether or not persons with different social class values have different conceptions of friendship, though this has been suggested (Ford, Young & Box, 1967). If this is the case, people of the same social class might be mutually attracted because of these similar conceptions of friendship in addition to their similarities with respect to other social values.

At the same time the selection of friends along social class lines may also be a function of a bias in the opportunity structure, since it may be that middle-class pupils are assigned disproportionately to the higher streams and working-class pupils disproportionately to the lower streams. Schools obviously show wide variations in this respect. In Ford's study such a bias is evident in all the schools, being particularly strong in the grammar school (90 per cent of the A stream being middle class and 80 per cent of the D stream being working class). In McDonald's (1969) research the bias is evident in the modern schools but non-existent in the grammar schools. In New's study there is the expected bias in the boys' grammar school but the *opposite* bias in the girls' grammar school. It is clear that the propinquity factor with respect to social class varies considerably from school to school and stream to stream.

Even in a school where every stream contains equal numbers of pupils from different social class backgrounds, a tendency to choose friends from members of the same social class cannot be ascribed to a similarity factor. Children from the same social class are likely to come from the same residential area. Such children have greater opportunity for interaction on the way to and from school, in local neighbourhood groups and institutions such as churches and youth clubs, and they will probably have attended the same nursery or primary schools. Friendships along social class lines may be more a function of past and current opportunity structures than of similarity in social values. In other words, social class may be merely a *correlate* rather than a *determinant* of friendship choice in school.

One American study by Gallagher (1958) has shown that pupils tend to choose as friends those pupils who live in close physical proximity to them, yet most other studies fail to take this factor into account. Many factors which can be shown to be associated with friendship choice, other than social class, might similarly be regarded as correlates rather than determinants. It is not easy to know how we can differentiate the one from the other. Unfortunately it is difficult to make sense of many of the available studies since most of them give findings on but a small number of the factors that are known to be associated with friendship choice.

The better researchers do demonstrate how the factors of propinquity and similarity are operating simultaneously. The study of Lundberg & Beazley (1948) shows that the friendship choices of college women were directed to those students with a similar social class background and with similar scholastic interests. It was also shown that students' choices were most strongly related to membership of the same residential house on the campus. One can account for studies such as this in terms of a limitation imposed by the opportunity structure within which similarity factors can operate. Thus members of the same house have high opportunities for interaction with one another. In addition, students who study the same subject(s) will have increased opportunity for interaction since they attend common classes together. In this case the opportunity structure reinforces the similarity of scholastic interest. It is most interesting to note that similarity with respect to scholastic interest was at its most powerful among science students who are more likely to spend time together in the course of their studies than are arts students because of their more frequent classes and practical sessions. Within the opportunity structure so created, the students can select their friends with respect to further similarities, less affected by the opportunity structure, such as attitudes and social values.

In schools the opportunity and similarity factors are particularly difficult to disentangle. Take, for example, the work of Norman Harrison (1970). The third-year pupils in this comprehensive school had been 'setted' for the basic academic subjects. Seventy-two per cent of the friendship choices of these pupils were directed to members of the same set; 66 per cent to members of the same previous primary school; and 69 per cent to members of the same social class. Each of these three elements—set, primary school and social class—is related to all the others. Fifty-eight per cent of the middle-class children came from one of the three primary schools and 80 per cent of the working-class children came from another primary school. Set was also related to social class, 59 per cent of set 1 being middle class. Since set and primary school are both related to social class it is not surprising to find that there is a relation between set and primary school: 73 per cent of the predominantly middle-class primary school are in Set 1. On the basis of these complex facts we can expect that the majority of friendship choices will show a similarity with respect to at least two of these three elements, set, primary school and social class. As can be seen from Table 8.2, 74 per cent of all friendship choices are directed to pupils who are similar with regard to two or more of the three elements, this figure exceeding what could be expected on the grounds of chance alone. Yet the factors of propinquity and similarity interpenetrate in such a complex way that it is difficult to assess the contribution of either separately.

It is important to recognize the potential significance of individual differences among the pupils. In Harrison's study it is clear that whilst the majority of the pupils appears to be influenced by the three

275

elements, some 6 per cent make friendship choices that are unrelated to the three elements. It seems that some friendship choices may be made on the basis of some obvious propinquity factor whilst others are not. In the same way some pupils may be affected by similarity of social class and other such similarity factors whilst other pupils are not. Even where the pupils appear to be so influenced it may be that in some cases the propinquity and similarity factors are simply coincidental—correlates but not determinants of friendship choice. From the available evidence we cannot tell. At present we know almost nothing about why and when particular individuals are influenced by certain sorts of similarity and propinquity but others are not. It is obviously a most important area for future research.

TABLE 8.2 *Friendship choice in a comprehensive school in relation to three factors*

	Observed frequency (%)	Chance frequency (%)
Same in all three factors	33	16
Same in two factors	41	41
Same in one factor only	20	34
Different in all factors	6	9

One interesting method by which we could consider the impact of similarity and opportunity factors would be through a more systematic investigation of the operation of similarity factors under different opportunity structures. In this respect the N.F.E.R. report, *Streaming in the Primary School* (Barker Lunn, 1970) is relevant, since it offers evidence on friendship choices in streamed and unstreamed schools. Children were divided into five ability groups on the basis of teacher assessments. In the streamed schools, as can be seen from Table 8.3,

TABLE 8.3 *Friendship choice in streamed and unstreamed schools*

Type of mutual pair	Streamed schools (%)	Unstreamed schools (%)	Chance expectation (%)
Same ability band	59	42	20
One ability band apart	34	33	32
Two ability bands apart	7	23	24
Three ability bands apart	—	2	16
Four ability bands apart	—	—	8

the vast majority of pupils choose their friends from a similar ability band, nearly two-thirds being within the same ability band. There are no mutual choices where the pupils are three or more ability bands apart. This makes sense, since as the pupils are streamed by ability, pupils who are greatly dissimilar in ability are most unlikely to be in the same stream. The findings are explicable in terms of the opportunity structure created by the streaming system. In the un-streamed school, however, no such limitation on opportunity is imposed by the organization for each class contains pupils of all ability levels. If friendship choices are unaffected by similarity of ability, then the distribution of friendship choices should be no different from those expected by chance. Although the spread of choices is wider in the unstreamed than in the streamed school, even in the unstreamed school the distribution departs considerably from that expected by chance. Children from the same ability band choose one another twice as much as can be expected by chance, and where the pupils are three or more ability bands apart mutual choices occur much less frequently than chance allows.

There are several possible explanations of this, though the report itself simply presents the facts. First, the opportunity structure within each class may not be as equal as the term unstreamed suggests. It may be, for example, that some teachers allow the pupils in each class to work in groups of similar ability, which would obviously create an additional opportunity for increased interaction among children of similar ability. Second, it may be that children in un-streamed classes choose their friends from a similar social class, either because of similarity of social values or because of common residential location. Since social class correlates with ability and attainment, pupils might appear to be choosing friends on the basis of similarity of ability when in fact their choices spring from a different source. If this were the case, then similarity of ability would be a correlate but not a determinant of friendship choice. (It is interesting that the report shows that some 80 per cent of the friendship choices were directed to children in the same social class or one level apart—and there were no differences in streamed and non-streamed schools.) Third, pupils may actually choose one another on the basis of similarity of ability quite independent of social class. This could have several sources such as common aspirations, achievements, rates of progress and so on. If this is the case, the findings suggest some interesting possibilities. In unstreamed classes the pupils rarely make friends if they are three or more ability bands apart; here the choices suddenly drop far below chance expectations. It is possible that at this point—three ability bands apart—that children erect a barrier of ability dissimilarity. In other words, pupils in the same ability band or one or two bands apart are perceived as falling within the area of similarity, but beyond this point they are perceived as being quite 'different'. If this were the case, any child's freedom of choice of friends is a function of his ability band. Pupils in the middle ability band can select their friends from all the other children in the class, for no pupil is more than two bands away from this central position. But for children at the extremes, in the top and bottom ability bands, choice of friends is restricted to three-fifths of the pupils in the class, for the remaining two-fifths of the pupils are in ability bands more than two categories apart from their own.

Although most of the research on friendship has been restricted to the impact of the factors of propinquity and similarity, and although we can come to no clear conclusions about the strength of these two factors relative to one another, our survey and analysis does warn against facile interpretations of data on the friendship choices of children in school. Take, for instance, the work on the friendship choices of pupils of different races. Lack of friendship choices among children of different races may spring from a particular opportunity structure. Sometimes this may be a product of the school's formal organization, as when the immigrants are segregated into separate classes or are disproportionately in lower streams. Sometimes this may be reinforced by external opportunity structures, as when children of one race live in one particular residential area or attend common neighbourhood institutions. On the other hand, children of the same race may choose one another as friends on the basis of a common language or of cultural similarities, even when the school removes its barriers to interaction opportunities. One cannot simply take the existence of a friendship cleavage between pupils of different races as evidence of inter-racial hostility or discrimination.

Again, some comprehensive schools seek to create friendships among children of different abilities and different social class backgrounds. As a minimum requirement they must provide pupils who differ in these respects with adequate opportunities for interaction. Any form of streaming will inevitably be inimical to such a goal, since it separates children of different abilities into different class groups, which often but not always have a biased social class composition. It is not enough for the headteacher and staff to claim that it is the tutorial group, fully heterogeneous in terms of ability and social class, which is the most important grouping in the school. However important it is in the eyes of the staff, it is unlikely to make any real impact on the friendship patterns of the pupils if the opportunity structure is weighted in another direction by other groupings which take up the majority of the school day, as the research evidence of Ford (1970) confirms.

The provision by the school of an adequate opportunity structure for children of different abilities and different social backgrounds will not of itself be sufficient to promote the development of friendship ties among such children. As we have already noted, pupils may

278

continue to select their friends according to certain criteria of similarity—and in this they may be encouraged by their parents—and so fail to avail themselves of the opportunities provided. It is here that the important experimental work of Sherif (1966) becomes relevant. In a series of experiments, a number of white, Protestant, middle-class boys aged twelve were taken to a summer camp. The boys were unaware that they were the subjects of an experiment, and were not aquainted before the experiment. For a while the boys lived together, during which period friendships developed. The boys were then divided into two groups, living in separate cabins, in such a way that the majority of each boy's original friends were now in the other group. Within a short time all the boys made new friends within their own group. As they lived, worked and played together, each group developed its own norms, jargon, jokes, secrets and nicknames, and also a status structure with leaders and lieutenants. At this point, by means of games and other activities, the two groups were given a competitive relationship. This became associated with marked in-group solidarity and hostility towards the other group. Sherif then tried to restore the two groups to a friendly and co-operative relationship. Several ploys, such as the making of appeals for co-operation or arranging conferences between the two leaders, and including attempts to maximize contact between the two groups, failed to achieve this goal. Co-operation and friendly relationships developed only when the two groups were brought together under what Sherif calls a superordinate goal, by which the groups had to co-operate if either group was to achieve its goal. For example, Sherif arranged an 'accidental' breakdown of a truck on a picnic and a failure of the water supply, problems which could be solved only when the groups co-operated. The result was a reduction in the hostility between the groups and, which is very relevant to this chapter, a restoration of friendship ties between members of the two different groups. The lesson for schools is obvious. If teachers want to promote friendships among children of different organizational groups or among sub-groups of friends with similar abilities and social background, then competition among such children must be avoided and co-operation in pursuit of superordinate goals must be actively stimulated. This is not an easy task for teachers, who too often are quite unaware of the friendship structure among the pupils, who also often believe that the provision of an adequate opportunity structure will solve the problem if it is recognized, and who in some cases believe that the social relationships and friendships among children of different abilities, social class and race are irrelevant to the school's functions as long as no major conflict becomes overt.

The perception of liking

Whilst social psychologists have devoted scant attention to some of the practical issues discussed in the last section, they have undertaken some fairly extensive investigations into other aspects of interpersonal attraction. Much of this work has been concerned with the conditions under which the individual responds with liking to another, and with individual psychological phenomena associated with feelings of liking. This is largely because such problems are more subject to experimental control than are friendships in real life. Indeed, some experiments do not involve a second real person at all. The subject is presented with information about another person, a 'stranger' who does not in fact exist, and is then asked to estimate whether or not he thinks he will like him and to what degree. In spite of the obvious artificiality of such experiments, some interesting and suggestive results have emerged. It is known, for example, that if Person perceives Other as similar to him, then Person will tend to like Other (Byrne, 1961c). Also if Person likes Other he will perceive Other as being similar to him (Broxton, 1963). Indeed, if Person likes Other he will overestimate the degree to which Other is similar to him, though this is true only in matters which Person regards as important (Byrne and Blaylock, 1961). Thus if Person has strongly held political views but is indifferent to religious affairs, he will perceive Other as being more similar to himself in political but not in religious attitudes.

Similar phenomena have been demonstrated with regard to reciprocity of liking. If Person likes Other, he will perceive that Other likes him (Newcomb, 1961); and if Person is given information that Other likes him, Person will tend to like Other (Backman & Secord, 1959). In the former case, Person again tends to overestimate the degree to which Other likes him (Tagiuri, 1958). It seems that we like those who like us and expect those whom we like to like us in return.

A few experiments have been conducted into other bases for attraction, though this research is still in its early stages. It seems that if Other evaluates Person favourably, then Person will tend to like Other, whether or not the evaluation is a correct one (Jecker & Landy, 1969). Perhaps flattery is more effective than the popular adage suggests. It has also been shown that if Person can be persuaded to do a favour for Other, then Person will come to like Other (Deutsch & Solomon, 1959). Presumably this is because normally we do favours only for those whom we like, so if we allow ourselves to be persuaded to do a favour for someone to whom we have neutral feelings, then we justify having done so on the grounds that we really like him. To make this justification we have to increase our attraction to that person. This suggests that we may be able to promote feelings of liking among children who are not similar—in ability, social class or race—by placing them in situations where they are encouraged by the nature of the situation to help and do favours for one another. The research supports our earlier suggestion for co-operation in relation to a super-ordinate goal.

Theories of attraction

A good theory is one which on the basis of a small number of postulates can predict a wide range of phenomena. Theories in social psychology, as in the human sciences in general, are relatively few or restricted in range, and do not usually meet the more strict criteria of a theory as used in the natural sciences. One of the major theories in social psychology, *balance theory*, was first formulated by Heider (1946, 1958) but has since been adapted and extended in a number of ways by different writers. One of the main advantages of the balance theories is that they can be usefully applied to several fields within social psychology. In the main they have been used most within the area of attitude change, but Newcomb, in his own version of balance theory, has shown that it can throw light on various aspects of interpersonal attraction. It is not possible within a short space to present Newcomb's extensive treatment of the theory as given in his book *The Acquaintance Process* (1961). We shall be concerned only with the most elementary form of the theory.

Balance theory analysis is concerned with two persons, Person and Other, and a third entity referred to as X which can be a third person, an object, an idea or an event. These three elements are linked together, diagrammatically in the form of a triangle, with positive and negative signs to indicate the nature of the relations between the elements (Fig. 8.1). There are two types of triangle, balanced and unbalanced. If there are three positive links $(+ + +)$ or one positive and two negative $(+ - -)$ then the relationships are balanced. If there are three negative links, or one negative and two positives, then the triangle is unbalanced.

Let us illustrate with reference to Diagram 8.1. P and O represent two schoolboys and X represents school. The positive and negative signs represent like and dislike respectively. In the first of the balanced triangles, P likes O and both P and O like school. This obviously 'feels' balanced in the sense that there is no strain among the elements. In such a relationship, all should be well. In the first of the unbalanced triangles, P likes O, P likes school, but O dislikes school. This obviously feels unbalanced since there is likely to be some strain among two friends who like one another but who disagree in their attitude to school.

Since there is tension in the unbalanced relationships, one might posit some force within the relationships towards the creation of

281

balance. (Any reader who is familiar with theories in general psychology will recognize that balance theory is a form of equilibrium or homeostatic theory.) Newcomb calls this force the *strain towards symmetry*. For example, in the case of our first unbalanced triangle, balance or symmetry can be restored by changing the nature of one of the links. Balance is achieved if O will like school (balanced triangle 1); if P will dislike school (balanced triangle 2); or if P dislikes O (balanced triangle 3). Balance can also be restored, of course, by the more complex task of changing two links rather than one.

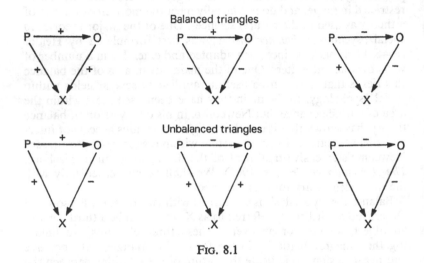

FIG. 8.1

X can also be made to represent a third person. The same processes of balance apply. We shall have a balanced relationship if Tom likes Dick, Tom likes Harry, and Dick likes Harry; or if Tom dislikes both Dick and Harry but Dick likes Harry; or if Tom likes Dick, but both dislike Harry, and so on. A lack of balance arises in such cases as when Tom likes Dick and Harry but Dick dislikes Harry. In these unbalanced triadic relationships there will also be a strain towards the restoration of symmetry.

Much of the earlier discussion on the importance of similarity of attitudes and values in friendship formation can be fitted into the theory, if X is made the object of some attitude of P and O. If two friends are dissimilar in certain values or qualities, or in their attitudes to a third object or person, then their friendship will be under strain and there will be a pressure to restore balance by developing a common attitude to X or by ending the friendship. The theory would thus predict the relationship between similarity and attraction.

282

Of course, in practice balance may be restored in other ways than by changing the signs of the links. For example, if X is a source of disagreement between P and O then the lack of balance can be removed simply by making X *unimportant* or *irrelevant* to the relationship. If an ardent socialist marries a committed conservative then we are likely to have the situation of unbalanced triangle 1, if X is made to represent the Socialist Party. The strain could be resolved in the usual ways—by P persuading his wife to become a socialist (balanced triangle 1); by O persuading her husband to become a conservative (balanced triangle 2); or by P disliking his wife and asking her for a divorce (balanced triangle 3). No doubt all three things happen in such marriages, though obviously if they married at all they were acting against theory by creating an unbalanced triangle. But given that they do marry, hopefully because they share many other common attitudes, they can restore balance to the political triangle simply by agreeing to make politics irrelevant to their marriage and deciding not to discuss politics at home.

Many common situations can be subsumed under balance theory which permits many predictions about the similarity of friends in their attitudes to a third element, be it a person, a thing or an idea or event. Unfortunately the theory has not in its present form as much predictive power as one would like. In particular, it is not possible to predict in which specific way an unbalanced relationship will change in order to achieve symmetry. Presumably it is the weakest link that will change. In our political marriage, if P and O are very much in love and thus reluctant to part, and O is a less committed conservative than P is a socialist, then presumably the O–X link is the one which is most likely to change. Is it only when all the links are very strong that an unbalanced relationship is likely to resolve itself by making X irrelevant? We do not know. But certainly here is an exciting field for an enthusiast to suggest the conditions and ways in which balance will be restored.

The theory as I have stated may not seem to get us very far beyond stating obvious facts in a new way. Two things must be remembered. Firstly the theory can be applied to many different areas within social psychology which without balance theory would have few theoretical connections. Secondly, Newcomb in his own book has shown how the theory can be used to predict whole sets of complex relationships in the area of friendship formation. If you wish to test the potentialities of the theory, try to apply it to some of the experimental results given in the previous section. With a little ingenuity and skill you will discover that the theory will correctly predict what has been demonstrated.

Newcomb's own major study was a real life experiment. Seventeen men, who were complete strangers to one another, were admitted as students to the University of Michigan. They all lived in one campus house free in exchange for filling in questionnaires regularly for Newcomb. He obtained some biographical and attitudinal information about them before their arrival at the University and tested them on their attitudes and friendships at regular intervals during the first term. Newcomb used balance theory to predict various aspects of the relationships among these seventeen men, with special reference to their feelings of attraction.

He found that some of the background variables such as age, college subject, rural or urban origin, correctly predicted some of the early friendships which developed in the first few weeks. Also important at this stage was the propinquity factor—who shared a room or lived in close proximity within the house. But agreement in attitudes tested prior to arrival was much less successful as a means of predicting friendships. In other words, similarity of values was not strongly linked to friendships during early acquaintance. However, by the end of the term seventeen weeks later, background and propinquity factors had become much less effective in predicting the friendship structure than were values and attitudes. This suggests that in the early weeks the students were attracted to one another on the basis of rather superficial similarities and propinquity. As time passed they came into interactional contact with those students who held similar basic attitudes and values and became increasingly attracted to such persons whilst withdrawing from their initial friends. When the period of exploration and discovery was over by the end of the first term, friendships were very much in line with the similarity–attraction hypothesis and with balance theory. It must be said that in some ways this is a disturbing study. Most of us like to think that we choose our friends freely. Newcomb's work shows that if one knows the basic attitudes and values of a person and of all the persons with whom he has high interactional opportunity, as it were his pool of potential friends, then whom he will in fact become friends with can be predicted with a striking degree of accuracy.

Although balance theory can be said to have 'explained' why persons with similar values and attitudes are attracted to one another in the sense that it can make a successful prediction from the P–O–X triangles, one might well feel that such an explanation is inadequate. An account which is rather more in line with the popular rather than the scientific conception of an explanation is available from exchange theory. On this view there are several reasons why similar people should be attracted to one another. In the first place, if two people share similar values then they are more likely to be able to chat agreeably about these values and allied matters and to share interests that are compatible with such values. In addition they may be able to participate in activities associated with the common values. For example, if two people share common aesthetic values, they can discuss the books they have read or visit art galleries and theatres together. In this way their behaviour is mutually rewarding to a high degree and both can make a profit. Further, in sharing these common values, interests and activities, they provide consensual validations for one another, for they will agree on many matters that are perceived as important by both parties. Not that in order to provide consensual validation they need to agree in all matters of importance. Often the mere agreement that a matter is itself of importance followed by disagreement about details is a sufficiently rewarding aspect of one's needs for consensual validation. In this sense agreement that poetry is an important art form can be more significant than agreement about the merits of a specific poet. As La Bruyère noted in the seventeenth century:

> The delight of social relations between friends is fostered
> by a shared attitude to life, together with certain differences
> of opinion in intellectual matters, through which one is con-
> firmed in one's own views, or else one gains practice and
> instruction through argument.

Another basic explanatory idea that can be derived from exchange theory is that persons with similar attitudes are likely to find one another relatively predictable. On the basis of shared attitudes and values each can within limits predict how the other will react in relevant situations. This reduces the costs of interaction since misunderstandings, disagreements, disappointments and rebuffs can be more easily avoided, with the effect that profits are maximized. And who will not respond with liking towards those who make our interaction with them pleasurable?

Sometimes it is possible to combine exchange theory and balance theory in explaining the facts. For example, one might argue that the weakest link in an unbalanced triangle is the one which involves the lowest cost to the relationship. Consider the political marriage we discussed earlier. P and O are in love but disagree in their attitudes to politics. Their disagreement here will not break the mutual attraction, for they are in agreement on so many other matters. The rewards they obtain in this respect greatly exceed the costs of political disagreement, so the marriage is not threatened and we can exclude a change in the P–O link. Yet they are equally fervent in their support for their respective political parties. The obvious answer is to make politics irrelevant and to decline to discuss them at home. Exchange theory, however, suggests alternative solutions. If during a political discussion it is more costly for O to take the 'utter rudeness' of her husband than it is costly for him to tolerate what he calls his wife's 'unintelligent repetition of illogical propaganda' because P quite likes a good argument but O prefers not to get involved in a row, then it is more likely that O will change her political allegiance to achieve the symmetry of marital peace and quiet than it is likely that P will change his. In other words, exchange theory leads us to consider the relationship between P and O in a more general way, and on the basis of the rewards and costs involved in the relationship as a whole permits us to make predictions about the restorations of symmetry in specific situations.

Finally exchange theory offers an interesting interpretation of a problem inherent in Homans's two propositions discussed above. If it is true that the more two people interact the more they will like one another and it is also true that the more they like one another the more they will interact, then why do these two people not find themselves locked in a permanent interaction of mutual liking? When, in other words, does the spiral stop? According to exchange theory, an interaction is likely to continue as long as it is highly profitable, more profitable than alternative interactions which must be given up if the first interaction is to continue. Our two fictional characters do not get caught in the spiral because at a given point the profits to be made in an alternative interaction with someone else exceed those which will ensue from a continuation of the first. Young lovers spend so much time together because their exclusive interaction is the most profitable available—evidently often by a very large margin. For the rest of us, a good estimate of the strength of our liking for another might be the amount of time we either do, or would like to spend with that person. This gives little consolation to the wife whose husband spends most of his evenings in the pub and comes home only to eat and sleep. But then she does not need exchange theory to tell her the score.

Varieties of liking

So far in this chapter I have, like most social psychologists, tended to assume that liking or attraction is a relatively simple and meaningful dimension, varying only in degree. In spite of our loose usage of the term—I can like jelly and I can like my wife—this is not a very reasonable assumption since there are good reasons for thinking that there are many varieties of liking between persons, even if we exclude sexual attraction. Once it is conceded that liking is not a unitary concept, which can be reduced to some 'affiliative tendency' or desire to associate with others, then we must attempt to distinguish the possible varieties of liking and then examine their reality in human experience, investigating the sources and consequences of such different forms. The most significant contributor in this respect is Newcomb (1960), who has suggested five basic varieties.

1   *Admiration*—when Person attributes to Other certain 'inherent' qualities that Person likes. For example, if Other is a jovial person and Person approves of joviality, then Person will like Other.

2   *Reciprocity*—When Person attributes to Other positive attraction to Person. As we have seen earlier, if Person believes that Other likes him, he will also tend to like Other.

The next three varieties concern the attribution to Other by Person of attitudes which are similar to Person's own attitudes. Person will find such support for his attitudes since they support his conception of himself and his conception of reality.

3   *Role support*—when Person and Other participate in activities that are rewarding to both, as when two people play golf together. This form of liking should be shown by the teacher towards the pupil who enjoys his role and supports the teacher's conceptions of the teacher and pupil roles.

4   *Respect*—when Person values Other's expertise and skill and knowledge. This might typify the pupil's liking for the teacher.

5   *Value support*—when Other shares values that are similar to those held by Person, thus supporting Person's conception of social reality.

My own experience suggests that we can add several varieties to Newcomb's list:

6   *Gratitude*—when Other has helped Person in a number of ways.

7   *Mutual obligation*—when Person and Other have helped one another and done favours on a mutual basis.

8   *Appreciation*—when Person finds Other inherently likeable.

9   *Acceptance*—when Person feels relaxed and able to be fully himself in the presence of Other.

These varieties may or may not be experientially distinct. My own crude attempt some years ago to distinguish a husband's liking for his wife from his liking for a close male friend in these terms was not successful. My hopes of making a claim to fame based on the Hargreaves Love Test were dashed, at least temporarily, since both relationships included all the above nine dimensions, though the husband–wife relationship was marked by a greater degree of intensity. Nor do we know at present whether these different varieties do have distinctive origins and distinctive consequences. It does not seem unreasonable to hypothesize that different liking relationships will consist of different complex combinations of several varieties, or that our conceptions of liking will change and grow increasingly complex as we pass from childhood, through adolescence, into adulthood. The youth of social psychology is betrayed by the fact that these interesting tasks have yet to be solved.

Recommended reading

F. HEIDER, *The Psychology of Interpersonal Relations*, Wiley, 1958.
T. M. NEWCOMB, *The Acquaintance Process*, Holt, Rinehart & Winston, 1961.

9 Groups

On several occasions it has been pointed out that within the field we call social psychology there is a constant tension between what might be regarded as the 'individual' and the 'social'. Sometimes the focus seems to be largely on the individual person; at other times it is the social relationship which is under analysis and the individual as such is much less evident. Up to this point in this book we have been mainly concerned with social psychological issues as they relate to the individual or to two persons (the dyad). The chapters on the self and on person perception, for example, gave major emphasis to the individual in his relations with others and to social psychological phenomena or mechanisms as they affect the individual. Later, in the chapters on role, interaction, and friendship formation, we began to change our perspective away from the individual and towards a more direct study of the personal relationships between two people. Even in the discussion of teacher–pupil relationships in the classroom the pupils were often treated as a unity and the analysis of relationships between teachers and pupils tended to follow a pattern set by dyadic relationships. Now we have arrived at the most 'social' area within social psychology, the study of relationships within groups.

Though by common consent the social psychological study of groups is referred to as 'group dynamics', it is surprisingly difficult to say precisely what is meant by the term *group*. It would not be very profitable to enter too deeply into the question of definition, but we must take the first few steps into this difficult terrain if we are to be clear what it is that social psychologists are seeking to study. One important kind of group is what might be called a sociological group or a *category*. We can group people together, in an infinite variety of ways, by members' common possession of a given characteristic. In this sense, all the following are groups or categories—males, adolescents, doctors, cricketers. Here people are classified according to a criterion of sex, age, occupation, and leisure activity respectively. But these categories are clearly not groups in the same sense of the term group as used in the following.

A group of friends met in the coffee bar.
A new cure for influenza has been discovered by a group of research scientists.
The pupils asked Mr Smith to be in charge of the Political Discussion Group.

These are examples of what the layman would call 'real' groups, whereas categories are logical groups. Sometimes they have a very real existence and members may be aware of their membership in a category, as in the case of doctors, but this need not be the case as in Sprott's (1958) example of 'red-haired archdeacons'. Groups of this latter variety exist only on paper, though these categories are often of major importance in sociological analysis. Social psychologists are little interested in categories as such, except in so far as membership in a category may affect the relations of members in a small group of interacting persons.

Groups as studied by social psychologists can also be distinguished from *social organizations*, such as a school, a factory, a village, a political party. These social organizations consist of complex networks of small groups. Sometimes they have a common spatial location (e.g. Dartmoor prison) but sometimes they do not (e.g. the Quakers). In becoming a member of a social organization one usually, but not always, also becomes a member of a small group. A teacher who joins the National Union of Teachers does not only belong to the massive social organization of thousands of teachers but he also belongs to a local branch and a small group of members within his own school. It is very difficult to say when a group develops into a social organization or to distinguish a large group from a small social organization. For the most part social psychologists have left the analysis of organizations to sociologists, though often the groups which social psychologists study are a part of a social organization.

Many writers have suggested ways in which different kinds of groups can be distinguished. Among the most famous is Cooley's (1909) distinction between 'primary' and 'secondary' groups. Primary groups are small, intimate groups of persons in face-to-face relationship, where there is a strong sense of unity and mutual identification. The nuclear family of parents and children is the archetypal primary group. In the secondary group, which tends to be larger, relationships are cool, impersonal, rational, contractual and formal. Although Cooley's two types of group have had considerable popularity within the social sciences, they have proved to be of limited value. For our purposes, the distinction does not help us to clarify what sort of group it is that social psychologists are concerned to study.

The group, in the sense used by social psychologists in the field of group dynamics, has five basic characteristics. Unless all five criteria are met, there is no group, and we might refer instead to a 'collectivity' or 'aggregate' or 'near-group'.

1 *A group is a plurality of persons.* That is, two or more persons are needed to form a group. Some authorities would insist that a group must consist of at least three persons, since there are very fundamental differences between the dyad and groups of three or more persons. The criterion of plurality has another limitation; it does not state the upper limit to the plurality. The importance of the upper limit is clear when we consider how size affects the number of member relationships. In a dyad there are only two relationships (A–B and B–A). In the triad there are six (A–B, A–C, B–A, B–C, C–A, C–B). In a group of twenty persons, there are 190 relationships. At some point, then, there is an upper limit to the group, partly because beyond a certain point it will become rather like a social organization with sub-groups within it, and partly because beyond a certain size the group will fail to meet the remaining four criteria. The size of a group is in fact of major significance, since size affects all the group properties which follow. The reader might like to suggest in what way.

2 *The members are in face-to-face relationship, and aware of their common membership.* This is really tantamount to saying that the members must interact, that is, their behaviour must be reciprocally contingent. It is not easy, though not impossible, for persons to interact if they are not in regular face-to-face relationship.

3 *The members have common goals or purposes.* Groups have one or more goals that are common to most members. It is the goal which unites and draws them together as a group. Of course, members often have additional and personal goals. An adolescent may join a youth club not only because he wishes to follow the group goal of enjoying his leisure in certain ways but also to achieve his private goal of becoming a powerful officer of the group or of keeping his Probation Officer quiet. In some cases, the member may not share the group goal at all, but may join the group merely to satisfy an individual purpose. An adolescent girl may join the youth club simply to satisfy her loneliness. However, where people join groups to satisfy individual purposes, they usually come to share the group goals for reasons we shall see in a moment.

Because the members share common goals they are attracted to the group. Groups obviously vary in their attractiveness to members. Groups which satisfy the members' needs will be very attractive. If the group fails to attain any of its goals or does not meet members' needs then its attractiveness will fall, and if the attractiveness declines below a certain point, members will leave the group where this is possible. Usually the attractiveness of groups is referred to by social psychologists as their *cohesiveness*. This is a good term since it suggests the degree to which members will stick together. Cohesiveness can be measured in a variety of ways, such as the resistance of members to leaving the group.

4 *Members subscribe to a set of norms*. Norms are standards of behaviour which specify the conduct expected of members. They determine what is 'proper' within the group. Norms are required to regulate the behaviour of members and one of the most important functions of norms is to allow the members to achieve the group goal. Behaviour which is in line with the norms is rewarded; behaviour which deviates from the norms is punished. One of the reasons why persons who join groups for an individual goal come to share the group goal is that the norms spring from the goal in order to facilitate its achievement. It is easiest to conform to group norms if one is committed to the group's goals. Even if a member does not share the group goals, he must act as if he does, for if he does not conform to the norms, he will be punished by, or expelled from, the group. Conformity to norms is related to other group properties. It has been shown, for example, that the more cohesive a group is, the more members tend to conform to the group norms.

5 *Members are differentiated into a structure*. The members are not completely homogeneous. Members develop expectations of particular individuals. In other words, members come to perform different *roles* within the group. Moreover, as we shall examine in detail, some members are liked more than others and some members exert more influence than others. Some members become what we call 'leaders'.

These five properties or criteria certainly help to clarify what social psychologists mean by the small group. They also serve to introduce us to the ways in which social psychologists try to analyse groups. But at the same time it must be admitted that the five properties fail to give a clear and adequate specification of a group. Let us consider a class of children who have just been admitted into a secondary school. A few of them know one another and were friends at a previous school; many are strangers. They have been assigned to the class randomly by the teacher. As they are, they are probably not a group in the social psychologists' sense, for they do not meet the five properties. If we look at this same class of children a year later, we may find that they have become a group in that they now meet all the five properties. At which point did they cease to be an 'aggregate', that is, a collection of persons in common spatial location, and become a group? The same problem occurs when we try to specify under what conditions we can regard the people standing in a bus queue or the patients sitting in a doctor's surgery waiting room as a group. These are questions that are fun and instructive to try to answer.

Although the five properties give us a rough definition of the meaning of the group as studied by social psychologists, it would obviously be useful if the properties could be used in a systematic way of classifying the many thousands of groups to be found in real life. Several approaches to dividing groups into different types are possible. One might, for example, classify groups according to how they vary with respect to each of the five properties. This would require us to analyse each property in greater depth. Take, for instance, the property of the group goal. Groups might vary according to the *clarity* of the group goal; the degree to which there is *consensus* among members about the group goal; the degree to which the goal is *specific or diffuse*. On this last point we might draw on March & Simon's (1958) useful distinction between non-operational goals where it is difficult to assess whether or not the goal is being achieved (e.g. the teacher goal of 'educating pupils to the full development of all their potentialities') and operational goals, where the goal is much more specific and its attainment is more open to accurate measurement (e.g. the teacher goal of getting pupils to pass national examinations successfully).

A similar approach would be to create a longer, more detailed list of group properties or dimensions along which groups can be expected to vary. Hemphill & Westie (1950) have suggested fourteen dimensions of group behaviour.

Autonomy—to what degree does the group function independently of other groups?

Control—to what degree is members' behaviour regulated?

Flexibility—to what degree are procedures rigidly structured?

Hedonic tone—to what degree is membership associated with pleasant feelings?

Homogeneity—how similar are members' characteristics?

Intimacy—how familiar are members with personal details about one another?

Participation—how much time and effort do members devote to the group?

Permeability—to what degree does the group provide ready access to members?

Polarization—to what degree is the group orientated towards a single, clear goal?

Potency—what significance has the group for members?

Size—how large is the group?

Stability—to what degree does the group persist over time?

Stratification—to what degree is the group structured into hierarchies?

Viscidity—to what degree do members function as a unit?

Although studies such as this have suggested ways in which the

293

dimensions are related (e.g. small groups tend to be more flexible than large groups; groups with high viscidity are high in hedonic tone, etc.) they have not provided a neat typology into which we can classify groups, so let us turn to the more productive aspects of the study of groups.

If we wish to study the behaviour of people in groups, we have to begin our investigation at some particular point. We cannot look at the whole group life at the same moment, no more than we can understand the life of an organism such as a frog unless we reduce the complex whole into smaller parts. However, in dissecting the group into convenient 'bits' which can be analysed separately, we run the risk of losing sight of the significance of the whole. For preference, then, we would make as few basic 'bits' as possible. In the biological sciences the basic division is between anatomy and physiology. For social psychologists, one of the most simple means of analysing group behaviour is to consider the group as consisting of a *culture* and a *structure*. Although this is a very useful analytical distinction, the two interpenetrate in very important ways.

Group culture

When we consider the culture (or ideology) of a group, we are mainly concerned with the fact that groups have values, beliefs and norms. The focus is on the *homogeneity* of the members; we are stressing what they share in common. A group's *values* are the overall guides to group behaviour, for it is the values which express what members regard as good, ideal and desirable. Group values are thus closely related to group goals. A religious group, for example, will have among its basic values the perfection and salvation of all its members and the conversion of non-members to these ideals. Among a group of teachers in school, basic values will be concerned with the intellectual, moral and social training of the pupils. These values will sometimes vary in different schools. Schools of a religious foundation, for example, tend to contain teachers with a higher valuation of religious and moral education than is the case in secular schools.

Values express themselves in the *beliefs* of members of the group. Beliefs are more specific than values. From the value 'religious education is desirable' flows a set of beliefs such as 'it is important for pupils to study the life of Christ', 'pupils should be taught the catechism' and 'prayers in school are an essential part of the educational experience'. When we remember that there are different religious groups, Christian versus Hindu, Roman Catholic versus Methodist, then it is clear that the same basic value—e.g. religious education is desirable—often leads to quite different sets of beliefs. Many teachers share the value that 'more education is desirable' but it is a smaller number of teachers who on the basis of this value will share the belief that 'children should be given a longer period of compulsory education'. I have said that in examining the group culture, the accent is upon the homogeneity of members. With respect to values and beliefs, members of a group tend to share the same or very similar values and beliefs. This is not to say, of course, that there will be no disagreement amongst members, but rather that such disagreement will be in matters of detail about accepted values, or it will be disagreement about values which are irrelevant to the group. There are several reasons why group members tend to share common values. In the first place groups are selective in their membership. 'Birds of a feather flock together'; people want to join groups which are known to possess values congruent with their own values. An ardent atheist does not usually want to teach in a Church school, and even if he does so, the school is unlikely to appoint him if there is a choice of candidates. Further, groups tend to be exposed to common information and ideas compatible with group values. In this sense the group acts as a filter on incoming information and rejects what is inconsistent. A Conservative Club does not usually subscribe to socialist periodicals. Even when information that is opposed to group values penetrates into the group, members interpret the information in the light of group values. A socialist pamphlet in a Conservative Club will be held up as an example of socialist 'errors' and 'lies'. It can be said, then, that there is a group interpretive scheme which is common to its members. When they are with the group or their membership in that group becomes salient, it is this interpretive scheme that will be activated. An individual's unique interpretive scheme is in part created and certainly supported by the (overlapping) interpretive schemes of the various groups to which he belongs. Once having developed a reasonably coherent set of interpretive schemes, the individual is then predisposed to favourable attitudes towards groups with compatible values and beliefs and interpretive schemes. The problems can become acute when involuntarily he is compelled to join a group that is incompatible in these respects.

Group *norms*, which flow from the group's values, are specifications of the beliefs, feelings, perceptions and conduct that are proper to group members. Norms are sets of shared expectations applied to the whole group and their purpose is to regulate conduct within the group so that the group goal may be achieved. Thus norms are a form of social control. Norms must be enforced, so they are backed by *sanctions*; rewards are given to members who conform to the norms and punishments to those who deviate from the norms. Sometimes the norms are formal, as when they are expressed in the rules and regulations of the group, but often they are informal. In the latter case the norms and sanctions play a subtle but important part in the life of the group and close observation is required before they can be specified.

Members of groups typically find it extremely difficult to explain to an outsider what the norms of the group are. This is partly because to members the norms are simply part of the taken-for-granted aspect of life in the group and partly because norms are often not stated explicitly. So if a teacher tries to 'observe' the norms of a group of pupils, a difficult and delicate task especially since the pupils may not behave normally if they sense they are being observed, then the teacher must take note of (i) how the norms are defined, what behaviours are expected or *not* expected, how the norms are expressed so that members become aware of them; (ii) how the members are constantly under surveillance by other members, how they monitor one another's behaviour to check on the conformity or non-conformity to the norms; and (iii) what the sanctions (rewards and punishments) are, how, when and by whom they are administered. This last point is particularly difficult for the observer, since the sanctions are very subtle. A smile or a nod from a leading member of the group can be most rewarding to other members, just as a slight frown or shrugging of the shoulders can be a signal of displeasure which warns the member concerned that he is departing from group norms. Extreme sanctions such as lavish praise, ridicule or ostracism are much rarer, though the observer can rejoice when he sees such reactions since they indicate that massive conformity or nonconformity are calling forth these sanctions. But in general it is much more difficult to observe and assess the significance of the smile, the frown and the mild teasing which are the constant feature of group life.

Most studies of life in groups do not give us a very extensive insight into the concealed intricacies of group norms. Rarely are we told of the extent to which members are aware of the norms; how much agreement among members there is about the norms; to whom and under what conditions and by whom the sanctions are applied; the degree to which the norms are rigorously enforced. However, one feature of norms which has received some attention is what has been called the 'latitude' of norms. Whilst norms prescribe the behaviour that is expected of members, they cannot specify in exact detail what is required. In other words norms can specify no more than what the Sherifs (1964) call 'the latitude of acceptable behaviour'. By this they draw attention to the fact that within a given norm there is a range of tolerable behaviour. There may be an ideal behaviour which is expected, but behaviours which fall short of the ideal may still lie within the acceptable range, though they will not result in as much reward or punishment as behaviour which falls nearer the upper or lower limits of the latitude of acceptable behaviour. A good example is based on work by Jackson (1960). Suppose some pupils are placed in a group to discuss a topic suggested by the teacher. A norm that members should contribute to the discussion will soon be evident, since if no-one contributes, there can be no discussion. The *amount* a particular individual speaks will be part of this norm. If the person says nothing, his behaviour may be disapproved. If he says little, the amount of disapproval shown to him will decrease, but in Figure 9.1

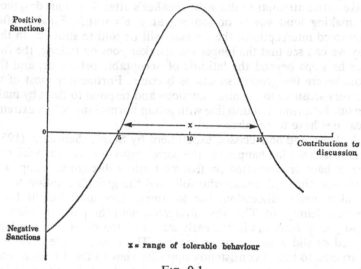

FIG. 9.1

—which is purely theoretical—the person who speaks for only 4 per cent of the discussion time is still not given any approval, on the grounds that he is not contributing a 'fair share' to the discussion. A person who speaks 5 per cent of the time has just entered the 'latitude of acceptable behaviour' and is treated with what can be called indifference, namely an absence of reward or punishment. From now on, the more he speaks, the more he is rewarded, until he reaches 10 per cent, which is the ideal behaviour and is the point of maximal reward. Speaking for more than 10 per cent receives less approval and such a person is seen by the members as taking more than his fair share in the discussion. Beyond 16 per cent, the speaker steps beyond the other limit on the range of tolerable behaviour and is likely to provoke various negative reactions from group members who see him as 'hogging' the discussion. Presumably most norms could be analysed in this way, though in practice it is very difficult to specify and measure the latitude of acceptable behaviour, though we do know that the more important an activity is to a group, the narrower will be the latitude of acceptable behaviour.

This last theoretical example of the group discussion also highlights the problem of observing sanctions. Just how do members react when a member talks too much in a discussion group? In my experience, the first negative reaction is to withdraw attention from the speaker, especially by declining to look at him. Later, members become more openly inattentive, by fidgeting or by looking elsewhere, such as through the window. If these sanctions are ineffective, members will make active attempts to draw the speaker's attention to his deviance, by making loud yawns or looking at one's watch. Finally, after attempted interruptions, the speaker will be told to shut up. In this way we can see that the longer the speaker goes on talking, the further he steps beyond the latitude of acceptable behaviour, and the more severe the group's sanctions become. Fortunately most of us are very sensitive to the mild sanctions and respond to them by making our behaviour fall into line with group norms long before extreme measures have to be taken.

A clever, and now classic, experiment by Stanley Schachter (1951) demonstrates what happens to the person who does refuse to fall into line. Schachter insinuated confederates into a discussion group. The *mode* was the confederate who followed the group's solution to the problem under discussion, the treatment recommendations for a juvenile delinquent. The *slider* disagreed with the position taken by most group members in the early stages of the discussion but later moved or slid with the majority view. The *deviate* confederate was instructed to take a consistently opposite view to the dominant one. When it was clear to the group members that the slider and deviate were failing to conform to majority opinion, they increased the amount of communication directed to them, presumably in an attempt to persuade them into accepting the dominant solution. When the slider accepted the persuasion, the amount of communication directed to him declined. The deviate, however, refused to slide. Still, more communication was directed to him but after a certain point, when the members concluded that he was not to be persuaded, communication to him ceased. However, it was clear from questionnaires given to group members after the discussion, that the deviate was not liked by the group members and was not regarded as a desirable discussion group member for future occasions. He was, in short, rejected by the group.

It is significant, however, that this clear rejection of the deviate was more readily expressed on paper than in the overt behaviour. I have already suggested that in observing real life groups it is rare to see the application of extreme negative sanctions. This is partly because there is great pressure exerted on members to conform and refrain from giving open expression to their deviant views or behaviour, and partly because the members who are likely to receive such sanctions, the non-conformists, do not remain long in the group. Whyte's (1943) description of life in a neighbourhood gang provides an excellent description.

In the spring of 1937 Nutsy was recognised informally as the superior of Frank, Joe, and Alec, but his relations with a girl had already begun to damage his standing. A corner boy is not expected to be chaste, but it is beneath him to marry a girl that is 'no good'. Nutsy was going so steadily with this girl that marriage seemed a distinct possibility, and, reacting to the criticism of his friends, he gradually withdrew from the gang.

In other words, expulsion of Nutsy from the gang was not necessary since the other members were able, by means of many less drastic sanctions, to make life sufficiently unpleasant for Nutsy that he voluntarily withdrew. Unfortunately Whyte does not tell us the specific content of the gang's 'criticism' of Nutsy.

It is very difficult to give an adequate definition of the term conformity and it is increasingly being recognized as an extremely complex concept (Beloff, 1958; Jahoda, 1959; Hollander & Willis, 1967). Certainly it is not a simple unidimensional continuum with conformity at one extreme and non-conformity at the other. In the popular view conformity is to some degree a pejorative term. The conformist is someone who follows the crowd because he does not possess the strength of character to do otherwise. So to be called a conformist is an insult since it implies moral weakness or failure. Conformity as used by social psychologists does not have these value-laden connotations, though it is still a term which is used very loosely and in different senses by different writers.

The first type of conformity to be distinguished is the consensus or agreement among group members in relation to a given value, belief or behaviour. For example, members of a group of clergymen are likely to believe in God. This is not surprising since they became members of the clerical fraternity precisely because they believed *inter alia* in God. In other words the uniformity of group belief arises from the selective nature of group membership. Persons who do not believe in God are not normally allowed to become clergymen. The uniformity is a form of conformity in that the clergymen's beliefs are similar or congruent. This uniformity, or *congruent conformity*, can be distinguished from the conformity which implies a change or movement in values, beliefs or behaviour. *Movement conformity* occurs when I hold one belief but then as a result of my membership in the group I come to hold a different belief. There may, of course, be a

299

variety of reasons why I have come to change my values, beliefs or behaviour owing to my membership in the group. It may be that I am anxious to gain acceptance or receive approval from other members and/or it may be that group members are exerting pressures on me to accept their influence towards change under threat of negative sanctions. Whatever the cause of my change, it is the fact of change which distinguishes movement conformity from congruent conformity—though in everyday life it is not easy to tell the one from the other.

One important form of movement conformity is called *compliance*, which arises when a person publicly changes his opinion or behaviour into line with the norms but privately holds a dissenting view. The man who tells his wife that she looks lovely when he is really thinking what an awful frump the boss will think he has married; the pupil who promises to do his homework more carefully when he knows that he has arranged to go to the cinema and will have to copy the work from his friend on the way to school next day; the teacher who agrees that a pupil should be expelled when he really feels that school ought to take responsibility for the child rather than passing the buck to some other school; all these persons are being compliant. They are giving in to the norm of expected behaviour in order to attain the rewards accruing from the deception or to avoid arousing the displeasure of others that will result from saying what they really think.

A second form of movement conformity occurs when a person changes in a given direction but genuinely believes in the rightness of the newly adopted values, beliefs or behaviour. This sort of movement conformity is usually called *internalization*, for the person has come to accept as his own the position he did not hold previously. Again, it is not easy in practice to distinguish the two forms of movement conformity. If all the members of the Conservative Club vote conservative in an election, it could be said that they are all conformists; but this would not distinguish the types of conformity involved or the motives of the individuals for conforming. We might have (a) those who vote Conservative because they have always been convinced, even prior to joining the club, that this is the best political party; (b) those who vote this way because they have been persuaded by their fellow members and now believe that this is the right way to vote; (c) those who so vote because it seems right to do so since they have enjoyed the social facilities of the club; (d) those who vote Labour but tell their friends they have voted Conservative so that they will not be forced to resign their membership. We have as yet no agreed system whereby these types, and the many others that can be postulated, can be clearly distinguished and categorized.

Persons who do not conform to group norms are usually termed non-conformists or deviants. Here again we have blanket terms that cover different types of deviance. First we have what might be called the *incongruent non-conformist*, the person whose values, beliefs or behaviour are simply not in line with the norms of the group of which he is a member. He is distinguishable in theory from the *independent* who does not follow the group norm in spite of group pressure, or perceived group pressure, or other forces directed toward changing his opinion. He declines to change and makes no secret of his non-conformity, though it is important to remember that non-conformity in this group may be conformity in another group. A third type, called the *counter-conformist* by Krech, Crutchfield & Ballachey (1962), is the person who compulsively takes the opposite view to the dominant one simply because it is the group norm. Like the movement conformist he is very susceptible to group pressure but in the opposite direction. Every school has its quota of counter-conformists among the pupils and staff.

There has been considerable research by social psychologists to examine the conditions under which people can be persuaded to yield to group pressure. Most of this research has been in rather artificial laboratory situations. The classic research on conformity is that of the American psychologist, Solomon Asch (1951). Nine students are seated in a semicircle as part of a psychological experiment. The experimenter presents two cards. On one card is a single vertical line, and on the other three vertical lines of different lengths, one of which is as long as the line on the first card. The task is to say which of the lines on the second card is of the same length as the line on the first card. The subjects announce their judgments in the order in which they have been seated. Since the judgments are very easy, all the students are unanimous. However, on the third trial one student near the end of the group disagrees with the other students. To this deviant the situation is very disturbing, for he does not know that the other eight students are confederates of the experimenter and that it has also been agreed beforehand that on certain trials the confederates will reach a unanimous but clearly incorrect judgment. The dramatic result of this experiment was that one-third of the 'naïve' subjects in this experiment followed the majority in making an obviously erroneous judgment in one-third of the trials.

Since one would expect the subjects to be aware of the gross discrepancy between their private judgment and the announced verdict, it is not surprising that in general they experienced considerable tension and anxiety. About one-quarter of the subjects remained independent throughout the experiment and refused to yield to group pressures. As Asch says of the independents:

The most striking characteristic of these subjects is the vigour with which they withstand the group opposition.

Though they are sensitive to the group, and experience the conflict, they show a resilience in coping with it, which is expressed in their continuing reliance on their perception and the effectiveness with which they shake off the oppressive group opposition.

At the other extreme is a group of subjects which yielded on most of the trials. Some of these were simply compliant; they followed the majority because of some overpowering need not to appear different from or inferior to others, but they still retained a discrepant private judgment. A few tried to rationalize their conformity by blaming one's own poor eyesight or by claiming that 'it looked different from where I was sitting'. In a few very extreme cases the subjects believed that they had made correct judgments and were unaware that their estimates had been distorted by majority opinion.

That intelligent university students should so frequently conform to majority opinion against the evidence of their own eyes has disturbing implications for the power of conformity in less dramatic situations in society at large. Further experiments were clearly needed. The most ingenious investigator to follow Solomon Asch has been Richard Crutchfield (1955) who ingeniously contrived an experimental situation in which all the subjects could be 'naïve' and thus avoided the expensive need for many confederates. Crutchfield's results confirm Asch's on perceptual judgments—about one-third of the judgments are conformist. However, Crutchfield also conducted some experiments with *opinions*. For example, when professionals and businessmen were led to believe that other subjects agreed with the statement *I believe we are made better by the trials and hardships of life*, 31 per cent followed the majority, whereas none of the control subjects—not under pressure to conform—agreed with the statement. Again, whilst only 19 per cent of students in a control situation agreed with the statement *Free speech, being a privilege rather than a right, it is proper for a society to suspend free speech whenever it feels itself threatened*, 58 per cent of those in a group pressure situation conformed. Army officers were presented with the statement *I doubt whether I would make a good leader*. Such a view is clearly incompatible with an army officer's position, so not surprisingly none of the control subjects agreed. Yet under group pressure where the officer was led to believe that other officers had agreed to the statement, 37 per cent yielded to group opinion.

The experiments of Asch and Crutchfield, though superficially very similar are in fact concerned with different areas of social influence. In Asch's experiments the subjects are pressured to accept the opinions of others as evidence of *physical reality* about which the subject has the direct evidence of his own sense organs. In Crutchfield's experiments the pressure is to conform in areas of *social reality*, that is of opinions and values, where the subject has no direct evidence about the correctness of his judgment. To find support in social reality matters one has to rely on argument or the agreement of others. So it is not surprising that persons tend to conform to group pressure in situations concerned with social reality to a greater degree than in the case of physical reality. Probably the most disturbing experiments on the induction of compliance are those of Stanley Milgram (1958). Subjects were required to give an electric shock to another subject ostensibly to study the effect of punishment on learning. (The 'learner' was a confederate of the experimenter and in fact did not receive any of the shocks.) The electric shocks ranged from 15 to 450 volts and this last point was marked DANGER: SEVERE SHOCK. Each time the 'learner' made a mistake on the learning process, the subject was asked to administer a shock, increasing by one step with each error. Although the 'learner' demanded an end to the experiment at 150 volts and claimed he could stand the pain no longer at 180 volts, no less than 62 per cent of the subjects administered the full 450 volts, and one-quarter of the subjects did so even when the 'learner' was in close physical proximity to them. Milgram was so disturbed by these results that he assumed his early subjects, students at Yale, must have complied with the experimenter's requests as a function of their subordinate status as students in the presence of a staff psychologist. However, later experiments with older male subjects from many different occupations produced substantially the same results.

The implications of the experiments of Asch and Crutchfield are highly disquieting, since they suggest that people can be persuaded to conform, even if in many cases this conformity is no more than an expedient compliance, to a surprising degree and against the evidence of their senses and on matters of opinion to which they do not individually subscribe. Milgram's experiments are even more distressing since they raise the spectre of Auschwitz, Buchenwald and the worst Nazi horrors that are normally regarded as exceptional phenomena. It is true that these experiments were not conducted in 'real' groups but in artificial laboratory conditions. It is difficult to experiment on conformity in real groups, but we have no reason to believe that degree of conformity indicated in these experiments would be reduced in a real group. Indeed, the evidence is to the contrary for it has been shown, though on less dramatic matters, that amongst highly cohesive groups (that is groups which are highly attractive to members) conformity tends to be much higher than in uncohesive groups. In the main, laboratory groups are uncohesive since they are formed simply for the experiment and have no real significance for their temporary members. In a cohesive group of friends the members have a great need to be accepted by their fellows, and the rewards for being allowed to remain in the group and the sanctions that can be brought to bear on deviating members are tremendously powerful. If this were not the case, the Nazi political and military machine would never have become a terrifying reality. More groups than we realize have greater powers over members than we imagine to pressure members in directions that go against individual opinions, judgment or conscience. Under skilful manipulation groups can always be used for evil purposes, or for political objectives as notably in China today. In the light of this the importance of 'education for independence' assumes an urgent significance.

Group structure

If the culture is one side of the coin in the study of groups then the structure is the other side. Whereas in group culture it is the homogeneity of group members that is emphasized, their common involvement in values, beliefs, norms and pressures to conformity, when we consider group structure it is the *heterogeneity* of members that is stressed. In other words it is the structure which differentiates the members from one another. The key notion in this process of differentiation is that members are ranked into a set of *hierarchies*, by which in certain respects some members are more valued by the group than are other members.

Before we can analyse this process in detail, it is essential to distinguish the formal structure from the informal structure. In many groups some members are assigned to formal positions. Clubs and societies frequently elect officers to execute special functions for the group, such as president, chairman and secretary. Sometimes the officers are appointed by an authority external to the group. In a few cases we have a mixed variety where members elect one officer who then has a right to appoint officers to serve under him. In many cases the roles of the officers are governed by a set of rules which expound their duties and rights. In large social organizations the formal structure is likely to be extremely complex. The rules and regulations controlling the behaviour of members will be minutely detailed. Patterns of authority, the chain of command, specialization of tasks and responsibilities, channels of communications, all these will be part of the rational plan or blueprint that is drawn up to determine the behaviour of members. Every social organization has its own version of Queen's Regulations.

Many small groups do not have a formal structure. But it is erroneous to conclude that such groups therefore do not have a structure. It is simply that the structure is *informal*. Some groups have a formal structure, others do not; but all groups have an informal structure. The distinction between the two types of structure is very important. Consider a class of pupils in a school. Not infrequently the school class has a formal structure. The officers, such as the class president or the form captain and various other prefects and monitors, may be elected by the pupils themselves or appointed by the teacher. When the officers are elected, it is likely that the formal and informal structures will overlap and interpenetrate, since in general the pupils will elect to the formal positions class members who are most valued by the group. If the officers are appointed by the teachers, then the two structures may no longer interpenetrate. The teacher will probably appoint pupils that he values, but these pupils may not be the ones who are valued by the pupils and who have a high rank in the informal structure. Indeed, the teacher may appoint pupils who are among the least valued by the class, in which case the formal and informal structures will be in conflict.

To recognize the important consequences of such relationships between the two structures, we must try to analyse in depth the dynamics of the informal structure. To the outsider—and in relation to the class the teacher is in many respects an outsider—the informal structure is concealed; its workings and often its very existence are shrouded in mystery. Yet its presence and its hidden processes are of crucial significance for the group members and thus cannot be ignored by the teacher.

The informal structure of a group is concerned with the differential ranking of group members according to the degree to which members are valued by the group. The product is what from now on I shall call a *status hierarchy*. Three main dimensions of the status hierarchy have been elaborated by social psychologists, though there is unfortunately no agreed terminology with which to describe them. In other words, there are three forms of status. The first is what is often called *sociometric status*. This dimension is concerned with the distribution of liking within the group, whereby some members are liked more than others. The members who receive most liking are popular; those who receive least are unpopular. This dimension is called sociometric status out of respect for Moreno, who invented a method of measuring the distribution of liking within a group which he called sociometry.

A sociometric test typically asks pupils to name three pupils with whom they would like to associate in three areas. For instance, pupils are asked whom they would like to sit next to in class, with whom they would like to work on a group project, and with whom they would like to share a holiday or play situation. Thus each child has nine votes at his disposal. When we consider the distribution of these nominations we find that some children receive many choices and others receive very few. The former, namely the popular pupils, are referred to by sociometrists as 'overchosen' or 'stars', and the latter are termed 'underchosen'. Children who cast votes to others but receive none are called 'neglectees' and pupils who neither give nor receive choices are called 'isolates'.

Sociometry has become a highly developed system with its own language and forms of analysis. Most teachers are now familiar with the most common form of representing the sociometric structure, namely the sociogram, but many more elaborate forms of measurement are readily available (Proctor & Loomis, 1951; Evans, 1962). One of the reasons why the sociogram has become popular is that it reveals the clique or sub-group structure within the larger group.

The sociometric structure arises out of the tendency of human beings to develop feelings of like and dislike towards others. It is because the evaluation of others in terms of liking is so basic to human interaction that it becomes one of the main dimensions by which group members can be differentiated. Whilst it is a relatively easy matter to measure the distribution of liking within the group, it is much more difficult to explain this distribution. Why is it that some members are popular but others are not? Social psychologists who have specialized in sociometry have conducted extensive researches into the correlates of high and low sociometric status within children's groups. In some school classes the popular children tend to be more intelligent (Bonney, 1944; Heber, 1956; Thorpe, 1955) but this is not always the case (Kelley, 1963). Sometimes popular children are disproportionately from the higher social classes (Bonney, 1944), sometimes they are not distinguishable from the rest in this regard (Dahlke, 1953). We shall examine the reasons for these inconsistent results later. Even where such factors as intelligence and social class are shown to be associated with popularity, it is not at all easy to explain why this should be the case. In other words, even where these factors are demonstrably *correlates* of popularity, it is difficult to explain for what reason they might be regarded as *determinants* of popularity. Attempts to relate popularity to personality characteristics have been more successful. Usually the more popular children are more cheerful, happy, considerate, fair-minded, sociable, extravert, socially sensitive and have a better sense of humour. Pupils of low sociometric status are often belligerent and quarrelsome. These findings confirm common-sense predictions. We can see very good reasons why one should be attracted to the former personality but be repelled by the latter variety. Northway (1944) has suggested three different types of unpopular child. The *recessive* child is listless and lacks interest in people or activities. The *socially uninterested* pupil is shy, quiet and retiring. He does have interests, but they are of passive nature, personal rather than social. The *socially ineffective* are noisy and rebellious, and lack the social skills necessary to make satisfying relationships. Because of their failure in this respect they make conspicuous but foolish and futile attempts to be recognized and accepted by others.

One common misuse of the sociometric test and the sociogram should be noted. The sociometric test is an excellent device for measuring the popularity status of individual children, but when these choices are charted diagrammatically in the sociogram it is not legitimate to use the sociogram as a representation of the clique or sub-group structure within the class. This is because the sociometric test elicits perceived as well as actual relationships. In asking children with whom they would like to sit or work, one is asking them to make ideal choices. Those who in fact have many friends will nominate these friends in the test. Others, however, are likely to make unreal choices in the sense that they choose persons they admire, but with whom they rarely actually interact. As Potashkin (1946) noted, the underchosen child

> though he may not be rejected, is not generally sought out as a companion by his class mates. His social aims and relationships are unsatisfactory and he aims to remedy this by making 'unreal' choices, by indicating as preferences classmates who for him are impossible. Instead of choosing from within his own experience, he chooses the stars or idols of the class, with whom he may have had little if any contact.

In consequence it is dangerous to regard the sociogram as a representation of the structure of actual relationships.

The second dimension of status within groups is concerned with *social power* of influence. We may define social power as 'the potentiality for inducing forces in another person toward acting or changing in a given direction' (Lippitt, Polansky & Rosen, 1952). A person with high social power possesses the ability to exert influence over another so that the second person will behave in line with the intentions or desires of the first person. As the sociometric structure betrays the unequal distribution of liking within the group, so the social power structure reveals the unequal distribution of influence within the group. Group members at the top of the social power hierarchy influence more than they are influenced and have a high resistance to being influenced; low social power members have a low resistance to influence from others and are influenced more than they themselves influence.

The interpersonal nature of power relationships is made clear by Emerson's (1962) analysis. A person can have power over another to the degree to which the second person is dependent on the first. In other words, the power of one person is definable in terms of the dependence of the other. This power-dependence relationship between two persons A and B can be represented as

$$PAB = DBA$$

A's power over B is equal to B's dependence on A. Whilst A may have power over B is one respect, it is quite possible that in other areas B has power over A. Two types of relationships emerge. In the first the power *as a whole* is shared equally, though in particular situations the power is sometimes exerted by A and sometimes by B. This relationship of balanced power can be represented as

$$PAB = DBA$$
$$PBA = DAB$$

If, however, A influences B to a greater extent than B influences A, then A can be said to be more powerful than B and B more dependent than A. This is clearly indicated in the diagram.

$$PAB = DBA$$
$$\lor \quad \lor$$
$$PBA = DAB$$

If we are to make an adequate analysis of power relationships we must try to distinguish different types of social power. Many writers have embarked on this task and the most notable contribution by social psychologists is probably that of French & Raven (1956), who propose five bases of social power. The first type is *reward power* which arises when Other perceives that Person has the ability to mediate rewards to him. The second type, *coercive power*, occurs when Other perceives that Person has the ability to mediate punishments to him. A teacher has both sorts of power over his pupils. He can reward, for example with praise or marks, the pupils who accept his influence and punish, with disapproval or other penalties, those pupils who do not conform to his requests. Reward and coercive power are similar in that Other's conformity may take the form of compliance, whereby Other accepts the influence merely to obtain the promised reward or avoid the threatened punishment. If this is the case Person will be forced to keep Other under constant surveillance, since Other is unlikely to conform in Person's absence when the reward is not available or the punishment cannot be imposed. All teachers are familiar with this phenomenon, and are aware of the limited nature of a teacher–pupil relationship which is founded exclusively on reward and coercive power. *Referent power* arises when Other identifies with Person or wishes to be like him. Teachers who are heroified or admired by their pupils acquire this form of social power. Whilst it is generally accepted within the teaching profession that it is dangerous to court popularity with pupils, a teacher whose behaviour leads him to be disliked by the pupils clearly cuts himself off from this sort of power. (Readers might like to speculate why pupils of high sociometric status can be expected to exert referent social power.) *Expert power* is based on Other's perception that Person has special knowledge of skills that is valuable to Other. For pupils who want to learn or to pass the examinations the teacher has considerable expert power. Where the pupils are alienated from learning or bored by the curriculum content, they are likely to despise the teacher's expertise for it is in areas not valued by them. In this situation the teacher has no expert power and because the teacher has to rely on other forms of social power to get the pupils to learn, teaching becomes difficult. *Legitimate power* arises when Other believes that Person has a right to prescribe behaviour for him. Most pupils believe that much of the teacher's power is legitimate, but if the pupils regard some of the teacher's power, whether reward or coercive power, as illegitimate, then they will not willingly accept the influence. In other words, it is when the teacher's reward and coercive power are perceived by the pupils as illegitimate that their conformity takes the form of compliance. If the pupils share the same values as the teacher, they are likely to regard his reward and coercive power as legitimate.

These five types of social power as defined by French and Raven take no account of the basic distinction between formal and informal structures. In giving examples of the varieties of social powers from the teacher–pupil relationship it is clear that we are basically concerned with a formal rather than an informal relationship. Some aspects of the teacher's reward and coercive power are formal and others are informal in that they are not formally prescribed rights. The teacher has no formal referent power and where he acquires it he does so on an individual and personal basis. In this chapter we are more concerned with informal relationships and informal status structures among the pupils. The reader should now check his understanding of the five types of social power by trying to give examples of how high social power pupils might acquire and exert each type of social power among the informal group of pupils.

One of the best experimental studies of social power among children is that of Lippitt, Polansky & Rosen (1952), whose subjects were boys at a summer camp. Measures were taken of *attributed* social power, how the children rated one another in terms of the ability to influence ('who is best at getting others to do what he wants them to do?'), and *manifest* social power, that is actual success in exerting influence as observed in the daily life of the camp by the researchers. There was a high level of agreement among the children about which members possessed high social power and the study shows that children who are high on attributed social power are also high on manifest social power. High-power children are more active, make more direct influence attempts and are more often successful in their influence attempts than are low-power children. Moreover high-power children are more often imitated, which is a form of unintended influence. Low-power children are deferential and approval-seeking in their behaviour towards high powerers, and if they do try to influence are more non-directive in their approach, that is, they are more likely to suggest 'Shall we . . .?' than to state 'We shall . . .' In short this study shows that children are aware of the distribution of social power within their groups and that this distribution has a marked effect on social relationships in the groups.

The third element which differentiates group members into a status hierarchy is *prestige*. As we shall see, prestige is normally the least important of the three dimensions. Unfortunately the term prestige is used by different writers in so many different ways that it is a highly ambiguous concept. In this book I shall be using it in a specialized and restricted sense. Whereas liking and power status are acquired by particular group members in their relationships with one another in the interactions of the informal group, prestige arises only within a *formal* structure. This is because prestige is attached to positions, not persons. All occupations have varying degrees of prestige and can be arranged into a prestige hierarchy. Some professions, such as medicine and the law, have high prestige status. The point is that the prestige is attached to the position. Dr Smith has high prestige because he is a physician, not because as John Smith he has particular qualities or behaves in certain ways. Often these positions have prestige symbols—what are popularly called status symbols—which make the position highly visible to others. Teachers wear academic gowns, soldiers wear uniforms, and in the Civil Service you can gauge a man's position on the basis of the size of the carpet in his office.

In the small group, prestige can become an important dimension of status where there is a formal structure. The group president or secretary are positions that may carry considerable prestige and for this reason members may aspire to these positions. If there is no formal structure, there can be no prestige structure. The possession of high prestige status may help a person to acquire high status along other dimensions, but this is by no means always the case. A headteacher has high prestige status, but if the teachers who serve under him do not like him and are little influenced by him then his sociometric and social power status will be low.

School prefects are a good example of the complex way in which prestige relates to other forms of status. The position of prefect carries great prestige. The prestige comes with the office and does not have to be earned. Thus one of the major problems of the prefect is to acquire sociometric and social power status that are congruent with the high prestige status. The situation is complicated by the fact that a prefect is delegated definite formal power by the teachers. Since he is expected to carry out supervisory tasks and enforce school rules and regulations he is given the power to punish pupils by means of detentions, impositions and sometimes even corporal punishment. However, a prefect may also influence the behaviour of pupils without resorting to his formal power and he is likely to elicit feelings of like and dislike from them. The interpenetration of formal status (prestige and formal power) with informal status (popularity and social power) can be clarified by considering different types of prefect. The 'hard' prefect believes in carrying out his duties to the letter. Whenever he discovers that the rules are being broken, he punishes the offender according to the book and does not easily listen to or accept excuses. He does not relish the task, but he does not shrink from it. He does what he considers to be his duty. In a more extreme form he becomes the 'tyrant' who fulfils his duties with ardour, actively seeking out offenders whom he punishes ruthlessly, even when the offence is minor or merely technical. He arrogantly abuses his power because he enjoys it. The 'hard' prefect and the 'tyrant' rely entirely on their formal power. They do not possess informal social power; they can influence people only under threat of punishment. Obviously they have low sociometric status. The 'soft' prefect is one who attempts to exercise his formal powers but finds that he cannot do so. His diffidence and lack of confidence prevent him from enforcing the rules and the pupils do not obey him. In extreme cases his threats based on his formal power elicit scorn and laughter from pupils. He may try to exercise informal power by making personal pleas or similar influence attempts but in this he is equally unsuccessful. Whilst the 'soft' prefect is not disliked, he is not liked either. He is despised. The 'irresponsible' prefect avoids the problems of his office by declining to exercise any power at all, either formal or informal. He shirks his duties and whenever possible remains in the sanctity of the prefects' room. He accepts and enjoys the rights and privileges of his position, but rejects his duties and obligations. In some cases this may be because he enjoys high status in other respects, e.g. as a star footballer, which may be jeopardized if he tries to enforce school rules.

The 'ideal' prefect, who like the 'tyrant' is so vividly portrayed in schoolboy stories, prefers informal to formal power. Whilst he is willing to use his formal power and does so successfully on occasion, he is generally reluctant to do so. He is willing to listen to excuses and explanations and to give offenders a second chance. It is no pleasure for him to punish. Because he exercises his formal power so rarely, and never in an arbitrary or unfair manner, he is liked and admired and respected. Pupils obey him and are open to his influence. Through his greater use of informal power he is able to influence pupils voluntarily to conform rather than to punish them for non-conformity. He transforms his high prestige and formal power into high sociometric and social power status. He is what we call a 'natural leader'.

Within the informal group, prestige status and formal power are of course absent by definition. In the informal group it is the relationship between sociometric and social power status which is problematic. Do these two forms of status tend to hang together or are they alternatives that do not reside in the one person? Before we answer this question we must consider the more fundamental issue of why some persons acquire either type of status. To show that informal groups develop sociometric and social power hierarchies does not tell us why they emerge in the first place. In other words, we must examine the reasons for the unequal distribution of liking and influence in the group. To do this we must introduce some new concepts. All group members bring to the life of the group certain *properties*. These properties may be a possession (e.g. a car or a cricket bat), a skill (e.g. the ability to fight or to play a game well) or a characteristic (e.g. a personality trait or an interest). Because groups have distinctive values and goals some of the many properties of the group members are more relevant to the life of the group than others. A property such as the possession of a cricket bat is more relevant to a group that is interested in cricket than to one that is not. When a property is valued by a group, because it is likely to make a contribution to the achievement of the group's goal, it is called a *resource*. When a member possesses a resource (a property that is of value to group members) he is able to mediate rewards to the group, and this ability gives him power over the group members. The group member who possesses the cricket bat has power to determine whether or not the group plays cricket, at least to some degree.

The concept of resource is important because it emphasizes the interpersonal nature of power. In Emerson's analysis of power it was made clear that power is not an attribute of the single person but something which arises in a relationship between two or more persons. The concept of resource is interpersonal in that a property becomes a resource only when it is valued by another person. One gains power by having a property that is valued by others. It is also an essentially economic concept in that the more scarce a resource is within a group the more it will be valued by the group members and thus the more power it will give to its owner. The possession of a cricket bat will give little power to its owner if all the members have a cricket bat—or if alternative sources of or substitutes for the cricket bat are available. The concept of resource clearly fits in with an exchange theory analysis of human behaviour.

The relationship between resources and social power has been demonstrated by Gold (1958) among children aged between five and twelve years. Gold gave the children two tests, one to find out which properties were valued by the children and which members possessed the resources, and the other to find out who exercised most influence. He confirmed that the group members who were attributed most social power also possessed more resources. In all the groups a helpful and friendly personality was regarded as a resource, but it was only in boys' groups that strength and fighting ability were resources. In the study of boys at a summer camp by Lippitt, Polansky & Rosen, discussed earlier, the boys who possessed the greatest social power were also superior in camp-craft skills, which would obviously be valued by the boys and thus contribute to status.

We can now return to the question of the relationship between sociometric and social power status. Several researchers have examined the problem and find that the two tend to correlate positively. The Sherifs (1964) found that in boys' groups the two forms of status correlate between 0·64 to 0·84 and Lippitt, Polansky & Rosen (1952) report almost exactly the same result, 0·63 to 0·82. In short, those who are more liked also exercise the most influence. The concept of resource helps us to understand this correlation, for the same resource may be the basis of both liking and power. In most groups a friendly personality is valued by the members since such behaviour makes life pleasant. Friendliness stimulates feelings of liking, leading to sociometric status, and also gives power as well. A similar situation arises with other resources, such as the ability to play a game well. In a football team the star's ability with the ball is highly valued by the members, who want to beat their opponents, and he thus acquires power. (In French & Raven's terms it is expert power.) But his skill and expertise arouse admiration, which is a form of liking. It is because one resource can bring attraction *and* power to its owner that sociometric and social power status tend to go together.

But there are exceptions. Some of the most important experimental studies of leadership in small groups have been conducted by Robert F. Bales (1952, 1953) of Harvard University. The students in Bales's experiments were put in a boardroom situation where they were given a human relations problem to solve through discussion. Bales found that certain members tended both to talk more than others and to receive more communications than others. It is with these members that we are concerned. At the end of each experimental session each member was asked to rate the others in terms of liking and of having contributed the best ideas. At the end of the first session the man who was seen as having contributed the best ideas was also the most liked.

This is in line with our previous correlation between influence and popularity. Yet after four sessions this effect dissipated. Now the most liked man was second or third on best ideas and the best ideas man became the target for feelings of dislike. When the content of the contributions of these two men is analysed it is clear that the best ideas man contributes in the task-related or *instrumental* area, that is, he makes suggestions, gives information and opinions, evaluates action and inhibits irrelevancies. The most liked man contributes in the socio-emotional or *expressive* area, that is, he is concerned with group maintenance, he encourages others to speak, boosts morale and relieves tension. Bales went on to suggest that groups throw up two sorts of leader, the instrumental and the expressive, which in our terms can be said to have high social power and sociometric status respectively. In only a few cases did the two roles combine in one person—Bales calls such persons 'great men'. In general, a bifurcation of the two forms of status occurred.

Since bifurcation of status seems much less common in natural groups outside the laboratory, as in the work of the Sherifs and Lippitt *et al.* discussed earlier, then Bales must be creating special conditions which make it difficult for the two forms of status to coincide. We have to remember, as Verba (1961) points out, that there was no appointed leader in these groups. Since the group was given a difficult task to solve within a limited period the emergence of a leader who would suggest ideas and control the group discussion so that valuable time would not be wasted was essential. Yet in pressuring the group members to focus on the central task and to make progress with the solution of the problem, such a leader would arouse antagonism and negative feelings from others, partly by his unofficial assumption of the instrumental leadership and partly by his demands for progress. An expressive leader was thus required to cope with these tensions. Only an exceptional person could both pressure the group to progress *and* manage the tensions which ensued. In a real life group, conditions would rarely be similar. Groups are not normally under such pressure to solve alien tasks and in helping the group to reach its goal a group member does not need to arouse the antipathy of others. Moreover real life groups are very much more cohesive than laboratory groups and Theodorson (1957) has shown that bifurcation of leadership is much less likely to arise in cohesive groups. It must also be remembered that one of the central tasks of a friendship group is the fostering and maintenance of pleasant personal relationships. Thus the member who is friendly and cheerful, who possesses the requisite social skills, is likely to become both a popular and influential member of the group. In most informal groups of friends, then, we should expect social power and sociometric status to go together.

The way in which the situation of the group can affect the nature of the resources and of the status structure is clear in Jennings's (1950) distinction between *psyche-groups* and *socio-groups*. When subjects made a sociometric choice on a leisure criterion, the group (psyche-group) was characterized by greater mutuality of choice than when the criterion was a working companion (socio-group). In the psyche-group we find a more even distribution of choices since members are opting for a mutually pleasant companion to spend time with. In the socio-group there is a greater concentration of choices on a few persons, thus creating a more clear-cut status structure, which suggests that fewer members are perceived as possessing the resources appropriate to the achievement of work goals.

We can now make sense of some of the earlier findings on the relation between popularity and group member characteristics. Some studies of children's groups revealed a correlation between popularity and intelligence, but others did not. This can now be explained in terms of the degree to which intelligence is a group resource. Where intelligence is valued by group members, then intelligent members will acquire informal status. In Hallworth's (1953) study of several classes in a mixed grammar school, popularity correlated with intelligence or attainment in some classes but not in others. We can explain this variation by considering the group's values. In one class the central value among the girls was academic success. From this fact we can predict that there will be a positive relationship between informal status and academic position. This is so, for there is a very high correlation between popularity and examination performance. Indeed, sociometric status gives a better prediction of academic success than does the intelligence test score! In short, if one wishes to examine the reasons behind the nature and the distribution of informal status within the group, one must consider the group's central values.

In our discussion of group structure we have been concerned with the unequal distribution of influence and liking among group members. It is not the case that one person does all the influencing and receives all the liking. It is a matter of degree. Some persons influence and are liked more than others and all members can be ranked into a hierarchy according to the extent to which they influence or are liked. How then does the concept of group *leader* fit into our conception of group structure and status hierarchies?

In earlier years leadership tended to be conceptualized as an all-or-none phenomenon. Either a person was a leader or he was not. When such a conception dominated in psychology—as it still dominates the popular view of leadership today—the main problem was to discover the characteristics or traits which distinguished leaders from non-leaders. After many years of investigation the result was a highly confused and inconsistent picture. For every study which appeared to demonstrate a clear relationship between leadership and a particular personality trait, another could be found which failed to confirm the first result. When Stogdill reviewed the literature in 1948 he was able to report several trends. It could be said that leaders tended, in the majority of studies, to possess the following characteristics to a greater degree than other members:

1 *Capacity*—intelligence, verbal facility and fluency, originality, insight, speed and soundness of judgment, height and weight.
2 *Achievement*—scholarship, knowledge.
3 *Responsibility*—dependability, reliability, trustworthiness, conscientiousness, self-confidence, self-assurance, persistence, industry, initiative, dominance, ascendance.
4 *Participation*—activity, sociability, extraversion, talkativeness, sense of humour, emotional control, co-operativeness, tact.

Yet the low levels of the correlations and the inconsistency of the experimental results marked the end of the trait approach. A different conceptualization of leadership was needed and began to emerge. As Stogdill puts it:

> The findings suggest that leadership is not a matter of passive status, or of the mere possession of some combination of traits. It appears rather to be a working relationship among members of a group, in which the leader acquires status through active participation and demonstration of his capacity for carrying co-operative tasks through to completion. . . .
> A person does not become a leader by virtue of the possession of some combination of traits, but the patterns of personal characteristics of the leader must bear some relevant relationship to the characteristics, activities and goals of the followers . . . It becomes clear that an adequate analysis of leadership involves not only a study of leaders, but also of situations.

Interest shifted from the study of individual personality towards a consideration of the leader's role relationship in interaction with other group members, the goals and activities of the group, and the current situation of the group. This is not to suggest that the leader's personality is irrelevant but rather to see personality in relation to other factors. As C. A. Gibb (1947) states: 'Leadership is both a function of the social situation and a function of personality, but it is a function of these two in interaction.' More recently Kelvin (1970) has suggested conditions under which the contribution of these two factors might vary.

Put very baldly, the relative influence of situation or personality will be a function of the extent to which the environment (or situation) is itself structured and predictable. The greater the degree to which the situation and its problems are defined, the greater will be the influence of that situation, and its problems and demands, in determining the individual who will lead. Conversely, the less structured the situation, the less the possibility to predict its demands, the greater will be the contribution of the individual leader's own personality.

Our analysis has conceptualized leadership in terms of the more general structural differentiation within the group. The term leader is applied to the member (or members) who emerges at the top of the social power and sociometric status hierarchies. This assumes that these hierarchies are fairly static, but it may be that one person may not exercise most social influence in *all* situations. If this is the case, then we shall have to say that the leadership shifts between different members in different situations according to which member is exercising most influence at a particular time. It is true that at times many groups may find themselves in situations where the person who normally exercises most influence loses his leadership role and gives place to another member who has more resources to cope with the novel situation and thus assumes the leadership. However, in real life groups such situations are likely to be rare. Most groups operate within a very limited variety of situations in which one person is in general likely to exercise more influence than other members. It is to such persons that we could apply the term leader. In Kelvin's (1970) words: 'The leader is not simply the particular individual who happens to influence the group at a particular moment; he is, rather, the individual who influences it most consistently over a more or less prolonged period.' At the same time we must remember that many groups, when asked, deny the existence of a leader. This may be because the group is unaware that one person is consistently exercising more influence than others or because they may resent the notion that one member is more powerful than others, even though such is the case. But it may also be true that among small groups of three or four friends it can justifiably be claimed that there is no leader. The differences in the amounts of influence exercised by individual members may be very small, or influence may be exercised by different members in different situations and on different occasions. Thus no one member may merit the title of leader. The exercise of influence is an inherent part of group life, but the existence of a leader is not.

Given that most groups do contain a general leader, we need to examine in further detail the behaviour of such persons that leads them to rise to the top of the status hierarchy. The most important element, as we have suggested, is the possession by the leader of resources that allow him *to make the greatest single contribution to the group goals*, be they 'instrumental' or 'expressive'. In this sense the structure of a group is closely linked with its culture. Structural differentiation is related to each member's ability to realize group values and goals. Hamblin (1958), in an ingenious laboratory experiment, showed that a person who is seen to be most skilled on the group task is perceived as the group leader, but when the task performance is manipulated so that the leader ceases to make the principal contribution to task achievement, he loses his position and is replaced. This phenomenon is readily observable in political life. Politicians and political leaders who fail to achieve the party's goals, including the basic goal of wooing the electorate, are dispensed with. In the British election of 1970 most people were expecting a Labour victory and criticism of the leader of the Conservative Party was at a high level. It seems very likely that if the Conservatives had lost the election, Edward Heath would have been under tremendous pressure to resign his leadership. It is probably the unexpectedness of the result which in no small measure allowed Harold Wilson to retain the leadership of his own defeated party.

The greater skill of the leader in tasks central to the group has been demonstrated experimentally by Carter & Nixon (1949) and by Carter, Haythorn & Howell (1950). Further, as Sherif, White & Harvey (1955) have shown, group members tend to overestimate the skill of the leader and underestimate the skill of low status members. The relationship between task skill and leadership is complex. As Whyte (1943) has noted, the leader need not be the most outstanding footballer or fighter, but he must have some skill in the group's major pursuits. The leader also promotes activities in which he is skilful, so that his high level performance is both a consequence of his position and an important support of that position. Should the group's central activity change due to external factors, then a change in the group's structure and leadership is possible. The Sherifs (1964) quote a case where a tough, lower-class group of boys, with the help and encouragement of an adult, developed an interest in basketball. After a period of training and coaching the original leader began to lose status and was replaced by another boy who was both a better player and more effective in team decisions.

Sometimes a group member may be more skilled at a task than his position in the status structure would seem to warrant. In such cases the group may pressure the member to display a task performance that is congruent with his status. In Whyte's (1943) study, Alec was extremely skilled at bowling, but his status in Doc's gang was very low. A competition between members of the gang was arranged.

318

Alec let it be known that he intended to show the boys
something. . . . After the first four boxes [frames], Alec was
leading by several pins. He turned to Doc and said, 'I'm
out to get you boys tonight.' But, then he began to miss, and,
as mistake followed mistake, he stopped trying. Between
turns he went out for drinks, so that he became flushed and
unsteady on his feet. He threw the ball carelessly, pretending
that he was not interested in the competition. His collapse
was sudden and complete; in the space of a few boxes he
dropped from first to last place.

Alec's failure to produce a performance congruent with his ability
seems to have two sources. Firstly, his knowledge of his low status in
the group undermined his confidence in his ability to beat a high
status member. Secondly, the others heckled him when he was performing too well. In fact Whyte himself won the contest.

Every corner boy expects to be heckled as he bowls, but the
heckling can take various forms. While I moved ahead as
early as the end of the second string, I was subjected only
to good-natured kidding. The leaders watched me with mingled
surprise and amusement; in a very real sense, I was permitted to win.

Another good example of the way in which the group situation can
influence the ability of certain members to contribute to the group
goal is available in the N.F.E.R. report on streaming (Joan Barker
Lunn, 1970). There was a greater tendency for boys of above average
ability to emerge as sociometric leaders and for boys of below average
ability to be neglectees in non-streamed schools than in streamed
schools. This finding can be explained in terms of the way in which
school organization affects the resources in the children's groups. In
the streamed school the pupils within any one class are very similar in
ability and are less likely to be given group work. In the non-streamed
school children of all abilities are represented in each class and group
work occurs more frequently. In the A stream of a streamed school
the fact that a pupil is of above average ability cannot serve as a
resource to aid the acquisition of status since *all* the pupils are of
above average ability and ability does not differentiate the pupils. In
the non-streamed school, however, the boys of above average ability
are in a minority and are able to make a greater contribution to the
achievement of academic goals, especially where pupils are working
in mixed ability groups within the class. The bright boy's ability
serves as a resource and grants him status. Significantly this greater
sociometric status of above average boys is evident on the 'work
with' criterion of sociometric status but not on the criteria of 'play

319

with' or 'best friend'. This is in line with our explanation since greater ability is a resource when the goal is academic ('work with') but in other situations ('play with', 'best friend') the possession of greater ability is irrelevant to group goal achievement. A related finding which is of considerable interest to us is that in non-streamed schools this effect was more powerful when classes were taught by the more traditional (Type 2) teachers than by the more progressive (Type 1) teachers. This suggests that the greater valuation of academic ability and the lower interest in the slow learners characteristic of Type 2 teachers was conveyed to the pupils and influenced the ways in which they valued one another. This is one of the rare pieces of evidence which suggests that the teacher's attitude can affect the informal status structure of pupils.

A second way in which the group's culture and structure are inter-related is the area of conformity to group norms. The members of high status, the leaders, *are the most conformist to the group norms.* Members who conform least have the lowest status. That conformity to group norms is one of the dimensions of structural differentiation is easily explained by the fact that lack of conformity to group norms clearly threatens the stability and cohesion of the group and also the achievement of group goals. The member who conforms helps the group and is valued accordingly. The leader, then, tends to be an exemplar of the group norms.

The fundamental importance of conformity to norms in the acquisition of status is demonstrated in the work of Merei (1949) on nursery children in eastern Europe. Merei, from his observation of these children, was able to distinguish those who consistently exercised most influence (the leaders) from those who were influenced (followers). These latter children he assigned to a number of new groups. After a time each group had developed a distinctive culture and set of norms concerning such matters as the seating arrange-ments, the division of toys, ceremonies, special language usage and so on. Then one of the 'leaders' was added to the group. The leader immediately tried to give orders but his attempted influence was not automatically accepted by the rest of the group. On the contrary, he was ignored or rebuffed. The leader had to learn the norms and traditions of the group and he then began to exercise influence in these terms. In other words he had to adapt to the group culture before he could assume leadership. One type, whom Merei calls the diplomat, followed this pattern but also began to introduce small changes into the group culture and became the leader of these new games. Later he was able to effect more radical revisions in the group's traditional games.

There is also evidence that the leader can deviate from the group norms more than others. If the group meets a crisis the leader may well need to do so in order to save the group or to ensure that the group can continue to achieve its goal even though the circumstances of the group have changed. Hollander (1960) has shown that those who eventually emerge as leaders of a group tend to be more conformist than others early in the life of the group, but that once the leaders have achieved their high status, deviation from the norms incurs less disapproval than does deviation by lower status members. The Sherifs (1964) have suggested that in matters of great importance to the group the leader is allowed less room for deviation from the norm but that in more peripheral areas the leader is allowed greater latitude for non-conformity.

Hollander (1958) has tried to resolve the paradox that the leader is both highly conformist and the most free to deviate under certain conditions by means of his 'idiosyncrasy credit' theory. Each time a member conforms he is rewarded with a credit. The more a member conforms, the more he accumulates a balance of credits. These credits can later be spent in idiosyncratic or non-conforming behaviour. The leader, owing to his early conformity, has the largest balance of credits. Thus he can afford to deviate because he has more credits to spend than other members. The limitation is that he must not spend so many credits that another member will have a larger number 'in the black' than he does. Homans's (1961) discussion of the relationship between conformity and status which is couched in exchange theory terms readily fits this model. The leader can afford to take the risk of deviating, because if he is wrong, he has relatively little to lose, but if he is right he further confirms his leadership. Similarly a low status member can afford to deviate, for if he is wrong he cannot be penalized by loss of status since he has none to lose, but if he is right, he has much to gain, namely a rise in status. The middle status member is in the least secure position. If he deviates from the norm and no benefit accrues to the group then he will lose status, which is dear to him in his position among the many middle status members. If he conforms, he has nothing to lose, for even if conformity leads the group into difficulty, then no one can blame him for he simply went along with the rest. The middle status member plays a safe bet and conforms more frequently than do his superiors or inferiors in status. Harvey & Consalvi (1960) have demonstrated this relationship between status and conformity among delinquent boys. In real life, of course, leaders are unlikely to go on a spending spree with their credit balance and may adopt various devices to minimize the risk involved in introducing innovation. As Whyte (1943) points out, leaders do not stake their status on a particular course of action without having first obtained clearance from other high status members.

The leader of a group *tends to be central in the communication structure*. He must have access to information about what is happening in the group and he must also be able to disseminate his views to other members. Bales, in his experimental laboratory studies, showed that the leaders not only initiate more communications than others, but also receive more communications. In a more natural setting, Whyte (1943) talking about the Norton Street gang, shows that the leader is the focal point of group organization. When he is absent the group breaks up into small cliques. When he appears, the discussion is co-ordinated and focused on him. Members do not speak unless they know that the leader is listening. Although the members thus direct many of the communications to the leader, Whyte reports that the leader does not always communicate with the group as an undifferentiated mass. Often when mobilization of the group is required, the leader communicates his views to the lieutenants first. In other words, communications from the leader often pass down the channels of the status hierarchy.

Because the leader holds a central position in the communication system, he knows more than other members what is going on in the group. From this we can predict that the leader should be able to estimate group opinion more accurately than other members. Chowdry & Newcomb (1952) and Trapp (1955) have shown that this is the case among student groups, though the leader's superiority in predicting group opinion is limited to opinions and attitudes that are relevant to group values and goals.

Whilst the status differentiation of members within the group structure is a relatively easy group phenomenon to demonstrate, it is much more difficult to examine the *process* by which this differentiation occurs. The Sherifs (1964) have shown that it is the extreme positions which emerge first, a process which they call 'end-anchoring'. Group members first identify the leader, who acts as a primary anchor, and then the low status members emerge and act as a secondary anchor. Group members of intermediate status are the last to be differentiated and in practice these members cannot be assigned an exact rank order position. It may be that they are never clearly differentiated but it is also possible that there are frequent minor changes of rank order among middle status members.

We can now examine some of the basic principles of group dynamics in relation to the life of pupils in school by considering my own study of the fourth-year pupils in down-town Lumley secondary modern school for boys (Hargreaves, 1967). I shall confine myself to a rapid survey of the culture, structure and leadership of the A and C streams. As we shall see, these two classes represent radically different perspectives, the members differing in their interpretations of life and reality, the meaning they gave to their experiences, their perceptions of events and people, their definitions of the past, their orientations to the present and the future, their doings, their feelings, their self-images—all of these both within and outside the school. Such perspectives cannot be adequately grasped and appreciated without meeting, watching and listening to the pupils themselves. Here I wish merely to give some indication of these different perspectives by highlighting some of the major differences in group culture and structure.

1   In the A stream the central value is hard work and academic achievement. It is normative to work hard and pay attention in lessons; messing or fooling about and copying other pupils' work are forbidden. These values and norms spring from the main goal of A stream boys which is to do well at school and pass the examinations. In the C stream, by contrast, the central goal is to 'have fun'. It is normative to avoid academic work, to 'mess about' and to copy. In the A stream 93 per cent of the boys 'like a boy who gets on with his work', but only 32 per cent in the C stream do so; 43 per cent of the A stream 'like a boy you can have a lot of fun with in lessons', but 68 per cent of the C stream do so. On the basis of teacher ratings of pupils' behaviour in school, 70 per cent of the A stream boys are above the median for the whole fourth year and 64 per cent of the C stream are below the median.

2   In the A stream smart appearance is valued. It is normative to wear trousers, sport jacket or sweater, and a tie, and to have short, well-groomed hair. In the C stream it is normative to wear jeans and to have long hair. The wearing of a tie is well outside the latitude of acceptable dress. When the teachers rated the pupils' appearance, 87 per cent of A stream were above the median and 73 per cent of the C stream were below it.

3   In the A stream it is normative to attend regularly. In the C stream there is no norm with respect to attendance; pupils are free to attend school or to play truant as they wish. The absence rate for the C stream is more than twice as high as for the A stream.

4   It is normative to smoke in the C stream, but amongst A stream boys smoking is frowned upon. Of the C stream boys 77 per cent smoke regularly, whereas only 13 per cent of the A stream do so.

5   In the C stream most of the boys are involved in delinquency. Two-thirds are currently involved in petty thieving and over half have appeared in the Juvenile Court at least once. For the A stream the respective figures are 7 and 3 per cent.

It is clear that the cultures of these two streams are in opposition. The A stream culture can be regarded as 'academic' or 'pro-school', the C stream culture as 'anti-academic' or 'anti-school'. However, these two cultures are not fully homogeneous. Within each stream a minority of boys have values which are against the dominant trend. In the A stream the anti-academic pupil is a deviant, whilst in the C stream it is the academic pupil who is a deviant. Since a measure of the pupils' social power status within each stream was taken, we can examine the relationship between status and values or conformity to the dominant norms more systematically.

1 *Work and behaviour norms.* In the A stream, status and examination performance correlate $+ 0·50$; in the C stream the correlation is $-0·15$. In the A stream the correlation between status and behaviour rating is $+ 0·18$; in the C stream it is $-0·58$. To put it more graphically, in the A stream no high status pupil disapproves of 'a teacher who makes the boys work hard', but 20 per cent of low status pupils do so. In the C stream this trend reverses; 73 per cent of the high status boys but only 36 per cent of the low status boys disapprove of such a teacher.

2 *Appearance norms.* In the A stream status and appearance as rated by the teacher correlate $+0·37$, but in the C stream they correlate $-0·41$. In the C stream all the high status boys think that 'boys should be allowed to wear jeans in school', whereas only 45 per cent of low status pupils agree. In the A stream 47 per cent of low status boys but only 33 per cent of high status pupils agree with the statement.

The relationship between the structure and culture of each class is particularly marked among the leaders. Adrian, the leader of the A stream, was the Head Boy of the school. He was also acknowledged to be the best dressed boy in the school. Clint, the leader of the C stream, was the school's leading delinquent and trouble-maker. Adrian was the most popular as well as the most influential member of the A stream. The resources he brought to the group—his formal position as Head Boy, his academic ability, his hard work (though he never obviously worked so hard that he could be called a 'swot'), his elegant appearance, his loyalty to the school, his exemplary behaviour, his outstanding ability at football, his friendly and cheerful personality, his sense of humour—all these combined to give him the highest sociometric and social power status.

Clint's leadership in the C stream takes a very different form. One of the principal values of this group was fighting ability. The boy with the greatest fighting ability was given the title 'cock'. Clint was able to maintain this position by his swaggering, threatening display as being 'hard' or tough and by the lack of challenge to his supremacy from other boys. Being the cock, Clint exercised the most influence. But although this valued fighting ability gave him social power, it did not elicit liking from other boys. Most boys secretly disliked him. Other resources were required for this second dimension of status and these Clint lacked. Indeed his cold, vindictive personality and his arbitrary aggressive outbursts alienated the boys who succumbed to his influence. Yet they dared not resist Clint's influence, for to do so would be to incur his hostility. They were influenced through fear. Thus, Clint's power, unlike that of Adrian, can be regarded, in French and Raven's terms, as coercive and illegitimate. His followers expressed this by calling him—though not to his face—a 'bighead'. Had he been challenged and beaten in a fight, the others would have been glad. Unfortunately the only potential pretender to the title of cock was Don, who being the sociometric star in the low streams, was aware of the difficulties in being a popular cock and never issued a challenge. It was thought that Don could beat Clint in a fight. Clint seems to have thought this too, for he scrupulously avoided antagonizing Don.

The bifurcation of social power and sociometric leadership is not inevitable in a group which values fighting ability. Doc, the leader in the Norton Street gang, was a popular best fighter. But as Whyte observes, considerable skill is needed to do this. The leader must be fair and cannot afford to bear grudges or in other ways arouse hostility towards himself. The group members turn to him for advice and encouragement. He must also make sure that he does not become heavily obligated to other members, especially low status ones. Consequently he tends to spend more money on the group members than they spend on him.

If norms are shared expectations common to all group members, then expectations applied to individual group members can be regarded as group roles. Traditionally in the literature it is the leader's role which is given by far the most attention and relatively little emphasis is given to other roles in the group. We have already noted that in some respects it is dangerous to regard the school class as a group, since often the class consists of several sub-groups or cliques with varying cultures and structures. In the Lumley study just discussed, both the A and the C streams consisted of three main cliques. In each case two cliques shared the dominant culture of the class and the third clique had a deviant culture. Further, the two main cliques can also be differentiated, one having higher status than the other. In other words, the school class tends to be divided into cliques which can be ranked into a clique status hierarchy: within each clique, members can be differentiated into an individual status hierarchy. However, the school class can be said to form a group in the sense that it is clearly distinguishable from other classes and in some cases all the cliques may share in a common class culture. The roles we shall now examine may be either class roles or clique roles, or both.

The comedian. The role of the comedian (clown, jester) is one of the most ubiquitous. It takes many forms, but in all cases the comedian is expected by others to be a source of humour. He achieves this by telling jokes, by a ready wit, and by impersonations, especially

325

of teachers. In the presence of the teacher, he is often noted for his smart asides, his backchat and repartee. Sometimes he is of high status and sometimes of low status among the pupils—the difference being whether the pupils laugh *with* him or *at* him. Sometimes a pupil assumes the comedian role in order to gain acceptance from others, especially where he lacks other resources that could give him status. Once assumed, it is a role that is not easily shed, especially when the teachers also treat a pupil in terms of this role.

The bully. This is another universal role. Whilst it is usually seen as a male pupil role, there is a female equivalent which is more subtle. The bullying can be physical and/or psychological; it is the former which is more easily noticed and so rigorously condemned by the teacher. Sometimes the bully has high status, as in the case of Clint at Lumley, but more often he is of low status. Psychologists usually ascribe bullying to feelings of rejection or inferiority. If this is so, then bullying usually has the reverse effect to the one intended, for the bully's status becomes further undermined. It is important to distinguish the pupil with fighting ability from the bully.

The scapegoat. This is a complementary role to the bully. Typically the scapegoat is a deviant from group norms and is thus attacked by the bully or by other pupils. Sometimes the scapegoat is picked on because he has some 'defect' which makes him stand out as 'different', e.g. he speaks with an unusual accent, has body odour, is effeminate. In my own experience the scapegoat usually both has a defect and is a group deviant. The scapegoat soon becomes the target of group hostility and blame, the butt of practical jokes and is regarded as 'fair game' to all.

The instigator. Such a person is brilliant in his invention of escapades but is equally brilliant in avoiding involvement and apprehension. Often he also manages to maintain an image of himself in the teacher's eyes as 'good' and protests his innocence impressively.

The fall-guy. This is a complementary role to the instigator. He is so gullible that he is easily dared or cajoled into mischief for which he invariably takes the blame. I once knew a brilliant instigator who would persuade the fall-guy into a prank and then call the teacher's attention to the misbehaviour. In this case the fall-guy was also the class scapegoat.

The teacher's pet. There are various forms of this role. There is the informer (or sneak) who reports all crimes to the teacher. The informer role is often combined with sycophant, who is constantly exhibiting his conformity to the teacher's requirements, offering his help to and generally ingratiating himself with the teacher. The teacher's pet is always rejected by other pupils, but not always by the teacher. There is also the inverted teacher's pet where the teacher reveals a preference for a particular pupil and discriminates in his favour. Pupils sometimes accept and sometimes reject this teacher-initiated pet role.

The swot. It is often thought that the swot is a pupil who simply works too hard. This is not so. Most pupils work much harder than they admit, for most pupil groups have a mediocrity norm with reference to work. If a pupil works very hard, then he threatens other pupils by setting high standards which the teacher may then wish to apply more generally. Most pupils so fear group disapproval for working too hard that they *conceal* their hard work. The swot, then, is not to be distinguished by his hard work but by his failure to conceal his industry because of his insensitivity to or lack of concern for the group mediocrity norm.

Many other informal class and group roles are well-known—the mascot, the womanizer, the beauty queen, the flirt, the lawyer, the group spokesman, the newcomer. Unfortunately there has been relatively little interest in a systematic classification or investigation of these roles.

Group dynamics and the teacher

What are the implications of group dynamics for the teacher in the classroom? Perhaps the most basic is that the class of pupils does not consist simply of either thirty or more separate individuals or a unified entity. It is true, of course, that each pupil is an individual and the teacher cannot understand or make sense of the pupil's behaviour without regard to his individual uniqueness. It is equally true that in many respects the class forms a distinct entity, even if only because it is so at an administrative level. As we have discussed before, the class will have an awareness of itself as a class and will develop a distinctive climate and 'reputation' within the larger social organization of the school. It is part of the teacher's job to be responsible for the academic, social and moral development of both the individual pupil and of the class as a whole and he responds to pupils at both these levels. Just as the teacher evaluates individual pupils in terms of conformity to the role requirement of learning and discipline, so also he evaluates classes. As he develops feelings of attraction or dislike towards individuals, so also he responds emotionally to classes. The categorization process operates at both levels. It is classes, as well as individuals, who are held to be 'good' or 'bad'. Where a teacher has special responsibility for a class, he tends to see the pupils as in some sense an extension of himself. The collective 'you' of the class becomes also a collective 'us'.

Although the vast majority of the teacher's classroom behaviour is directed to one of these two levels, the individual pupil or the

327

collective 'team' of pupils, it is too simple a view of the life of the class. Group dynamics draws our attention to the intermediate level, the existence of sub-groups or cliques within the class. Very little of the teacher's behaviour is directed towards these cliques as such. The evidence suggests that teachers vary considerably in their ability to predict the sociometric structure of the class but in general cannot do so with much accuracy (Bonney, 1947; Gronlund, 1955). In a recent British study (Evan Wong & Bagley, 1970) there was no correlation at all between the class leaders as nominated by teachers and the actual sociometric leaders. Part of the explanation is that teachers do not respond to children in terms of their friendship groups and see little need to do so. It is also true, as we have seen earlier, that teachers tend to evaluate pupils in terms of the degree of their conformity to the pupil role as conceived by the teacher. Obviously this will often not coincide with the dominant values and norms of a clique of friends. Many teachers, the evidence suggests, find it difficult to perceive and evaluate pupils in other terms than the pupil role. Because the teacher perceives the child as a leader, he assumes that other pupils will also see him as a leader. The teacher cannot or does not make the imaginative leap necessary to perceive a pupil as other pupils see him.

Once again we have to recognize the disjunction between the teacher and pupil perspectives. Thus a teacher may decide that a pupil should be transferred from one group or stream to another in the interests of the pupil's academic development. For instance, it may be the teacher's view that in his present stream a pupil is 'out of his depth' or is 'not being sufficiently stretched'. To the teacher these serve as adequate grounds to justify a transfer from one stream to another. The teacher is, however, devoting his exclusive attention to the *formal* grouping. The pupil, on the other hand, is likely to be more concerned about the *informal* ties that have grown up within his present class. He knows that if he is moved into another stream his opportunity for interaction with his friends will be severely diminished and that he will be forced to try and maintain his current friends in out-of-class contexts and activities. There is a high probability that he will have to change friends. Further, the values and norms of pupils in the new class may not be congruent with those of his former friends, and if this is the case he will have to conform to such values and norms to gain acceptance and acquire status, or remain an isolate, clinging to his old friends whose group life continues unchanged except for his own constant absence. He will have to come to terms with his new social surroundings or play the role of the unwilling prisoner who dreams of the time when he can be with his real friends again and frets about the joys of group life that he cannot share. Obviously transfers between formal groups are not always as disturbing to a pupil as this, but they may be. The point is that the teacher does not always recognize the importance of these matters to a pupil, nor does he always take them into account when making an 'academic' decision.

Teachers are aware of the deleterious effects that a change of formal grouping can have on informal groups, for they know that a group of 'troublemakers' can often be effectively broken up by assigning the members to different classes. But then the deleterious effects are in the teachers' interests. Yet teachers do not always bear these effects in mind, nor do they always have much knowledge about these informal groups. In the Lumley study, teachers correctly perceived that an upper stream pupil could be induced to work harder under threat of demotion to a lower stream for this was a threat to the pupil's self-image as 'bright'. They did not recognize that there was an additional incentive arising from the informal group life, whereby upper stream pupils did not want to be transferred into a stream dominated by boys with incompatible values and norms. With respect to lower stream pupils, the teachers believed that promotion to a higher stream was a reward for academic merit. They did not recognize that most of the lower stream pupils did not want to be transferred into a class with incongruent values and norms or that the boys would intentionally underachieve in examinations in order to avoid such a transfer. The dimension of the informal groups was hidden from the teacher who had to account for the poor examination performance of such pupils in other terms, such as lack of ability, which was in some cases quite erroneous.

If the teachers did recognize and understand the group dynamics among the pupils not only would their perceptions be changed but they would also be constrained to alter some of their behaviour. If, as in the Evan Wong & Bagley study, the pupils the teacher thinks are leaders are in fact below average in popularity, then the teacher's attempt to appoint popular leaders for work groups will be unsuccessful. At Lumley, to take another example, the teachers of low streams would have realized that in approving the behaviour of the 'good' pupils they were in fact rewarding the low status pupils and reinforcing their low status. Further, in punishing the outstandingly 'bad' boys, the teachers were not simply disapproving unacceptable classroom behaviour but also confirming their high status in the dominant anti-school groups. A third example concerns the 'ripple effect' in classrooms (Kounin & Gump, 1958). When the teacher disciplines a particular child, it is often done in the presence of other children. Although the teacher's disciplinary action may be directed at the one pupil, there is an effect on the other pupils as well as on the target pupil. It is like the ripple effect caused by dropping a pebble in a pond—the whole water surface is affected, not just the point of impact. Gnagey (1960) showed that the peer group status of the target child in a disciplinary incident can influence the way in which the other pupils, who are subject to the ripple effect, perceive the teacher. When the target pupil submitted to the teacher, the teacher was seen as more powerful than when the pupil defied the teacher, but the effect was much stronger when the target pupil was of high rather than low status. When the high status pupil submitted, the teacher was also seen as more fair and more expert and the pupils actually learned more from the lesson, than when the high status pupil defied the teacher. There were no such differences between the submission and defiance conditions when the target pupil was of low status among the pupils.

Group dynamics suggest that we need to know the group affiliations of persons to understand their behaviour fully. The teacher needs to know the ways in which the class is structured into various cliques. These cliques may vary in size from a pair to a group of ten or more pupils. Some groups are highly cohesive; being in the group is important to its members, who will be fairly resistant to outside forces that threaten the existence of the group. Other groups are much more fluid, with a constantly shifting membership. How many teachers know (a) how many cliques there are in the class (b) whether or not some or all of these cliques are closely connected with other cliques and (c) the attractiveness of the cliques to their members?

To understand the individual within the group, we need to know the values and norms of the group. Do all the cliques in the class have similar values and norms, so that we might legitimately speak of class values and norms? Or do the cliques vary in their values and norms? Do the values and norms of different cliques within the class differ so much, as in the Lumley case discussed earlier, that there is a cultural clash and consequent inter-clique conflict? What is the structure within each group? Who are the leaders and what sort of leadership is being exercised? Are the cliques structured into a hierarchy of cliques? What of the pupils who do not belong to any clique? Are they aspirants to a clique but who for some reason lack the qualifications for acceptance? Are they rejected by other members? Or are they isolated within the class because being members of groups outside the class, they have no wish to join in any of the class cliques?

These questions are, to anyone with a knowledge of group dynamics, fairly elementary. They are important initial questions if one accepts the assumption that we cannot understand the individual pupil unless we know something of the nature of the groups and the group processes within the class. Thus it would seem to be a sensible idea for teachers to study group dynamics. At the end of a suitable course the teacher will be in a position to ask the elementary questions posed above. Yet presumably in giving teachers a course in group dynamics we do not merely want to provide them with the *intellectual* knowledge and skill required to ask the questions. We are, presumably, aiming beyond this, towards *sensitizing* teachers to *recognizing* the problems in the group behaviour of children and to asking the right questions at the right time, and towards providing them with methods whereby they can also *answer* the questions and *apply* what they know usefully within the classroom situation. If we cannot do this, then teachers can with justice claim that the knowledge is useless. I have met many teachers with training in the social sciences applied to education who believe that this training made them more sensitive teachers. I do not know to what extent they are right, but in general I am sceptical. Earlier I have mentioned what I call the Sociological Myth, where I suggest that informing teachers of the powerful influence of home environment on school performance may lead teachers to feel and believe that they can do little to mitigate the effects of environment on socially deprived children. In consequence the teachers lower their expectations so that the children do indeed reach the low levels of attainment predicted by sociological research. If there is any truth in the Sociological Myth, then it is the reverse of what training in sociology of education was intended to achieve. I am likewise sceptical that a knowledge of group dynamics will in itself sensitize a teacher to the group dynamics in the classrooms. Some teachers do have a 'natural' understanding of classroom group dynamics, though they cannot usually couch their understanding in technical language. Training in group dynamics might help them by allowing them to conceptualize their 'natural' knowledge more adequately and even to extend their skilled use of this knowledge. It may, on the other hand, have the reverse effect, as in the proverbial case of the centipede and his legs. To assume that a training in group dynamics for teachers without this 'natural' knowledge will in some way make them 'better' teachers seems quite unwarranted.

Let us illustrate the problem with reference to an example of the sort of knowledge that will be available to the teacher who has some understanding of group dynamics.

It is pointless to expect the leader of an anti-school clique to adopt pro-school behaviour unless this new behaviour brings greater rewards and satisfaction to him than does his leadership of the anti-school group.

In the light of our knowledge of group dynamics this seems an eminently reasonable statement. The question is how valuable is this assertion to the classroom teacher? That the leader is anti-school in his attitudes and behaviour will be obvious to the teacher. But how can the teacher discover his leadership position within the clique?

Further, whilst the statement gives us a general condition under which we can expect his behaviour to change, no precise detail is given. On the assumption that the teacher is aware that he is the leader of the group, it is obvious that the leader will not be allowed by other group members to deviate from central group norms without losing his leadership position. The teacher with a knowledge of group dynamics knows, in theory, that it will probably be ineffective to lecture the pupil in private and counsel him to mend his ways, for group pressures to conformity will soon re-assert themselves against any resolutions made by the boy, however, sincerely, to be a better pupil. The teacher can recognize that although the boy's anti-school behaviour seems irrational to him, to the boy it is rational, given that he wishes to retain his membership in and leadership of the group. A good example is given in Evan Hunter's *The Blackboard Jungle*. The teacher, Rick, is producing the school's Christmas show. In rehearsal, he finds that the leading anti-school pupil, Gregory Miller, is surprisingly co-operative. In the following English lesson, Rick finds that Miller has reverted to his more usual pattern of behaviour.

> Rick blinked at Miller, not able to understand the change in the boy. Was this the same helpful co-operative kid who'd worked out the speech rhythms in the auditorium just a period ago? This wiseacre who had just now initiated a new series of jibes against the teacher? He couldn't believe it, and so he let it pass because he couldn't understand it. But in the days that followed he learned a basic fact and he also learned to live by it.
>
> He learned that Miller formulated all the rules of this game, and that the rules were complex and unbending. And just as Miller drew an arbitrary line before the start of each fifth-period class—a line over which he would not step—he also drew a line which separated the show from anything academic.
>
> It was a confusing situation. It was confusing because Rick really did get along with Miller at rehearsals. The student–teacher relationship seemed to vanish completely. They were just two people working for a common goal, and Miller took direction and offered helpful suggestions, and stood by shamefacedly whenever Rick blew his top about a bit of stage business or a fluffed line. Rick valued the boy's participation in the show, and most of all the way Miller led the sextet, helped Rick mould it into a unified, smoothly functioning acting and singing machine.
>
> And then rehearsal would be over, and in class Miller drew his line again, and he pushed right up to that line, never

stepping over it, always baiting Rick so far, always annoy-
ing him until Rick trespassed onto Miller's side of the line
and Miller was faced with the choice of retreating or shov-
ing over on to Rick's side of the line, and that he would
never do . . .
    The English class was another matter. The other boys in
the class considered English a senseless waste of time, a
headless chicken, a blob without a goal. Miller may have
felt the same way, though it was impossible to know just
what he felt. But he sensed that approval lay in disorder,
that leadership lay in misbehaviour. And so he drew his
line, and he drew his second line, the line that told Rick,
'The show's one thing, Chief, but English is another. So
don't 'spect me to go kissin' your ass in class.'

This teacher discovered a group dynamics principle for himself. But
how could Rick change the situation? He did not find a solution, nor
does our present knowledge of group dynamics suggest any easy
solutions to problems such as this. In short, a knowledge that the
leader of a group is unlikely to adopt behaviour that is the opposite
of that prescribed by group norms is largely *negative* knowledge—
knowledge which tells us where solutions are *not* to be found, rather
than where they are to be found. This may be no mean contribution
of the group dynamics approach when one considers the amount of
time spent by teachers in trying to solve problems by methods which
are almost certainly doomed to failure from the beginning. But,
quite rightly, the teacher wants positive solutions to his problems.
In general such positive solutions are not available. Most research in
group dynamics is not based on children's groups in school but on
students in psychological laboratories and on small groups in
industrial organizations. Whilst much of this research on other groups
has yielded principles of group dynamics which can be applied to
classroom groups, there is a desperate lack of research geared to
analysing classroom problems with a group dynamics component, to
examining the range of possible solutions to such problems, and to
assessing the relative difficulty, merit and success of such solutions.
    The other problem for the teacher is that of how he can discover
the nature of the dynamics of the children's groups. Research reports
often make the evidence look so easy and obvious, but the researcher's
only function in the classroom is to discover the group dynamics and
he is able to use a variety of very sophisticated techniques in the
service of his goal. To the practising teacher, who is very busy with
the myriad problems of doing his job to the best of his ability, the
group dynamics of the class are not obvious as he looks at the thirty
or so faces before him in the classroom or on the corridor or on the

playground. He has very little time to devote to systematic investigation of group dynamics. Moreover many of the measuring devices —sociometric and social power tests, questionnaires, interviews— are not always easy to use, analyse or interpret. So far as I know there is no consultative function, exercised by researchers or educationists, parallel to consultants in industry, who will perform these tasks for the teacher's information. In general the teacher will have to rely on spasmodic observation, which can be extremely misleading—though with practice in looking at the right things at the right time there is more to be observed than one might imagine, as John Holt has demonstrated—and the occasional use of the test, if he knows how to use it properly.

I look forward to the creation of such a pool of knowledge about classroom dynamics and techniques designed for use by teachers to explore these dynamics, but its existence would not in the least guarantee its usefulness to teachers. Rick discovered a principle of group dynamics the hard way, after considerable puzzlement and confusion and after a long person-to-person discussion with Miller. Only when Rick was able to see the situation from Miller's perspective as well as from his own teacher perspective was he able to make sense of the boy's behaviour. In the chapter on teacher–pupil interaction, emphasis was placed on the importance for the teacher of gaining the *general* pupil perspective that is common to the majority of the pupils, and the *individual* perspective of the unique individual pupil. Group dynamics emphasizes the *group* perspective, common to the group's members, and the concepts of group culture, values and norms throw light on this group perspective. Group dynamics also emphasizes the *individual* perspective of a particular member of a group, and the concepts of structure and status help us to understand the perspective of the individual-pupil-within-his-group. Thus one of the major values of a knowledge of group dynamics for the teacher is that it suggests concepts which can be used to analyse an aspect of pupil behaviour about which, as I said earlier, the teacher is not directly pressured to be aware. The teacher is in a position to make better sense of a pupil's behaviour once he recognizes that the pupil is a member of groups, both formal and informal, which contribute to his over-all pupil perspective. Many teachers, like the teachers in the Lumley study do not recognize the impact of the group upon the pupils and in consequence their appreciation of the pupil perspective is seriously diminished.

Learning the principles of group dynamics does not in itself give the teacher the capacity to capture a pupil's perspective as a group member. In my view the ability of a teacher to capture the perspective of the individual-pupil-within-his-group is essentially the same ability as that required to capture any aspect of the pupil perspective, namely empathic understanding. If this is so, then it suggests that if teachers are to be taught the principles of group dynamics an essential part of their education would be training in empathy or sensitivity.

At present psychologists have done relatively little in the area of sensitivity training. The most important exception is the T-group approach, where the group studies itself in order to improve the sensitivity and social skills of its members (Bradford, Gibb & Benne, 1964). Some attempt has been made to organize T-group courses specially for teachers (Ottaway, 1966). A second exception is role-playing, where persons play the roles of others with whom they are likely to interact in order to improve sensitivity. Although this technique has been incorporated into a few conferences for teachers and headteachers, it is still not widely used nor has its effectiveness for teachers been assessed. A third approach, of which we are likely to see more in the coming years, is typified by the research of Jecker, Maccoby & Breitrose (1965) who have shown that teachers can be trained to judge accurately the facial expressions of pupils who are learning what they are being taught. A fourth development is the attempt by interaction analysis experts to use their methods to help teachers to achieve insight into and to obtain feedback on their own teaching behaviour (Amidon & Hough, 1967, Part III).

Unfortunately there are many trends within teacher education against such innovations. Colleges of Education, especially since the introduction of the B.Ed. degree, and University Departments of Education are still predominantly concerned with academic or sub-ject training rather than the social skills of teaching. Even in 'educa-tion' courses, the stress is on the traditional educational disciplines of history, philosophy, psychology and sociology. Further, advanced diplomas and Master's degree courses in education are even more concerned with these and associated disciplines. It is the teachers who take these courses who tend to become the next generation of teacher trainers in the Colleges of Education and thus the academic tradition is perpetuated at the expense of the development of tech-niques designed to improve the social skills of teaching, which are assumed to be acquired through experience. This is, in my view, a frightening abrogation of responsibility, yet University Departments of Education are under no pressure to change, for their clients are either students in initial training, who are not sure what is the most appropriate training for them, or experienced teachers (intending to leave schools for posts in Colleges of Education) competing for a place on the over-subscribed advanced courses whose academic content meets the needs of future lecturers in academic subjects. Teaching is a social process, yet we give so little guidance to teachers during training in the necessary social skills, and once they enter the profession they are left to achieve success in the isolation of the classroom on the basis of their 'natural' insight and skills, or, as in too many cases, without the necessary help being either available or acknowledged as important or essential. It is time that teachers began to recognize their own needs in this respect and relinquish their belief that every teacher is and should be self-taught—and that if he seems to be unable to teach himself then he is unteachable and must be tolerated as an unsuccessful member of the profession. If they could take this step, then perhaps teacher trainers, the organizers of courses and conferences, and the staff of University Departments of Education would pay some attention and try to meet this need in addition to continuing their essential task of stimulating the academic study of education and curriculum development.

Recommended reading

D. CARTWRIGHT & A. ZANDER, *Group Dynamics*, Tavistock, 1963 and subsequent editions.

D. H. HARGREAVES, *Social Relations in a Secondary School*, Routledge & Kegan Paul, 1967.

J. KLEIN, *The Study of Groups*, Routledge & Kegan Paul, 1956.

M. OLMSTED, *The Small Group*, Random House, 1959.

M. & C. W. SHERIF, *Reference Groups*, Harper & Row, 1964.

W. J. H. SPROTT, *Human Groups*, Penguin, 1958.

For a quite different approach to group life see:

B. R. BION, *Experiences in Groups*, Tavistock, 1961.

10 Youth, youth culture and the school

It is a bold author who is prepared to write about youth, for it is a subject in which we all, and especially educators, claim to be experts. It is even more dangerous to write briefly about adolescence, since one cannot hope to say very much that is new about such a vast topic within a few pages. Again it is often the novelist, rather than the professional social scientist, who has best characterized adolescence. Carson McCullers' *The Member of the Wedding* (1946) is an outstanding example. What social scientists have had to say about adolescence has not usually been so illuminating. Too frequently they have been reporting the obvious in jargon terms, measuring the irrelevant, or making claims of extraordinary incredibility or obscurity. There have, of course, been exceptions. Few social scientists have displayed the good sense, insight and humanity of J. B. Mays in his book *The Young Pretenders* (1965), which admirably surveys the field in a highly readable form. In this chapter we can deal with but a few problems rather superficially, taking those which are of particular relevance to education.

The contemporary study of adolescents in Western countries has been considerably influenced by the work of anthropologists on preliterate or 'primitive' societies. The pattern of adolescent development is now known to be much more profoundly affected by the culture of the society than we had previously assumed. In some societies the transition from childhood into adulthood is a period of relatively smooth psychological and social development (Margaret Mead, 1928, 1930). Some societies facilitate the transition between these two states by means of *rites de passage* or initiation rites. These are ceremonials or rituals by which a change from one position to another is effected. Weddings in our own society mark the sudden change in position from being single to being married. Funerals mark another sudden position change, though here the new position that is entered is rather more nebulous. Some preliterate societies have

337

initiation rites for the transition from childhood into adulthood. The rites are often very dramatic and imbued with deep emotional significance. For example, the young boy is withdrawn from the tribe and left alone, naked, in the forest to fend for himself. On his return he is subjected to elaborate rituals, often involving quite terrifying experiences, in which he is symbolically reborn and given a new name. Circumcision or other forms of bodily mutilation are not uncommon. He emerges as a full adult member of the tribe. Such societies have effectively abolished adolescence in the Western sense of the term. Childhood, with its lack of responsibility and emotional and economic dependence on parents and other adults, gives way within a very short space of time to full adult status.

In our society we have no such *rites de passage* for the adolescent. It is true that the first job, reaching the age of eighteen or twenty-one, or getting married, have been regarded by some as the point at which one becomes an adult. In former days national service for males and 'coming out' balls or being presented at Court for upper-class females were somewhat akin to a *rite de passage*. Yet there is no consensus about the point where the transition is complete. The end of childhood and the beginning of adulthood are very blurred. There is, to use Ruth Benedict's term, a *discontinuity* between childhood and adulthood, with adolescence falling uneasily between the two. There is thus considerable ambiguity in the position of the adolescent. He is no longer a child, but not yet an adult, suspended between the two without a clear social role.

We have come to accept adolescence as a difficult time. This is hardly a new phenomenon. As the shepherd remarks in *The Winter's Tale*:

> I would there were no age between ten and three-and-
> twenty, or that youth would sleep out the rest: for there is
> nothing (in the between) but getting wenches with child,
> wronging the ancientry, stealing, fighting . . .

Recently we have come to speak of a *youth culture*. It is one of those sociological terms which has passed through the Sunday colour supplements into common parlance. It is a conception we shall have to examine carefully. The background is relatively well known. In contrast to life in earlier ages, the young are now all compelled to attend school for a minimum of ten years, a growing minority of older adolescents continuing for between one and ten years more in further or higher education. One of the main effects of schooling is the segregation of the adolescent from parents and small children for several hours each day. Since schools are age-graded, adolescents spend a tremendous amount of time in school with persons of their own age, and this spills over into out-of-school associations. In addition, the adolescent has in the post-war era become an important consumer. He has money to spend and a whole economic market has grown up around him to cater for his needs—or in some cases to promote needs which can then be catered for.

In spite of the popularity of the term youth culture, it is not easy to find a careful definition. Perhaps the most famous is that of Coleman (1961) who suggests that the adolescent

> is 'cut off' from the rest of society, forced inward toward his own age group, made to carry his whole social life with others of his own age. With his fellows, he comes to consti- tute a small society, one that has most of its important inter- actions *within* itself, and maintains only a few threads of connection with the outside adult society. . . . Our society has within its midst a set of small teen-age societies, which focus teenage interests and attitudes on things far removed from adult responsibilities, and which may develop standards that lead away from those goals established by the larger society.

Coleman uses two pieces of evidence to support this claim about youth culture in American society, both of which are extremely dubious. He finds that a very small proportion of boys want to follow in their father's occupational footsteps and uses this as an index of the apartness between the generations. This evidence is much more easily interpreted as a function of the changing occupational and educational structure of modern America. His second piece of evidence concerns the answers to the following item on his question- naire.

Which of these things would be hardest for you to take:
 —your parents' disapproval
 —your teacher's disapproval
 —or breaking with your friend?

Slightly over half the subjects said parental disapproval and just over 40 per cent nominated the friend. Coleman suggests that this indicates the peer orientation of many adolescents. Unfortunately the item is not balanced; the disapproval of parents is pitted against *breaking* with a friend. As Epperson (1964) has shown, when adolescents are faced with a choice between the disapproval of parents or friends, 80 per cent find parental disapproval harder to take, which hardly indicates massive peer orientation.

It will be easier to analyse the concept of youth culture if we can find a more adequate definition than Coleman's. Most are as vague as that of Coleman, but a recent British writer, Sugarman (1967) has summarized the two principal elements in earlier definitions as, first,

339

'the existence of distinct values and norms among youth which conflict with some of those held in adult society' and, second, 'strong pressures among the young enforcing conformity to the norms of youth culture in preference to those of adults.' This is still rather vague. The values are alleged to conflict with *some* of those held in adult society, though these are not specified in detail. Similarly Coleman asserts that youth *may* develop different standards from those of adults.

It may help to clarify some of the issues if we introduce the concept of *reference group*. The term was first introduced by Hyman (1942) and has subsequently been developed by social psychologists (notably Newcomb, 1950; Sherif, 1953; Shibutani, 1955) and sociologists (notably Merton, 1957). Traditionally, a reference group is a group in which a person seeks to attain or maintain membership or in whose terms he evaluates himself. Kelley (1952) has suggested two main functions of reference groups. The first function is *normative* and is specified in the definition just given. Here the person wishes to become a member of the group or to maintain his membership in it. To do this, he conforms to the group's norms, adopts its values and evaluates himself in these terms. The members of the group become his significant others. With respect to the normative function a reference group can be negative. A negative reference group is a group of which a person would like to cease being a member or of which he would have definite antipathy to becoming a member; he would not evaluate himself in the group's terms. The second function of reference groups is *comparative*. In this case a person uses the reference group as a standard of comparison with which to estimate his own position. For example, as a university teacher I might use schoolteachers as a comparative reference group when they are given a pay rise. I may have no interest in them normatively, i.e. I do not want to become a schoolteacher or to conform to their norms, but I may take their pay rise as a basis for estimating my own financial position. By comparing my position with theirs, I might feel economically deprived. Thus groups, whether or not one is a member, can be used as reference groups with one or both functions.

To make the concept clear, we can consider the data of the Lumley study (Hargreaves, 1967) described earlier. The low stream pupils used the high streams as a comparative reference group. They complained that the high stream pupils were given privileges denied to them. They felt deprived relative to the upper streams. Yet at the same time they used the upper streams as a negative reference group with respect to the normative function. The norms and values of the upper streams were the inverse of their own; they did not want to be transferred into the higher streams and intentionally under-achieved in order to avoid transfer.

The concept of reference group is a useful analytical tool in the study of the processes of social influence and in the explanation of the uniqueness of individuals. A person does not merely take to himself the dominant attitudes of his culture and subculture (Mead's 'generalized other'), but from all the possible others available to him selects some others as significant others. Moreover, the reference group which is taken as a set of significant others need not be a group of which the person is a member. Social influence is not confined to those others with whom fate has put a person in immediate contact, nor even to those who are living, for reference groups can be taken from the past or from unborn future generations. As yet there is little research on the determinants of the selection of particular reference groups, but it is clear that we cannot account for individual action without considering both the groups of which a person is a member and the groups of which he is not a member, and then that person's orientation (positive, negative, neutral; normative, comparative) to these groups. The individual's perspective, with its complex processes of meaning interpretation through which action is constructed, is related to that individual's significant others. The concept of reference group can assist us in this endeavour.

The adolescent is not, by definition, a member of the adult group. He is a member of the adolescent group as a whole and he is a member of one or more particular friendship groups within the totality of his age-mates. All these groups can form reference groups, both normative and comparative, to the adolescent.

Comparative reference groups

The adolescent resents the fact that he legally cannot drink alcohol until he is eighteen.

The adolescent complains to his parents that he is not receiving enough pocket money. He argues that nowadays all adolescents get more than 15p a week to spend.

The adolescent feels badly treated because all his friends are allowed to come home late on Friday and Saturday evenings.

In these three examples the adolescent is using adults, all adolescents and his particular friendship group respectively as comparative reference groups.

Normative reference groups

(a) *Positive.* An adolescent grows his hair long because all the members of his friendship group do so and he wishes to conform to this norm in order to be accepted.

An adolescent refuses to grow his hair long because he knows that

341

his parents and other adults disapprove of the style and he wishes to be accepted by them.

(b) *Negative.* An adolescent grows his hair long because he knows that his parents and other adults disapprove. He does not wish to be an adult or to acquire their norms and values which he despises.

The adolescent grows his hair short because he wishes to dissociate himself from his adolescent friendship group where long hair is the norm.

Thus an adolescent may determine his hair style in response to various social influences. The concept of reference group helps to clarify this social influence. Hair style can be determined in relation to adult or adolescent reference groups, positive or negative. It may represent conformity to a positive reference group, adult or adolescent, or a reaction against a negative reference group, adult or adolescent. It may of course, be a combination. For example, he may grow his hair long in order to gain acceptance in his peer group (adolescent positive reference group) *and* in order to reject the values of adults (adult negative reference group). A third possibility is that the adolescent may be caught between two positive reference groups. He may be pressured by his friends to grow his hair long and by his parents to keep it short. Presumably the final length of his hair will indicate which is the stronger of the two positive reference groups.

We can now set up three basic reference group situations.

Type A   Peers form a positive reference group.
Adults form a negative reference group.
Type B   Peers form a negative reference group.
Adults form a positive reference group.
Type C   Peers form a positive reference group.
Adults form a positive reference group.

The proponents of youth culture seem to be arguing that the adolescent is basically in a Type A situation. The argument of the first part of this chapter will be that this is an erroneous view, based on a superficial interpretation of the facts and an inadequate conception of social influence during adolescence. The argument maintains that Type B and C situations are very common for most adolescents and that very careful research is needed before we can explain an adolescent's actual behaviour as belonging to a particular Type. As the illustration of length of hair indicates, it is very easy to mistake a Type A situation for a Type C situation.

If the majority of adolescents are basically in a Type A situation, if they possess values and norms which conflict with those of adults, then we should expect relationships with parents and other adults to be generally rather poor. Much of our present evidence seems to suggest that in general adolescents have favourable attitudes to parents and to adults. In the work of Musgrove (1964) the average percentages of favourable attitudes to fathers, mothers and peers were 49, 78 and 49 per cent respectively. Only about a third of their statements about adults in general were wholly or mainly critical. Musgrove concludes:

There was little support in this inquiry for the often alleged 'solidarity' among adolescent males. Their approving statements about their peers were no more frequent than their approving statements about adults in general, less frequent than their approving statements about mothers, and up to the age of 15 less frequent than their approving statements about fathers.

Wilmott (1966) shows that between 40 and 60 per cent of his adolescent boys in a poor district of East London think that father understands them very well. The study of the Eppels (1966) suggests that there is a generally high commitment to accepted social values by adolescents, though the sample here is somewhat atypical. Other studies (e.g. Morris, 1958) also throw doubt on the notion that the generations are marked by hostile relations or sharp value differences.

What is much more striking in the research is that adults tend to be highly deprecating of adolescents. In Musgrove's study no less than two thirds of the adults were wholly or mainly critical of young people. Why should adults be more hostile to young people than young people are to adults? Perhaps in part adults suffer from a certain envy of the young. Certainly they often behave as if they are envious of the freedom and the relative wealth of young people today. They are constantly reminding the young, 'When I was your age I had been working for years and my pay was only . . .' Their complaints against the young imply that if they had had such freedom and wealth they would have used it much more wisely—which is easily claimed from a middle-aged standpoint.

It seems reasonable to suggest that adults interpret adolescent behaviour as deviating from accepted social values and standards much more radically than is really the case. It would be an excellent rationalization of their hostility. The generation gap would then be perceived as much more pervasive and fundamental than it really is. Certainly the evidence does not point to such a pervasive disjunction in values. One of my own students, Niles (1968), following the pioneering work of Brittain (1963) in the United States, put adolescent girls in a situation where they had to choose to accept the influence of parents or peers in the solution of certain problems. In matters of taste or personal appearance, peer influence was preferred to that of parents. For instance, 65 per cent of the fifteen year olds would change their hair styles to please friends rather than parents and

76 per cent would adopt a dress colour approved by friends rather than parents. But in more important matters, especially those involving moral questions or long-term decisions, parental influence tended to be accepted. For example, when faced with the problem of which of two boys she liked equally well she should date, only one-third accepted the recommendation of friends rather than parents. When pressured by friends to stay late at an exciting party at the cost of worrying her mother, less than one-third actually stayed.

In one of the best recent studies of adolescence, Douvan & Adelson (1966) suggest that, whilst the peer group does become more important in adolescence, many writers have over-emphasized the potency of peer group norms and the discrepancy between parental and peer standards. They suggest that many of the conflicts between parents and adolescents are over trivial matters rather than over basic values and norms. They cite the case of bitter family disputes about popular music as being typical. They suggest that many adolescents make mock revolts and that parents and adolescents enter into a tacit understanding to disagree only over 'teenage' matters, thus allowing this pseudo-rebellion to forestall serious questioning of basic values.

Putting the matter in reference group terms, it seems that in matters of taste and personal appearance many teenagers are in a Type A situation. Peer group influence is accepted and parental influence positively rejected with a certain amount of defiance. Occasionally, when very serious long term or moral decisions are at stake, a few teenagers are in a Type B situation, where adult influence is accepted and peer influence rejected with suspicion. For the most part, however, the adolescent is in a Type C situation, susceptible to the influence of both parents and peers. Generally speaking, when the problem is relatively inconsequential, peers will win; when the problem is of deeper significance, parental influence will be accepted (though the adolescent may often try to conceal his acceptance of parental influence, since he wants to be, and to be thought of as, an independent agent). Many decisions are naturally not of this simple either/or variety. The conflict can be resolved by compromise between the two positive reference groups. Good evidence for this can be found in Musgrove's (1964) work. The self-expectations of adolescents with respect to such items as 'Behave sensibly and generally "act your age",' 'Be respectful, courteous and polite to your parents and elders', and 'Be quite free to stay up late, or to stay out late, if you wish' fell almost without exception somewhere between the perceived expectation of adults and the perceived expectations of friends. And in eight cases out of twelve the compromise was closer to the perceived expectations of adults than of friends. The research of one of my students, Winder (1970), shows that the educational values of adolescent boys are a compromise between the perceived educational values of friends and parents. Other studies (e.g. Riley, Riley & Moore, 1961) fit this notion of an adolescent compromise between two essentially positive reference groups.

Of course, many of the proponents of youth culture do not seem to accept such an interpretation; Sugarman (1967), for example, states

> This repudiation of [parental and adult] standards extends
> beyond matters of taste and fashion to a repudiation of
> other adult standards, so affecting the behaviour of the
> young over a wide area.

Unfortunately he does not bring any evidence to support this contention.

It is clearly important to try to explain the negative attitudes of adults to young people. Part of the answer may be that adults are victims of the mass media view of the young, for it is the delinquents, drop-outs and other deviants who are so persistently news. Perhaps this biased presentation of the behaviour and values of adolescents has been believed by adults to the point where they now accept further news as further confirmation of their belief in deep inter-generational conflict. If so, they are in the company of several social scientists.

Perhaps, too, the adolescents have come to accept this picture of themselves, despite their occasional protestations that they are grossly misrepresented by the mass media. The researches of Musgrove and Winder indicate that many adolescents perceive and believe that their peers have standards, values and attitudes which diverge markedly from those of parents and adults *and* from their own actual values and attitudes. We obviously need more research on this question, but it does seem that whilst a small minority of adolescents do diverge in fairly fundamental ways from accepted values, adolescents in general have come to believe the myth that this dis-junction in values is much more widespread than is in fact the case.

The mass media and social scientists are not, of course, entirely responsible for the myths about adolescence. That would be ascribing to them a quite unjustifiable influence on public opinion. It is easy to see how, in everyday experience, adults are led to make inferences about deep value differences on the basis of what are relatively superfical differences. Take the case of manners. Many young people in my experience pay relatively little attention to what we have traditionally regarded as good manners. Like many other adults I applaud their rejection of elaborate etiquette and formality in social relationships. But like many other adults I also find myself distressed by their rejection of good manners defined as a basic consideration

345

for other people's pleasure and a respect for other people's opinions. It is difficult for an adult not to be intensely irritated by loud talking and shouting in public places such as buses and theatres, and by the denial of free speech to those whose views differ radically from their own. But we have to take great care that we do not allow these irritations to lead us into imagining that a whole generation is rejecting the fundamental assumptions of our society. Negative halo effects are, as we have seen, potentially very dangerous.

Often I suspect that the conflicts which do arise between adults and young people are not so much rooted in real value differences but in the fact that adolescents behave in adult ways which adults themselves regard as *inappropriate* to adolescents. Sex is a typical case. The sexual problems of adolescents are partly induced by our culture. We make inconsistent and impossible demands on adolescents with respect to sexual behaviour. We know that the vast majority of adolescents are physically mature by the age of fifteen. Not unnaturally they experience sexual urges. We adults try to insist that sex is good and natural, but we stipulate that this is so only within marriage. This means that the adolescent will have to remain celibate for between five and fifteen years after reaching puberty before he can practise sexual intercourse. This in itself poses a major problem for the adolescent. Matters are made worse by the fact that during childhood and early adolescence we have shielded children from sexuality as if it were a nasty part of reality. The work of Schofield (1965) has shown that 62 per cent of boys and 44 per cent of girls get most of their information about sex from their peers. Thus sex soon becomes associated with sin and the dirty joke. This is contrary to the current view that sex is not inherently bad at all. It appears that we have thrown off our Victorian heritage only at a superficial level, though the coming generation is having more success in this respect. Finally, we subject adolescents to massive sexual stimulation through films, books, newspapers and advertising. The choice for the adolescent seems complex. We demand that he abstains from sex, though he is constantly bombarded with titillating invitations to its delights. We expect him to struggle with his impulses as best he can. On the other hand, if the adolescent gives in to his desires, we expect him to feel guilt and shame at his self-indulgence and are even more scandalized when he fails to give evidence of such feelings.

The work of Kinsey, and of Schofield in Britain, shows that we are not terribly successful in damming up teenage sexuality, though we do manage to ensure that in the male most of the outlets take the form of masturbation, which we also tend to condemn. Schofield's figures show that by the age of eighteen one boy in three and one girl in six have experience of full sexual intercourse. It is impossible to say whether or not this is an increase in teenage sexual behaviour. I suspect it is, largely because the Victorians were at least more consistent in their attitudes to sex.

Sex illustrates my point that it is not sexual practices in themselves which adults condemn, but rather that they wish to preserve them as a privilege of the married adult. Extramarital sexual relationships are much more easily tolerated in the adult. The conflict is not about values or behaviour as such. The conflict is that the adolescent adopts adult beliefs and behaviours at a point which adults consider to be premature. The adolescent has become an adult too soon. Taking the view that Iris Murdoch is the Jane Austen of this century, it is clear that adults in all sections of society are rather fascinated by sex, but it is a fascination that adolescents are not expected to develop.

With respect to sexual matters some writers seem to imply that once again adolescents are in a reference group Type A situation, with adults forming a puritanical negative reference group and peers forming a permissive positive reference group. Whilst this may be correct in some cases, I doubt if it is generally. Given the great changes in attitudes to sex in recent years, changes which can hardly be said to be confined to the young, the surprising thing is the large areas of agreement between the generations in sexual matters. In line with my previous argument, it does seem that some adolescents treat adults as a *comparative* rather than a negative normative reference group, feeling that they are unjustly treated with respect to sex.

The sexual problems of adolescence must be seen in the wider context of the adolescent task of preparing for marriage. Since marriages in our society are not arranged, but freely entered after choice, the adolescent period is used for acquiring the experience that is necessary for making that choice. As Wilmott (1966) has shown, the normal pattern is to belong to a group of friends of the same sex, then to move to a group of mixed sexes, and finally to an attachment to a single member of the opposite sex. (After marriage there is a curious tendency to return to the first stage.) Dating, in pairs or with a mixed group, acts as a means whereby the adolescent can come to test feelings about and explore relationships with the opposite sex. After 'going out' with one or more persons in an uncommitted way, a couple begins to 'go steady' and becomes 'serious'. By progressive steps they are led to engagement and then marriage. It is a path of progressive involvement and commitment (Waller, 1937). The problems posed for the adolescent become clear when we contrast this situation with a society where marriages are arranged by parents, as in traditional India. Our society makes the route to marriage like a visit to the supermarket. One is expected and quite free to look around the shelves for a suitable and pleasing purchase. One is quite free to walk around at leisure and to take one's time in making a choice. One is free to put a ware into one's basket and then change one's mind and return it to the shelf—though it is not advisable to do this too often. There is no real commitment until one reaches the cashier, and once past this point the choice has been made. But the choice is complicated, since in contemporary society many adolescents are shopping in a hurry and have not time to look round the store. In addition, parents and other well-meaning adults are constantly at the adolescent's side, giving full advice about the best buy. In a society where marriages are arranged, getting married is like working through a mail order firm whose catalogue is unavailable except to parents.

Nor is this the end of the problem, especially for girls. Western society has what has been called the Romantic Love Complex. One is expected to 'fall in love' with one's marriage partner before marriage, not afterwards as in societies where marriages are arranged. This Romantic Love Complex consists of three elements; an exclusive and deep emotional attachment to the other, a feeling that the relationship was destined, and a tendency to idealize the other. It is a notion celebrated by the Hollywood Doris Day–Rock Hudson films, by magazines for teenage girls, and by pop songs. The difficulty is that the reality of married life, the routine, the hairpins and the nappies can come as a severe and disillusioning shock. One sometimes finds, to return to our supermarket analogy, that the advertising has been very misleading. It is the lucky ones who are pleased with their purchase or grow to like it and find that they have an unexpected gift of Green Shield stamps. The Romantic Love Complex can inhibit rational choice or the paying of attention to important matters such as common interests. The disillusioned wife comforts herself by indulging more than ever in romantic fiction.

One form of adult resentment against the young may spring from the fact that adolescents express some social values more openly, with less guilt, shame or ambivalence. One of the dominant values in our society is hedonism—having a good time. In my view this is not, as some have argued, a value confined to youth. We all seem to be scrambling to make money, and more money, in order to worship the gods of entertainment, food and drink, holidays abroad and the rest of the pantheon. Bingo and football pools enjoy an unprecedented boom among adults. It is known that whenever there is a very large win on the pools, large numbers of irregular punters fill in their coupons with renewed vigour and hope. Indeed, the points system for the Treble Chance was changed in order to increase the chances of a single person scooping the pool.

Not surprisingly many young people come to worship at the same or similar shrines. The difference is that the young do not have the responsibilities or the restrictions of adults and can thus indulge themselves rather more thoroughly. The adult stresses the so-called 'balanced view'. He says, 'I'm not against your having a good time, but remember . . .' and the 'but' is always restrictive. But the adults would probably follow the same path, if only they could. Indeed, much of the entertainment provided by television, films and books or magazines seems to be aimed at giving adults vicarious pleasure by recounting the experiences of real or imaginary persons who can.

The other difference is that adults are caught in the midst of a changing and progressively more affluent society. With more wealth, leisure and opportunity many adults are beginning to shed the Protestant and puritanical ethic of the middle classes (and perhaps the working classes) of the past. As a result they tend to be genuinely ambivalent in their values and attitudes. They espouse the current hedonism but try to articulate it with the more traditional values of hard work and deferred gratification, and a traditional sexual morality. As a result they have one foot in the past and one foot in the present and are not sure whether to more forwards or backwards. They tend to be inconsistent and at times hypocritical. The woman who has nothing but contempt for the young but spends a fortune on making herself look as young as possible represents a whole genera-tion of confused adults who expect young people to do as adults say, not as adults do.

Adults tend to judge young people in terms of the more traditional values rather than the current ones which they partially espouse. Kingsley Davis (1940) has said:

> Since the parent is supposed to socialize the child, he tends to apply the erstwhile but now inappropriate culture con-tent. He makes this mistake, and cannot remedy it, because, due to the logic of personality growth, his basic orientation was formed by the experiences of his own childhood. He cannot 'modernize' his point of view, because *he* is the pro-duct of those experiences. He can change in superficial ways, such as learning a new tune, but he cannot change (or *want* to change) the initial modes of thinking upon which his sub-sequent social experience has been built. To change the basic conception by which he has learned to judge the rightness and reality of all specific situations would be to render subsequent experience meaningless, to make an empty caricature of what had been his life.

There are some adults of whom this is still true. They are the ones who abide by all the old values and resist change. But for many adults things have changed since 1940 when Davis wrote. They are partially committed to newer values, but tend to evaluate youth by the older values in which they were socialized when they were young. Their ideals, created in their youth, are beginning to crumble in their own lives, so they are displaced on the young. They set standards for the young which they cannot themselves achieve. And when the idealism of contemporary youth—of which there is ample evidence—goes against their own assumptions, it is condemned as a product of the permissive society.

The concept of a distinctive youth culture begins to look rather thin. It is not at all clear, when one looks below the surface, what is indeed so very distinctive about the young in the sense of values which conflict with those of adults. Perhaps in the end what is alleged to be youth culture is no more than pop culture. There is no doubt that young people are frequently influenced (and perhaps exploited) in the matter of styles of appearance, of clothes, of musical taste and dancing. It is hardly surprising that new trend setters have been found for this very important commercial market, even to the point at time of superseding the hegemony of the Parisian *haute couture* salons. But again these differences are hardly differences in values, for many women have always wanted to follow the fashion and so also, though perhaps to a lesser degree in the Victorian era, have men.

We also need to remember that pop culture is not exclusively a concern of youth. Adults have tended to follow, after a while, most of the youthful fashions. Women shortened their skirts following the mini-skirt craze; men have let their hair grow somewhat longer than was required by the traditional short-back-and-sides. The middle-aged have absorbed the passion of youth for vivid colours in dress design for women and psychedelic ties for men. Even the musical market is not consumed exclusively by the young. It may well be true that the vast majority of pop records are bought by teenage girls. But through the country hundreds of thousands of women between twenty and fifty listen to the same records on the radio. The BBC would not have so radically transformed its radio programmes in response to the threat of the pirate stations like Radio Caroline if pop music had only teenage girls for an audience. Today's pop stars, idolized by screaming teenage girls, have a curious habit of becoming the heroes and heroines of many adults tomorrow, as in the case of Cilla Black and Cliff Richard.

The second element in the definition of youth culture asserts conformity among the young to the values and norms of youth culture in preference to those of adults. Since there is, in my view, little evidence of distinctive values and norms, the assertion about conformity to them seems superfluous. However, it might also be claimed that young people are conformist to the norms of the peer group. This would not be surprising, yet some of the writers on youth culture have spoken of *compulsive* conformity among the young. This implies two things; that such conformity is undesirable, and that young people are more conformist than are members of adult groups.

The first is a value judgment and is not open to verification, but the second can be put to the test.

All cohesive groups put members under great pressure to conform to the norms, so there is a sense in which the young can be regarded as highly conformist if it can be shown that their groups are more cohesive than those of adults. It might also be possible to show that young people are less resistant to social influence under conditions of equal pressure. I know of no satisfactory evidence to settle either of these two issues. There are, however, certain grounds for believing that the contentions may to some degree be true. The adolescent is likely to spend more of his time in intensive interaction with his age-mates than do many adults. He is often segregated into a friendship group for a large part of the school day and he is likely to spend much of his spare time with his friends in the evening and at the weekend. The male adult is likely to belong to a work group, but in many cases the opportunity for intensive interaction will be small. The average adult woman is a housewife and again is not likely to spend a large portion of her time in the company of friends and neighbours. Both male and female adults spend much of their evening and weekend spare time at home within the family, rather than in the company of work-mates or friends. So there are grounds for believing that many adolescent groups may be more cohesive than adult groups simply because the members interact much more frequently and mix with other groups or individuals with less frequency. If the groups are highly cohesive then we should expect a high level of conformity to group norms. We have not, however, any grounds on the basis of this evidence alone for believing that the level of conformity would be any higher than that in an adult group of equivalent cohesiveness.

It is on other grounds that it can be argued that the adolescent is particularly susceptible to group pressures. One of the central tasks of adolescence is the gaining of independence from parents. He has to abandon childhood dependence in favour of a greater degree of personal autonomy, both in his goals and in his ability to achieve them. In this process the adolescent tends to be ambivalent in his attitudes and inconsistent in his behaviour. At times he will struggle fiercely to obtain his independence and be resistant to adult control. To the annoyance of parents he will reject their advice, for he wishes to experience life directly rather than accepting vicariously the experience of adults. At other times his fear of life, of the unknown, of the complex may cause him to revert to a child-like dependence on parents. He will swing between being sociable then lonely, submissive then rebellious, ascetic then self-indulgent, enthusiastic then apathetic, altruistic then self-centred and so on.

Inevitably parents find this rather tiresome and difficult to cope with. Some overprotect or dominate (placing the adolescent in a permanent Type B situation). More typically, they also respond ambivalently. In one situation they claim, 'You can't do that—you're not old enough.' In other situations the adolescent is greeted with, 'Why on earth don't you behave like a grown-up?' When I was an adolescent, if I stayed at home during the evening for a whole week, my parents claimed that it was unhealthy and that I ought to go out more. If I then went out every night the following week I was told that I was using the house like a hotel.

Tragically, conflict between parents and adolescents can arise from the inability of each to take the perspective of the other. Take the case of the mother and daughter arguing about the time at which the daughter should return home in the evening. The mother insists that the daughter should be home by eleven o'clock. She imposes this limit since she is anxious that her daughter does not find herself in difficulties late at night and that she gets a good night's rest. The daughter, however, perceives this time limit as a restriction, motivated by a lack of trust in her. The daughter's angry reaction is then interpreted as a rejection of her mother's care and affection. Soon a full-scale row is in mid-flight, leaving both in a state of emotional distress. A great deal of parent–adolescent conflict may have its roots in situations such as this.

The effect of such conflicts in the home can be that the adolescent turns to his peers for help, comfort and understanding. If he believes that only his friends really understand him, then he may be particularly open to their influence. The peer group does serve an important function in helping the adolescent to emancipate himself from his parents, but it is possible that part of the price for achieving this independence is that for a period the adolescent will tend to be highly conformist to peer group values and norms. Fortunately, of course, many of these values and norms are ones which support those of the home.

More important, the peer group tends to be a main source of status to the adolescent. In society as a whole the adolescent has very little status. Indeed, the word adolescent is often used pejoratively by adults. Within the peer group the adolescent can work out a new status that is independent of his parents. As a child, status is *ascribed*; the child's significance accrued largely from the significance of his parents. During the adolescent period the adolescent is seeking to *achieve* status for himself. So in addition to compensating for a lack of status elsewhere, the peer group provides a medium in which he can achieve status in his own right. For this reason, peer group status is an important area of concern to the adolescent. A teenage girl is soon reduced to tears if she cannot have the right dress for a party, simply because she feels that her status with her friends may in part depend upon meeting the dress norm, at least minimally. Actually, she often hopes to exemplify the norm in order to maintain or enhance her status.

On grounds such as these it can be argued that group conformity in adolescence is particularly high. Fortunately in most respects there is little cause for alarm, partly because it is a natural part of growing up and partly because the pressure to conform is in quite innocuous areas. Where drug-taking or delinquency is the norm, there is cause for concern. There are no easy answers to problems such as these with which parents can suddenly find themselves confronted, but the avoidance of misunderstanding and conflict in quite superficial areas by parents might effect more than we could reasonably hope. Unfortunately parents find it very difficult to sort out the wheat from the chaff during a time when they are struggling to recognize that their child is growing up and to accept the fact.

We must now turn more directly to the relation between youth and the school. Without doubt the most important and influential study has been Coleman's *The Adolescent Society* (1961). As I have noted earlier, Coleman and others have used this work as evidence of a general youth culture. The book itself suggests that Coleman himself set out with the intention of collecting evidence to support his belief in a distinctive youth culture, though it is possible that he came to this conclusion when he had to interpret the material gathered in relation to his central objective, namely that of exploring the values and social system of youth in the American high school. His study is based on the questionnaire responses of over eight thousand pupils in ten high schools in the state of Illinois.

Coleman argues that the values and social system of youth in the high school are directed away from the major social goals of adults in contemporary American society. The evidence he produces is substantial, though some of it is of a dubious nature. For example, one item asked:

If you could be any of these things you wanted, which would you most want to be?

Boys
Nationally famous athlete (37%)
Jet pilot (31%)
Atomic scientist (26%)
Missionary (6%)

Girls
Model (32%)
Nurse (29%)
Schoolteacher (21%)
Actress or artist (18%)

The percentages of pupils choosing each response are given in parentheses.

It is very difficult to know how to interpret the answers to this question. Both the wording of the question and the alternatives allowed implicitly suggest that a fantasy rather than a realistic career choice is being probed. Very few of the boys are likely to become any of the choices offered to them, and only two of the girls' choices, schoolteacher and nurse, are likely to be open to realistic aspiration. (Even the nurse, with its Dr Kildare overtones, can easily acquire a fantasy quality.)

A second item asked:

If you could be remembered here at school for one of the three things below, which one would you want to be?

Brilliant student (boys 31%, girls 29%)
Most popular (boys 25%, girls 36%)
Athletic star (boys 44%)
Leader in activities (girls 35%)

This indicates that academic brilliance is a less attractive image than athletic brilliance in the case of boys, and than popularity or activities leadership in the case of girls.

When the parents of these pupils were given a very similar question, a different picture emerged. No less than 77 per cent of the boys' parents selected the brilliant student, as did 55 per cent of the girls' parents. The evidence suggests that there is a marked divergence in values between parents and their children. However, Coleman asked two further relevant questions of his pupil subjects. In one situation the pupil was asked by the teacher to act as the assistant in the class; in the other, the pupil became a member of the basketball team (boys) or the cheerleader (girls). The pupils were asked to report how proud their parents would be of these achievements. The boys report that their parents would show greater pride in athletic rather than academic achievement, and the girls similarly indicate greater parental pride for the non-academic achievement. What, then, are the parents' real values? Are the adolescents projecting their own feelings onto their parents? Or is it that the parents, when asked directly, simply produce an answer which they regard as socially desirable, namely academic brilliance? Perhaps the parents hold both sets of values. They want their children to do well at school, in order to gain entry to higher education and prestigious occupations, and they also want their children to shine in non-academic areas. Perhaps the average American parent has more than he would care to admit in common with Willy Loman in Arthur Miller's brilliant commentary on the American way of life, *Death of a Salesman*. It is in Willy

Loman that the American dream of money, fame and popularity finds its quintessential expression, both in his own failure to reach the dream and his attempt to conceal that failure, and in his desire to realize the dream through his son, Biff. When Biff goes off to an important match where he will become captain of the All-Scholastic Championship Team of the City of New York, Willy sees this as an event of major significance, for it will help Biff to be *liked*. In consequence the universities will beg for him and the doors of the great American business world will be opened to him. The dream is about to be realized.

Similarly, American parents are ambivalent towards their daughters, wanting them to be good students with good grades, but not at the expense of the traditional feminine role of glamorous and popular non-studiousness (Komarowsky, 1946). In Wallin's (1950) study, more than half the female students had been advised to 'act more feminine', and the main source of such advice was parents, especially the mother. Coleman himself acknowledges that parents help to reinforce the values of the adolescent culture, but he claims that this is not because they share the same values, but because they want their children to be successful and esteemed by their peers. Obviously were Coleman to concede that they share the same values, then his contentions about an autonomous adolescent culture would collapse.

Yet there is much in *The Adolescent Society* which stresses the similarity of values between adolescents and adults. Coleman writes:

> Cars are an important matter to a teen-ager ... *As cars have become more important for adults*, they have become more important for their adolescent children—as parents of adolescents will quickly attest. Consequently, as a boy reaches the age when he can possess a driver's license ... his pressure upon parents for permission to have a car of his own becomes extremely great. Parents find it hard not to give in, at least by letting him buy a car if he can afford to pay for it. (Italics added)

By the end of the fourth year in high school, almost half the boys possessed a car (range 17 to 81 per cent for different schools), and boys with high informal status were more likely to own a car than others.

In their study of adolescent males, the Sherifs (1964) note the same finding.

> Automobiles were a regular preoccupation in every group —one way or another. Whether they owned cars or not, these American boys discussed, compared and admired cars.

355

> Those who did not own a car knew what kind they wanted,
> and frequently faced problems of having access to one—
> in order to go some place or to take out a girl. Those who
> did have cars spent an amazing length of time in and around
> them with their pals . . .
> In our research, the intense interest in cars is also re-
> vealed by the adolescent's ability to specify the make,
> model, year and even the colour and engine type of the car
> he desired. Such specificity is impressive, particularly when
> compared with the general nature of preferences for occu-
> pations. . . . The findings almost force one to conclude that
> the automobile is a major cultural goal to American youth.

The Sherifs also report that in neighbourhoods of middle or high social rank, accessibility to a car is almost a 'must' for those boys with high informal status.

Now these facts must be interpreted in the light of the fact that cars are a major preoccupation of most adult males in our society, and, though perhaps to a lesser degree, an important source of status. We need to speak of a car culture, in which adolescent males share, rather than an adolescent car culture.

The Sherifs do not speak of an adolescent culture. Their study emphasizes the involvement of youth in, and their commitment to, the dominant American culture.

> There is one clear and striking generalization about the
> high school youth which holds in all areas and despite their
> differing backgrounds: their values and goals earmark them
> all as youth exposed to the American ideology of success
> and wanting the tangible symbols of that success. There were
> no differences between the youth in different areas with
> respect to desires for material goods. In addition to com-
> fortable housing, the symbols of success for these adolescents
> included a car in every garage, a telephone, a television set,
> transistor radios, fashionable clothing, time to enjoy them,
> and money to provide them . . .

Whilst it is possible to reject Coleman's claim to demonstrate the existence of a distinctive adolescent culture in the school, it is essential to recognize that he makes a major contribution to our understanding of the social system of the adolescents in school. In particular, his analysis of the 'leading crowd' or the high status pupils in school is both important and interesting. In every school the male members of the leading crowd were involved in athletics to a striking degree. Girl members of the leading crowd were highly involved in the school's extracurricular activities; indeed, such participation is a *sine qua non* for membership of the leading crowd. These facts are not very surprising, since if involvement in athletics and activities are part of adolescent values and norms in school, then we would expect the high status pupils to exemplify these values and norms to a greater degree than other students.

Having high status, athletes have more friends and are more popular with both sexes than are other male pupils. However, it is also important to note that the male scholar also has more friends and is more popular with both sexes than the rest. Coleman's finding is not that all informal status goes to the athlete rather than the scholar, but that whilst both have higher status than the rest, it is the athlete who constantly outdistances the scholar. Of course, the athlete and the scholar are not mutually exclusive categories; some 1·3 per cent of the student body are both athletes and scholars. Significantly, the athlete *doubles* his status, by the measures used, when he is also a scholar (Table 10.1).

TABLE 10.1 *Four measures of pupil status* (Data taken from J. S. Coleman, *The Adolescent Society*, Free Press, 1961)

	(a)	(b)	(c)	(d)	N
Athlete-scholar	9·9	12·5	7·1	4·9	54
Athlete	4·6	6·6	5·9	2·5	218
Scholar	1·9	3·1	4·4	0·5	224
All other boys	0·4	0·8	2·9	0·2	3,598

(a) Nominated under 'I would like to be like . . .'
(b) Nominated as member of the leading crowd.
(c) Number of friends.
(d) Nominated under 'the girls go for most'.

The problem is to explain these interesting differences. First, Coleman makes it clear that the athlete is more *visible* than the scholar. There is high consensus about the identity of the athletes because they are out there regularly on the sports field for all to see. They are constantly under a public spotlight in a way that the scholar is not. The athlete is much more available as a focus for adolescents and as a potential model. Secondly, there is the actual structure of school activities. In games the athlete is doing something for the school (and the community) as a whole; his victory is the school's victory. By contrast, the scholar's success is purely personal and is in a sense gained at the expense of his classmates. In Coleman's words:

> The scholar's efforts can bring glory to no-one but himself, and serve only to make work more difficult for the others.

But the athlete's achievements occur as part of a collective effort. He is working for his school, not merely for himself, and his extra efforts bring acclaim from his classmates, while loafing brings rebukes—all a consequence of the structure of activities in a school, which allocates interscholastic games to athletics, and allocates interpersonal competition (in the form of grades) to academic work.

Doubtless there is much truth in this assertion, and it could be used as further evidence against competitive evaluation of pupils in school. But it is obviously not a full explanation, since it does not account for the fact that the scholar has more not less status than ordinary pupils, not does it account for the spectacular gain in status of the athlete–scholar over the athlete. It seems that it is not academic brilliance as such which is frowned upon by the pupils, but the *studious* pupil. As Tannenbaum's work (quoted by Coleman) shows, the brilliant student is preferred to the average student, but the non-studious student is preferred to the studious. The reason is that studiousness threatens the mediocrity norm to which we have referred earlier, the norm which forbids very hard work. Brilliance (being intelligent or academically gifted) does not in itself threaten the mediocrity norm, which is about effort. If the brilliant pupil works only for as long and as hard as other pupils, then the other pupils will not complain; they will somewhat enviously admire. Indeed, the brilliant student is defined as brilliant because without special or devoted effort he can still make outstanding academic achievements.

Coleman rightly notes that the scholar is often met with ridicule, and is sometimes defined as deviant, being called a 'swot' or a 'grind' or a 'curve-raiser'. Unfortunately Coleman does not recognize that it is the studious student, not the brilliant one, who breaks peer group norms by his overt academic efforts. His category of the scholar does not distinguish between brilliance and studiousness. It is possible that the brilliant but non-studious scholar might have a status which is very similar to that of the athlete, whilst the brilliant but studious scholar is of markedly lower status. It is possible that the athlete-scholar has such a strikingly high status because his involvement in athletics (demanding, as it does, considerable amounts of time for practice, etc.) is taken by his peers as evidence that he is not studious. Athletic involvement might be regarded as proof that he works no harder than the rest.

Coleman believes that the problem will not be solved by the abolition or reduction of competitiveness in school, for he seems to regard competition as an inherent part of our society and of human nature. Rather he wishes to change the form of academic competition, away from competition between pupils in the same class, to inter-group competition. In other words, by institutionalizing scholastic competition between schools, in the form of science fairs and other types of academic contests, he seeks to give academic affairs a structure similar to that of games. This, he believes, would bring greater rewards to the scholar since he would be serving the school as well as himself. The pupils would have an incentive to work hard academically and be orientated towards the scholar image since there would be peer group rewards for such strivings. It would also, in Coleman's view, relieve the onus on the teacher to evaluate the pupils academically.

Coleman may be right. It is possible that under a different structure, where high academic attainment brings glory to the school and attracts peer group rewards, that the scholar image would improve in its acceptability to adolescents. It might to some degree help to destroy the norm against very hard work that arises in most classes. However, Coleman does not take sufficient account of certain basic differences between academic and athletic concerns. Not all students are expected to join in athletics. If one has little ability or interest in athletics one can withdraw since it is a voluntary optional extra. When a boy lacks the ability or interest to make the team, he is quite free to cease striving. Yet all pupils are expected to involve themselves in academic work, even if one has little ability or interest. This is a compulsory aspect of school from which one cannot withdraw. In other words, the structural change advocated by Coleman can be effective only if *all* the pupils are to a high degree involved in the proposed inter-school academic contests, and in all those aspects of the total curriculum which the school imposes. If there is little hope for a pupil of representing the school in a contest, then there is little incentive for him to strive academically, or at least no greater incentive than at present. Coleman's change might improve the incentives to academic achievement for a small minority, but it would be at the expense of the rest. He would not, I presume, be content if the majority of the pupils became passive spectators of outstanding contestants in academic games.

Coleman is dealing with the symptoms of what is wrong with schools, rather than the disease itself. He is concerned with improving external, social incentives to hard work and academic attainment, rather than trying to alter the structure so that pupils commit themselves to learning without the need of external pressures. Naïvely, he suggests that all pupils will become more involved in academic work, even though they do not represent the school, on the grounds that currently many pupils who are not on the athletic team happily spend their lunch break and other leisure time playing the same games as the team members. He seems not to realize that football and other games are perceived by boys as inherently pleasurable in a way that academic work is not. I fail to see why the creation of inter-school academic contests will cause a sudden rise in motivation to those pupils whose strivings will never get them onto the team. His proposed changes do nothing to alter the motivational problem of the vast majority of pupils in school. My own view, developed earlier in this book, is that relatively little success in trying to get pupils to work hard in school can be expected from a system based on the manipulation of social rewards and incentives. It is a structural change, which will draw on the child's inherent desire to learn and which leads the learner to recognize the intrinsic rewards of learning, that is needed. Coleman shifts the focus and form of the competition within school. It is my belief that the lack of interest in academic work and the norms against hard work can be undermined only when the pupils are encouraged by the school structure and by the teachers' attitudes and behaviour to recognize that, in spite of the undeniable competitive ethos of our society, real learning is not best fostered in a spirit of competition.

One interesting aspect of Coleman's study that has not been mentioned so far is the effects of the social system on the pupils. In particular, Coleman considered the relation between a pupil saying that he would like to be someone else, a measure signifying low self-evaluation, and the degree of informal status. The members of the leading crowd, the athletes, the scholars, the activity leaders, the girls who were popular with boys, all were less likely to want to be someone else than were the rest of the pupils. Those pupils who were not in the leading crowd were divided into three categories: those who wanted to be in; those who did not care; and those who did not want to be in. Those in the first category regard the leading crowd as a positive reference group and those in the last category regard the elite as a negative reference group. As might be expected, it was the pupils who took the leading crowd as a positive reference group but who were not members who most wanted to be someone else, and this was particularly true where such pupils had a small number of friends. In other words, self-evaluation is markedly low among those pupils who are not in the leading crowd, but would like to be, and who have few friends; it is they who lack status in the social system and the social support of friends. Coleman infers that such pupils feel hurt psychologically and need to take action to repair their self-evaluation.

It is possible, of course, that it is not the failure to make the leading crowd which induces low self-esteem, but low self-esteem which inhibits entry into the leading crowd. On this interpretation those pupils who want to be someone else, are not in the leading crowd and have few friends would simply be those pupils who had low self-esteem in the first place. The low self-esteem would be a cause of their social rejection, not a reaction to such rejection. Data from Rosenberg's (1965) study of the self-image of adolescents in high school offers support for such an interpretation. Rosenberg shows that pupils with low self-esteem tend to be more sensitive to criticism; to feel awkward with others; to find it difficult to make friends; to feel a need to put up a façade; to feel detached and isolated from others; to have less faith in people. The effect on social relationships is quite striking.

it is difficult to escape the conclusion that people lacking self-respect do not stir up much of a social breeze in the high school. They participate in fewer extracurricular activities (especially clubs); spend much less time at such activities; are less often elected as president, chairman, or other officer of clubs or school organizations; are less often elected as homeroom officers; are less popular with their classmates (as determined by sociometric methods); participate less frequently and actively in casual, informal conversations with schoolmates; are less often opinion leaders; are less likely to 'take a position of leadership in a group you are with'. . . .

Whether the group is formal or informal, voluntary or involuntary, the person with low self-esteem tends to be a relatively impotent social force.

Rosenberg shows that one of the reasons why such pupils fail to participate is because of their attitudes towards others and the impression they make on them. They do not have a high opinion of others, and they are not respected in return. They are submissive and unassertive in their dealings with others. It is impossible to know whether Coleman or Rosenberg is right, though in this matter Rosenberg's evidence is the more impressive. In the event, it seems likely that there is an elaborate interaction between the school's informal social system and the personalities of its members.

Coleman suggests that those who fail to gain status in the adolescent social system react by withdrawing from it and by turning to reading, listening to records and watching television as a compensatory substitute. He considers the differences between schools in the amount of status given to athletes and scholars. Where these elite boys are the recipients of very high informal status, they turn to the mass media to a smaller degree than do the athletes and scholars where the status given to athletes and scholars is much lower. (One is forced to wonder why Coleman fails to present any evidence on the obvious hypothesis that it is the pupils with the lowest self-evaluation, namely those who have few friends and who would like to be in the leading crowd, that should show the highest consumption of the mass media.) Coleman claims that 'when the adolescent is in a system that fails to give him status and allow him a positive self-evaluation, the adolescent often escapes to a world where he need not have such a negative evaluation, the world of the mass media.' Actually, Coleman's own data do not fully substantiate this claim. It would be more accurate to state that elite boys in schools where peers give high status to athletes and scholars tend to use the mass media less often than elite pupils in schools where peer group status for athletics and scholarship is lower and less than that of pupils in general. This difference is very important. It is possible to explain the lower consumption of the mass media by some boys in terms of the amount of time they have to devote to their status giving activities. Non-elite boys simply have much more free time to devote to the mass media. Also in schools where there is least informal status given to athletes and scholars, such boys have less incentive to devote large amounts of their spare time to these activities since greater devotion will not improve their informal status. If this is the case, it is misleading to claim that the non-elite boys turn to the mass media as a substitute for status; it is the elite boys who turn to the rewarding elite activities as a substitute for the normal amount of mass media consumption.

Much space has been devoted to Coleman's study, though only a small fraction of his findings have been presented. This will, I hope, indicate what an important book this is. I have offered some criticisms of the book in order to demonstrate how complex research in the area of the youth and the school really is. We know relatively little and the evidence that we do have is often ambiguous and open to a variety of interpretations. *The Adolescent Society*, like all good books, raises more problems than it solves, but inspires and illuminates the way to further research.

There is no British equivalent to Coleman's work. In this country researchers have been more interested in the relation between school and social class, for example the fate of the working class pupil in grammar school (Jackson & Marsden, 1962), or comparisons between pupils in different types of secondary school (Musgrove, 1964). Yet it is impossible to transpose the findings of Coleman and other American researchers to Great Britain. Not only are the cultures different, but so also are the educational systems and the internal organization of schools, though with the recent comprehensivization of our secondary schools the two systems may in some respects become more similar.

One obvious difference is the place of athletics in the school system. In this country physical education is as strongly associated with gymnastics and keeping fit as with games such as football and rugby. The P.E. teacher is not at all like the American coach, whose pay is often much higher than that of other teachers and who is often hired or fired by results. The good American coach can bargain for his salary because he is highly valued by the school. These differences represent differences of orientation here and in the United States. In both countries the pupils enjoy their games, but there is an important truth in the assertion that in the United States one gets fit in order to play the game, whereas in Britain one plays the game to keep fit. Moreover the large numbers of pupils, staff, parents and other members of the local community who attend sporting events at an American high school stand in marked contrast to the handful of pupils and parents who attend inter-school matches in English secondary schools. No doubt many headteachers and teachers would like much more extensive support for school games, for they regularly try to induce pupils to attend, but their attempts are usually in vain. The relative lack of enthusiasm to support sporting events in British schools is somewhat curious, especially when we remember that football and other professional sports form a major part of adult male interest and conversation.

From a British point of view it is clear that Coleman tends to underestimate the place of athletics in the *formal* school system. Coleman's work has sometimes been used to demonstrate the disjunction between adult, teacher or official school values and those of adolescents. With respects to athletics, this view is easily overstated. The remarkable thing about Coleman's study is in fact the degree to which the formal goals and values assume cardinal importance in the values and status system of the adolescent pupils. The adolescent informal status structure reflects the official and formal high valuation of athletics by parents and teachers alike. Coleman's work is basically in line with the tradition of Newcomb, who in the Bennington study (1943) showed how the politically and economically liberal views of the staff, in a college for upper class women of conservative background, deeply permeated the informal value system and status structure of the students, high status becoming associated with attitude change to non-conservatism.

Another important difference between Britain and America is that whilst the majority of American pupils over the age of fifteen stay in school for several more years, it is a minority who do so in this country. Because the senior pupils at school here are such a small number, or much younger than the seniors of the American high school, combined with the fact that many schools are single-sex schools, the dating phenomenon and cross-sex relationships tend to have a relatively smaller place in the informal system of the school. Further, the lack of a clear leading crowd in English schools—Coleman showed that there was an extremely high degree of pupil consensus about the membership of the leading crowd—means that we are unlikely to meet with the sort of pupil strivings that typify the American high school. This extract, taken from Gordon's (1957) study of a single high school, where the social system of the pupils was dominated by 'big wheels' (male athletes) and the 'Queen of the Yearbook', seems foreign as well as disturbing.

> A girl friend and I set out to be popular. This friend and I gave pyjama parties to help make the girls like us. When the day was over, we would get together and add up our progress. Such things were included: what boys had asked us for a date, or had talked to us? If any popular boy in our class talked to us or acted interested, what things could we do or say that would attract favourable attention from him? When we did get dates, with older boys particularly, we tried in a subtle way to have them ask us to go steady. We also tried to be friends with [the girls of high status]. Then we tried to join every club we could, so it would look like we had done a lot for the school . . . We had as many dates as we could get with boys of our own age who were the school's heroes. We were continually trying to impress and we flirted with any boy that might help us on the Coronation Court.

The lack of a leading crowd in British schools seems to signify that school plays a less central role in the life of the adolescent and that there is no dominant, pervasive and elaborate informal social system among the pupils. Certainly the teachers often stress the importance of the extracurricular life of the school as well as academic matters. They value the 'all-rounder' who is good at his studies, a keen and skilled games player, and an active participant in the school's clubs and societies. Yet the sports teams are less important to, and influential upon, adolescents in general, and the clubs and societies are smaller in both number and size as well as more intellectually orientated than seems to be the case in the American high school. True, in some schools there develops an 'establishment' of sixth formers or senior pupils, composed of the school's intellectuals, sportsmen and club officials. Many of them are prefects also because of their commitment to the goals and life of the school. Thus it tends to be the middle-class and upper stream pupils who become both the prefects and the members of school clubs and sports teams (Start, 1966; Holly, 1965; Hargreaves, 1967). Yet this rarely seems to take the form of a leading crowd; more commonly there are overlapping groups of sportsmen, club members and prefects.

Because there is no overall social system of adolescents in British schools, there is also no clearcut overall status structure among the pupils. The social life of the English pupil is typically centred on a small group of friends; the same is true of American pupils, but there

364

each clique has a known place in the general status hierarchy of cliques. Of course, *within* each clique there is a status structure, but the form that this takes depends on the values and norms of the group and the resources each member brings to that group. It is for this reason that the dynamics of friendship formation and the dynamics of small groups are potentially more important areas for the analysis of the informal life of pupils in British schools. A social systems approach becomes more useful when we want to examine the effect of the formal structure on pupil values and friendship groupings—thus the interest of British researchers in the social effects of streaming.

It has been suggested by some that as the number of older pupils in British schools rises, partly as a function of the raising of the school leaving age and partly as a function of the trend towards longer voluntary staying on at school, then we shall develop a social system which is much more akin to an American social system as analysed by Coleman and Gordon. Once the social life of older adolescents becomes focused so heavily on the school, then we shall follow the pattern set in the American high school. Whilst this may be true in some respects, such as the growth of school-based cross-sex relationships, the general thesis seems rather doubtful to me. This is in part because the community in Britain does not have the control, influence or interest in school that is typical in the United States and in part because I can see no reason why athletics should suddenly acquire the formal or informal importance it has in American schools. Indeed, I would go much further and suggest that despite the larger numbers of older pupils in school we are not likely to develop a situation where the major aspects of adolescent social life are school-based.

One of the most striking social changes in this century has been the change in parent–adolescent relationships. The central aspect of this change is the shift in power between the two. The adolescent has made marked gains in power, increasing his rights and reducing his obligations *vis-à-vis* his parents. It is not long since we were being harangued by public figures about 'the distressing abdication of parental authority'; currently we seem to take it more for granted. We do not expect, for example, our adolescent children to seek our permission in areas where it would have been normative to do so fifty years ago. Adolescents no longer need our consent about the ways they spend their money or their lesiure time, or about the friends they wish to go around with. The parent who does try to intervene in such matters is quite often told that it is none of his business. In the interest of good relationships parents concede their former rights and powers, though often after fighting a brief losing battle that precedes the reluctant retreat.

The remarkable aspect of this change is its impact on the school. All teachers accept that relations between teachers and pupils have become much more informal and friendly in recent years, even in the very traditional grammar schools. Yet—and this is the point—this has not been accompanied by a marked power shift. Pupils today have hardly any more rights than they had half a century ago, nor do the teachers want to concede such rights. In this respect the school has lagged behind the family. Not unnaturally, adolescents often— but by no means as a general rule—expect an increase in rights to match the growth of informality and the power shift that has taken place at home. As a result teachers tend to be ambivalent in their feelings and behaviour towards pupils. They approve of the informality, within limits, but when the pupils take things 'too far', when they ask for or demand rights that are not traditionally theirs, the teachers tend to swing back to the old authoritarianism. The school is normally much more restrictive and autocratic in structure than is the home, despite the general air of informality and friendliness that prevails in so many secondary schools. The difference between reality and appearance becomes obvious when new rights are desired by some pupils. Heads are content with school councils, as long as the function is advisory only, subject to the veto of the staff or head. Once the council wants power in areas which the staff regard as of more than peripheral importance, the proposals are rejected as unthinkable. More traditional headteachers try to preserve their conception of school in spite of the social changes elsewhere in society. Pupils who do not accept their assigned role, who fail to obey the school rules, even on such relatively minor matters as the styles of clothing and length of hair or jewellery, are punished or even expelled where this is possible. 'Pupil power' is a frightening word to many teachers and headteachers; they try to eliminate such dissenting groups as militant products of a permissive society who are potential sources of conflict and contamination within the school. (This remarkable over-reaction becomes evident when one reads the documents of the Schools' Action Union, which are very moderate in their demands and reveal a deep and idealistic concern with education. It is no less than tragic that the schools cannot tap this idealism rather than attempting to reject or suppress it.) The teaching profession has failed to learn a lesson from the universities, for it is not realized that in effect the school is setting itself up as one of the few remaining authorities against which the adolescent can rebel. Curiously, too, many teachers do not seem to recognize that much of their behaviour seems hypocritical to the pupils. The teachers preach democracy and talk about education for independence and creativity. What they mean is their own style of democracy which excludes the school itself. Independence and creativity must operate within the teachers' assumptions. If the pupils develop an independence or creativity which actually questions or rejects the teachers' assumptions, they are ruled out of court. It is easy for the teachers to explain things away in terms of a minority of troublemakers or 'outside influences' or 'militant anarchists'. It is much harder to recognize the deeper and more general social changes that are taking place in society and in the school and to accept that today's difficult minorities have a curious habit of becoming tomorrow's moderates.

Let us now examine in further detail some of the growing problems and conflicts between teachers and pupils. Smoking is a traditional school problem with boys, though recently the problem has grown more acute. In 1970 a report by the Tobacco Research Council showed that in the seven-year period since 1961 fifteen-year-old boys increased their average weekly consumption of cigarettes from 13·4 to 19·2, and girls from 4·1 to 7·8. There is evidence, too, that pupils are beginning to smoke at an earlier age—7 per cent of boys beginning to smoke at the age of six or under. Young people can now afford to buy more cigarettes and are now frequently smoking with parental consent. Yet very few schools allow their pupils to smoke in school. This is obviously a very difficult issue, but we cannot escape the fact that the school is in this area being more restrictive than the home, forcing the pupils to do their smoking illicitly in lavatories and other such places.

Drink is a relatively new problem, partly affected by the adolescent's ability to afford the cost of alcoholic drinks and partly by the greater social acceptability of drinking in the older adolescent. In some areas it is now customary for young people, not always over the age of eighteen, to meet socially in pubs rather than coffee bars. Teachers are having to face the problems of pupils arriving at school functions 'under the influence', and have to police the pupils to prevent illicit drinking at school dances. I know of one school which has abandoned its annual dance for this very reason. In this school the teachers dare not go into the pub near the school after a school function since they know it will be full of their own senior pupils.

Drugs present quite a new problem for schools, especially when we remember that almost all adults have neither current nor past experience of them, unlike the case with cigarettes and alcohol. Because drug-taking is not socially acceptable among adults, teachers tend to regard it as a much more serious 'crime' than smoking or drinking. Yet this is a problem which is likely to grow during the next decade, with respect both to the smoking of cannabis and to the taking of the amphetamines, barbiturates and opiates. There is a widespread fear and ignorance of drugs among teachers as among adults in general. The reaction of many headteachers to finding one or more cannabis-smokers or drug-takers among the pupils does not suggest that they will be able to deal with such cases rationally or sympathetically, or in a way that will inhibit the spread of drug-taking and the development of a drug 'underground' among senior pupils either on or off the school premises (McAlhone, 1970).

Sexual problems, too, are making a greater impact on the life of the school. Conscious of being to some degree *in loco parentis*, and fearing that sexual relationships between pupils may take place when they are in the school's care, teachers feel a need to supervise pupils very closely in this respect on school holidays and at other school functions. Teachers know that if the pupils commit what most adults would regard as sexual misdemeanours when they are in the school's charge, then the school will be harshly judged by parents and public. It is an unenviable position for teachers. When sexual scandals or pregnancies do arise, the offending pupil(s) is expelled or quietly extracted from the school. There remains a latterday form of being sent to the far reaches of the Empire. In effect, schools are reluctant to face sexual problems, even though they are aware of their existence among the older pupils. They are unwilling to encourage an open discussion of these problems and do not take the initiative in encouraging a dialogue between pupils, teachers and parents. They prefer not to see, taking action only when a crisis situation affords no alternative, such action dealing with the symptoms rather than the root of the problem.

Most important of all, the social life of the adolescent in school has, from the teenager's perspective, a definite old-fashioned air. The school offers activities of an intellectual quality (debating and scientific societies), sport, and the sort of activities traditionally associated with the church youth club (table-tennis, country dancing). Most teenagers do want these activities, but increasingly some also want the sort of activities provided by commercial enterprises but for which the school is generally unwilling to cater. Even the school parties traditionally provided by teachers for the younger pupils are often regarded by these pupils as 'square' or insufficiently sophisticated.

All these factors suggest to me that the school is likely to play a less important part in the social life of the pupils. The teachers are unwilling to adapt to the changing needs and expectations, in this field as in others. Many adolescents will continue to accept the school's facilities—leading teachers to believe that all is well—but they will also demand activities that are provided, or will be provided, elsewhere, either by commercial undertakings or by the pupils themselves. More and more the gap between the school's extracurricular social life and the other leisure activities of many adolescents will widen. In turning to commercial enterprises for leisure activities or in creating their own, such adolescents are in a vulnerable position with respect to commercial exploitation and are likely to reduce the amount of informal contact with responsible and caring adults.

368

Tragically, it is the pupils who are in most need of adult help and care who will most readily turn away from the school's extracurricular social life. The pupils most affected will be those from working-class homes, those who are less committed to traditional middle-class values and to the intellectual life of the school, those with the lowest intellectual attainments and behaviour ratings, as the research of Sugarman (1967) shows. Sugarman himself believes that it is the pupils who are alienated from school who turn to teenage culture— listening to pop records, going to dances and coffee bars, following adolescent fashion in clothing and hair styles, smoking and dating— as an *alternative* to the pupil role. (This is clearly an argument that is similar to Coleman's.) Whether it is more a question of such pupils doing badly at school because of their home background and/or because of their commitment to pop culture, or vice versa, there is undoubtedly a major problem posed for the school, a problem which defies easy answers but which must be tackled if we are not to be overtaken by events. Certainly matters are not helped by the inclination of many teachers to deride pop culture openly, rather than recognizing it as a useful basis for the education of discrimination and artistic appreciation. How few music teachers, for example, know anything about contemporary pop music, which is so familiar to and so enjoyed by most adolescents. If their own musical taste is ridiculed or despised or treated with condescension by the teachers, is it so surprising that these attitudes are reciprocated to the teachers' musical preferences, effectively blunting the pupils' musical sensibilities and interests for many years to come? In ways such as this the teachers often enlarge the generation gap and reinforce the adolescent's commitment to pop culture. The teacher's avowed intention to educate often boils down to a demand that the pupils unilaterally and unconditionally accept the teacher's valuations. Any movement on the teacher's part (in our musical example, permitting the pupils to play pop records on the last day of term) takes the form of a half-hearted concession to valuations of a lower order.

Finally, in a chapter on youth, we must consider the truly deviant minority among adolescents. It is, of course, extremely difficult to distinguish clearly between the 'deviant' and the 'normal', between the temporarily deviant and the permanently deviant. This latter point is particularly important, since if all the official delinquents become adult criminals we would have criminal statistics of unbelievable dimensions, and there is an interesting tendency of today's skinheads to become the solidly respectable working-class Conservatives of tomorrow. Yet there is a minority of young people who reject some of our common social values and standards of conventional society. If we apply the rather crude labels, they are the delinquents, the vandals, the Hell's Angels, the hippies, the drug addicts. It is probably a growing minority, since in recent years it has increasingly involved the middle-class adolescent from the 'good' home and 'good' school in receipt of all the privileges and opportunities afforded by our society. We still know very little about the causes of these forms of deviance, though it is clear that many varieties of social deprivation as well as frustrated ambitions and ideals are involved.

In schools we can see most of these forms of deviance in an embryonic state. It is a constant temptation for the teacher to judge and to blame rather than to help, and in this sense the teacher may unintentionally promote the deviance of which he so strongly disapproves. It is beyond the scope of this chapter to deal with the complex question of the relationship between adolescent deviance and the school. What I wish to emphasize is that the teacher should not be side-tracked into perceiving deviance as being 'caused' by various external factors, 'controlled' by law enforcement officers, and 'cured' by various treatment agencies, but rather that he should perceive his role in terms of his being in a position to understand and appreciate the perspective of young people, and through this, to offer a relationship in which young people can grow.

This is not an easy task for the teacher, who, despite his occasional protestations to the contrary, was usually a highly successful and conformist pupil and then student, and is now a respectable member of society, accepting the majority of our major assumptions, values and conventions. Yet if we try to stand outside ourselves for a moment, we may capture something of the perspective of these young people, for we contain within ourselves the seeds of most forms of deviance.

In the first place, much of our culture is highly materialistic and individualistic. We encourage the young to 'get on in life', to find good jobs which are well paid, secure, allow long holidays, and offer attractive pension schemes. Good jobs mean a high income, giving success and social status and permitting conspicuous consumption. Yet this is a rat-race that relatively few can win.

Second, education is one of the principal paths to occupational, social and monetary success. The main value of education is regarded as instrumental, rather than an end in itself. If education is not very exciting or interesting in itself then that is simply the price one has to pay. Even the teachers, realizing that many pupils do not see education as intrinsically interesting, offer the enticing incentives of examinations, qualifications and their relationship to future occupations.

Third, many of the occupations entered by a very large number, perhaps the majority, of young people are repetitive and tedious, giving very little personal satisfaction. Many involve tasks which make little use of skills and abilities, either known or potential; give

370

scope for minimal autonomy or responsibility; are quite unrelated to a meaningful end-product; make little obvious contribution to the well-being of society. The relevance of school education to such jobs is often minimal, nor has formal education contributed much to the adolescent's ability to use his leisure time in ways that bring high personal satisfaction.

Fourth, the pattern of adult life is for the most part constraining and unattractive. The normal prospects for the young are: have a good time whilst you are young, get married and have children (in either order), struggle to buy a suburban semi-detached or wait for a pleasant council house, and then save up for a car in order to escape the routine of one's job or one's home dominated by young children and a television set.

Fifth, as an adult one will have relatively little power or influence over what happens in society, in industry or in local and central government. One must accept one's relatively powerless position of being governed by people whose values and decisions one frequently dislikes or despises.

From such a perspective the future for an adolescent looks rather dismal. If this picture is even approximately close to the adolescent perspective of some, then it is hardly surprising if some of them rebel. Some do not, especially if they are likely to be successful in adult life, whatever the measures of that success are. Some simply adjust, scaling down their former ideals or coming to terms with their former frustrations. To me the great mystery is not why so many rebel, but why so many adjust. If we could only explain why so many do 'settle down' then we might be in a position to understand why a minority does not.

As teachers we take for granted the essential rightness of the 'good' or acceptable pupils, who make little trouble, who do not seriously threaten the system or our assumptions and values. Yet it may be that many deviants become deviants because of a profound sensitivity—to values, relationships and institutions—which is constantly frustrated or wounded in almost unimaginable ways. Currently we are writing off too many adolescents under collective and pejorative deviant labels and are failing to recognize that many young people are trying out new experiments in living for both the good of the individual and the group of which he is a member. The communes that have grown up among the young in the United States are a case in point. It is no coincidence that the greatest amount of deviance among the young, whether we define it as positive or negative deviance, is occuring in the United States, for this is the Western civilization which is nearest to the point of collapse. It may be that it is the deviant pupil or adolescent who is less sick and more human than the 'good' pupil or adolescent. For the average adult and teacher this is the hardest perspective of all to take, for it means shedding, at least temporarily, the fundamental meanings of our own existence.

Recommended reading

J. S. COLEMAN, *The Adolescent Society*, Free Press, 1961.

E. DOUVAN & J. ADELSON, *The Adolescent Experience*, Wiley, 1966.

C. W. GORDON, *The Social System of the High School*, Free Press, 1957.

J. B. MAYS, *The Young Pretenders*, Michael Joseph, 1965.

F. MUSGROVE, *Youth and the Social Order*, Routledge & Kegan Paul, 1964.

M. & C. SHERIF, *Reference Groups*, Harper & Row, 1964.

11 Changing attitudes

Education is concerned with the changing of attitudes. Whilst most discussions of education implicitly acknowledge the fact, to speak explicitly of the nature of attitude change within educational contexts creates a certain uneasiness. Perhaps it is because the term 'attitude change' evokes images of American soldiers being brainwashed by Chinese communists, or of psychologists ingeniously probing our attitudes and selling their findings to commercial organizations who can then manipulate us with their advertising campaigns. These are distasteful spectres that we do not wish to associate with teaching. It is foreign to what teachers think they are doing.

In this book we have also been implicitly concerned with the development and change of attitudes; it is a thread which has passed through most of the discussion. Attitudes are part and parcel of relationships and thus of education. In his instructional role the teacher is concerned with the formation and change of attitudes to academic matters, to the subject, to ideas, concepts and facts. In his disciplinary role he is concerned with the formation and change of attitudes to the teacher, to other pupils, to the moral and social rules that govern relationships in the classroom, the school and the wider society. Given that attitude formation and change are a central feature of educational enterprises we must now consider the matter in a more explicit way. Some authorities assert—wrongly in my view —that the concept of attitude is the most central and fundamental in social psychology. Certainly this is the branch of the human sciences traditionally most concerned with the nature, formation, change and measurement of attitudes so we have a further reason, if one were needed, for devoting a section of a book that purports to deal with social psychology and education to the concept of attitude.

It is possible—though unlikely in the extreme—that a teacher

373

might turn to a social psychologist and say, 'I am very worried by my pupils' attitudes. Why are they like this? And what can I do to change them?' These are two very large questions. The first is concerned with the origins and development of attitudes and the reasons for their persistence and stability. The second is an appeal for advice about how to change them. By and large social psychologists much prefer the first question; in this chapter we shall be concerned with the second. Of course the two are interdependent, and one cannot hope to understand how attitudes might be changed without a full appreciation of how they got there in the first place. Bearing in mind the obvious risk, we shall concentrate almost exclusively on the process of change.

An essential preliminary task is to distinguish the different approaches of social psychologists and teachers to attitude change. The social psychologist, recognizing that people's attitudes are remarkably stable and yet do sometimes change, seeks to investigate and classify the conditions under which attitudes change. He aims to systematize such knowledge and to construct theories that can explain the research findings. The teacher is less concerned with such general processes and conditions and theories of attitude change. His concern is essentially practical. Given that John Smith has the 'wrong' attitude to French, or to his teacher, or to his classmates, the teacher wants to know *how* this attitude can be changed. He is asking for techniques of attitude change which he can use effectively. The social psychologist is also interested in techniques and their relative effectiveness under different conditions but he tends to relate such questions to the wider, and perhaps less practical, issues of attitude theory and research. The teacher is asking narrow questions which cannot be answered except in relation to wider social psychological issues, so the social psychologist is unlikely to be able to provide the simple answers that the teacher expects. Such, unfortunately, is the nature of social science.

The teacher's search for appropriate techniques of attitude change immediately raises a number of *moral* issues. If the social psychologist can offer the teacher certain techniques for changing Smith's attitude, the teacher then has to decide whether or not a proposed technique is morally acceptable and in line with his philosophy of education. He might, for example, conclude that certain techniques involve an unacceptable manipulation of the pupil. This is not so much a problem for the social psychologist, not because he is morally defective (though this is arguable), but because his conception of his task excludes many of the moral issues. The social psychologist is seeking to clarify the conditions under which attitudes change and is more concerned with the effectiveness of certain techniques rather than their morality. His moral problem is with the form the experiments take, e.g. how much deception is justifiable, but not with the morality of the techniques which others might use in non-experimental situations. With respect to techniques, the social psychologist tries to remain morally neutral, and as long as he stays within the confines of his laboratory he seems to do so with some degree of success. The main moral questions arise when one tries to apply given techniques in real life situations, which is what the teacher is trying to do. At this point most social psychologists will claim that this is not their problem and in consequence the teacher can expect little moral guidance from his social psychologist mentors.

Because the social psychologist has been concerned to create a scientific understanding of attitude formation and change, most of his work has been conducted in experimental laboratory settings where he can, or believes he can, exercise greater control over many complicating variables with greater ease than is possible in real life situations. This means that it may not be easy to transplant laboratory findings into real life situations. Real life has a habit, very irritating to psychologists but consoling to the rest of us, of being very unlike life in the laboratory.

Some of the major differences between laboratory and natural settings have been expounded by Hovland (1959). In laboratory studies *all* the subjects are exposed to the influence that is intended to change their attitudes. They dutifully pay attention to what is presented to them. In natural settings only a limited number of persons are exposed, and even where they are exposed it is much easier to 'switch off' or in other ways minimize or avoid exposure. (Think what you do when the commercial breaks come on the television screen. How desperately the advertisers try to stop you going to the kitchen or the lavatory!) In laboratory studies the subjects are carefully segregated from competing influences or 'counter-communications', e.g. discussion with other persons, whereas in real life we are constantly subjected to these competing influences. In laboratory studies the experimenter tests the effectiveness of his influence at once before the competing influences have time to intervene, which means that the experiments are concerned with short-term effects. In real life it is the long-term influences which count. (What is the point of an impressive advertisement unless the customer remembers it when he next goes shopping? Of what consequence is an impressive party political broadcast by Harold Wilson if I still vote Conservative next day?) The subject matter used in laboratory studies tends to be of peripheral importance to the subjects. There are good reasons for choosing relatively unfamiliar and innocuous subject matter, perhaps the most important being that it minimizes differences in pre-existing attitudes among the experimental subjects, but it also has the effect of making the subjects most susceptible to change. In real life we are often more interested in changing attitudes which are of considerable importance to people. These, and other factors mentioned by Hovland, make it clear that in laboratory studies we are much more likely to succeed in inducing attitude change than will be the case in natural settings. It is essential to bear this in mind when reading books on attitude change.

Before we turn to attitude change research and theory and their educational implications, we must clarify briefly what we mean by attitudes. It proves to be a surprisingly difficult concept to define. The most famous definition is probably that of Allport (1935) who stated that an attitude is 'a mental and neural state of readiness, organized through experience, exerting a directive or dynamic influence upon the individual's response to all objects and situations with which it is related.' This is a clever definition, but not a simple one for the novice. It is probably easier to grasp what is meant by an attitude if we look at the generally agreed components of an attitude. The three components are *cognitive* (knowledge, beliefs), *affective* (feelings, emotion) and *action tendency* (predisposition to behave in a certain way). Kelvin (1970) puts this with great simplicity when he says that 'my attitude towards an object is a compound of what I know or believe about it, how I feel and what I am inclined to do about it.' We can illustrate with an easy example, a pupil's attitude to French.

Cognitive—he thinks French is dull and a waste of time.

Affective—he dislikes French and French lessons.

Action Tendency—he avoids French where possible and hardly ever does his French homework carefully.

Usually, as in this example, the three components tend to be congruent with one another. Certainly this is the assumption of most attitude tests, which usually measure only one of the three components. But this is not always the case. For example, Merton (1949) recognizing the difference between prejudice (cognitive and affective components) and discrimination (action tendency) points to the distinction between the prejudiced non-discriminator (e.g. 'I dislike coloured people but I can't keep them out of my restaurant without breaking the law') and the non-prejudiced discriminator (e.g. 'I think coloured people are like everybody else, but I daren't serve them in my restaurant or my white customers won't come here any more'). Here action and attitude are in conflict.

It is clear from our example of the pupil's attitude to French that attitudes are *evaluative*, that is, they involve judgments of the favourableness/unfavourableness, like/dislike, approach/avoidance in relation to a given object. Attitudes also possess three important properties. First, attitudes vary in their *multiplexity* that is, the number and variety of elements that make up the attitude. A religious man (and perhaps an atheist) is likely to have multiplex attitudes towards religion—cognitively, elaborate knowledge and beliefs about the Church, the person of Christ, the Bible and so on; affectively, complex sets of feelings in relation to each of these elements; and behaviourally, predispositions to act in various ways such as church attendance, regular prayer and evangelistic work. The agnostic, by contrast, does not in his attitude to religion make such differentiations —religion simply does not mean very much to him. Whilst we can say that the agnostic has a simple generalized attitude to religion, the religious man's attitude to religion really consists of a complex cluster of related attitudes. Second, attitudes vary in their *connectedness*, that is, the degree to which they are related to other attitudes. I have known some fundamentalist Christians who are also scientists. They maintain beliefs and attitudes, which to other men would seem to be in conflict, by declining to connect them. Their attitudes to religion and science are compartmentalized and can be operated quite independently. For other scientists, their attitudes are more closely connected and they may well reject certain religious beliefs because they are incompatible with their scientific beliefs. It is because the two sets of attitudes are connected that a change in one set causes a change in the other. Third, attitudes vary in their *centrality* or their importance to the person. When we say that this man is religious we are saying by definition that his attitudes to religion are important to him and play a central role in his orientation to life. We define and categorize people partly by those attitudes which are known to be central to them. We can now see more clearly why multiplex connected and central attitudes are particularly resistant to change: a change in one attitude or part of an attitude will create a need for appropriate changes elsewhere. Such changes involving radical restructuring throughout the attitudinal system do occur occasionally, as in a religious conversion, and the necessary reformulation of former attitudes is accompanied by a subjective feeling of rebirth and reformation.

Having said a little about the nature of attitudes, let us take a particular case of an attitude change in an educational setting. Let us take the not uncommon situation where a headteacher and his staff are disturbed by the amount of smoking among senior pupils. They are likely to be aware that simply forbidding smoking in school merely sends the problem underground. Suppose, though this is much less common, that the teachers agree to look for certain methods whereby they could cure the problem by persuading the pupils not to smoke. Before we can consider in detail any specific techniques which they might adopt, we need some model of the attitude change process to serve as a frame of reference in which we compare different techniques.

The process of changing attitudes can be conceived as including four possible elements.

1 *A communicator*, a person who is attempting to persuade.
2 *A communication*, the information or message by which it is intended to persuade.
3 *A subject*, the target of the persuasion attempt.
4 *A group*, of which the subject is a member.
Some techniques involve all four elements, others do not. Research by social psychologists has looked at each element in isolation and in various combinations with other elements.

With this model in his mind, our teachers can now begin to consider appropriate techniques. It is, however, important that as an initial step certain general considerations be taken into account.

Preliminary considerations

1 How many of the subjects share the attitude that it is intended to change?
In many natural settings not all the subjects share the attitude or do not share it to the same degree. In this case not all the senior pupils will be smokers. One might arbitrarily divide the pupils into four groups: non-smokers, light smokers, moderate smokers and heavy smokers. It may be that different techniques will be more appropriate to certain groups rather than others.
2 What sort of attitude is it?
We need to know how multiplex, connected and central the attitude towards smoking is in the various subjects. The favourable attitude towards smoking is likely to be more multiplex, connected and central among heavy as opposed to light smokers. However, it is also likely that some of the smokers have unfavourable attitudes towards smoking i.e. their cognitive and affective components may be incongruent with their behaviour. These subjects need not be persuaded that smoking is undesirable, but need help in giving up a habit of which they disapprove. Other smokers, however, may need persuading that smoking is undesirable. And what of the non-smokers? It is dangerous to assume that they have negative attitudes to smoking. They may have favourable attitudes to smoking but do not smoke for fear of parental disapproval. If so, they too may need to be persuaded about the evils of smoking in order to inoculate them against smoking at some future date.
3 What sort of attitude change is planned?
To use the terms of Krech, Crutchfield & Ballachey (1962), is the proposed attitude change congruent or incongruent? A congruent change is in the same direction, e.g. a change from being mildly unfavourable to being highly unfavourable towards smoking. An incongruent change is a change in the opposite direction, that is,

making the strongly favourable attitude towards smoking less favourable or distinctly unfavourable.

The importance of these preliminary considerations is clear when we recognize the most fundamental finding of attitude change research, namely that *it is incongruent changes of multiplex, connected and central attitudes that are the most difficult to effect.* Yet for a fair proportion of our pupils, the heavy smokers, the proposed attempt at attitude change is of precisely this type. Research, which is not typically cautious in its claims, suggests that our chances of success are unlikely to be high.

Techniques of attitude change

Bearing these general considerations in mind we can now examine various techniques in the light of our simple model. In trying to change attitudes by means of a persuasive communication our attention is drawn to the communicator and the communication. *Who* is going to say *what* to the subjects?

Whether the communicator makes a persuasive appeal (the 'sermon' so dear to teachers and headteachers) or gives relevant information, or uses a combination of the two, one most important variable is going to be the *credibility* of the communicator. The subject's reaction to the communication is determined not only by its content but also by its source. The more credible the communicator in the eyes of the subjects, the more likely attitude change is to be effected. Hovland & Weiss (1951) showed that the same communication about the atomic submarine was more influential on students when it was attributed to Robert J. Oppenheimer than to *Pravda*. The subjects of Kelman & Hovland (1953) listened to a speech in favour of lenient treatment for juvenile delinquents. When the speech was purportedly that of a juvenile court judge it was more effective in inducing change towards the advocated position than when the speaker was allegedly a former juvenile delinquent currently on bail.

A second aspect of the communicator concerns his intentions. The less the communicator is perceived by the subjects as intending to change their attitudes, the less resistant they will be to change. Walster & Festinger (1962) showed that subjects who 'accidentally' overheard a communication were more influenced by it than subjects in conditions where it was obvious that the communication was an intentional attempt at persuasion.

Teachers and headteachers are not, in my experience, always aware of the level of their credibility to pupils. Because they feel they ought to be credible, they assume that pupils regard them as such. If the headteacher has the honesty to recognize the possibility of his own lack of credibility he may well look for a communicator with greater credibility, such as a doctor. But these choices are not easy, since it is difficult to know whom the pupils will regard as credible. If, for instance, the headteacher chooses a member of his staff, should he choose a smoker, a non-smoker or an ex-smoker? Further, it will require considerable ingenuity if the chosen communicator is to avoid giving the impression that his intention is to change attitudes to smoking.

The significance of the content of the communication, whilst obviously crucial, is difficult to investigate since it will vary enormously to the attitude concerned, the type of communication and the personality of the communicator. However, one feature is common to many persuasive attempts, namely that a certain degree of change is advocated. It is known that the greater the amount of change advocated, the greater the amount of change effected (Hovland & Pritzker, 1957), but this is true only where attitudes are not central, multiplex and connected. In such cases the *reverse* is true. Hovland, Harvey & Sherif (1957) investigated attitudes to prohibition in Oklahoma. Subjects were presented with communications in favour of prohibition ('dry') or in favour of repeal ('wet'). On such an important matter the subjects tended to have a clear position prior to the persuasion attempt. It was found that subjects whose position was only slightly discrepant from the position advocated were influenced much more than those subjects whose position deviated markedly from the communicator's. For example, when a wet position was advocated almost a third of the middle-of-the-roaders changed but only four per cent of the drys did so. It is also interesting to note that where the discrepancy between the communicator's and subject's position was small, the communication was perceived by the subject as fair and factual. When the discrepancy was large, the communication was perceived as unfair propaganda. We also know from several experiments that when the communication is highly divergent from the subject's position, the persuasion attempt may produce a 'boomerang effect', that is, the subject moves in the opposite direction to the one advocated and becomes even more firmly entrenched in his original position.

A few experiments have investigated communicator credibility and the degree of change advocated at the same time. Aronson, Turner & Carlsmith (1963) had college girls rank stanzas of modern poetry in order of merit. Each girl then received a communication about a stanza she had ranked low. The communications were varied in the degree of discrepancy from the subject's position, and the source's credibility was varied in that in half the cases it was allegedly written by T. S. Eliot and in the other half by a student. The results showed that when the discrepancy between the subject's and communicator's position was large, the girls were much more likely to be influenced when the source had high credibility.

It seems clear that our communication about smoking must not advocate a change that is highly discrepant with the subject's position unless it is certain that the communicator's credibility is very high. Yet in schools it is the heavy smokers who are most frequently apprehended for smoking and sermonized by well-meaning teachers who both lack credibility and advocate the extreme position of total abstinence. Should we then be surprised if our exhortations cause a boomerang effect?

One important form of the communication is the emotional appeal, especially when it is designed to arouse fear. We are all familiar with such fear-based appeals on hoardings and they have an obvious relevance to smoking. Essentially such appeals arouse a fear by indicating the dire consequences of a given course of action and by recommending a course of action which will avoid this situation. The most famous experimental investigation of a fear-based appeal is that of Janis & Feshbach (1953). Three sets of high school pupils were given a lecture on dental hygiene involving three levels of fear-arousing material. The strong fear appeal emphasized toothache, painful dental treatment, mouth infections and secondary diseases including blindness and paralysis. In the moderate appeal the stress was on mouth infections, tooth decay and cavities. In the minimal fear appeal no photographs of diseased mouths were shown and only tooth decay and cavities were emphasized. Obviously it was the strong fear appeal that produced the greatest worry in the pupils. However, actual change in the direction of accepting and adopting the dental recommendations of the lecture was greatest in the case of the minimal fear appeal and least in the strong fear situation. It is difficult to explain these findings. It may be that the strong fear provoked aggression in the subjects who then rejected the communicator and his communication. Or it may be that the intense anxiety aroused by the strong fear appeal made the subjects feel very defensive, leading them to avoid or ignore the threat and the lecturer's recommendations. In short, the study suggests that an appeal based on a high level of fear may lead the subjects to resist the communicator's arguments, conclusions and recommendations.

Subsequent research has shown that the effects of fear appeal are much more complicated than the early studies seemed to suggest (Higbee, 1969). Indeed, there is some evidence that high fear can be more effective than low fear when the communicator is perceived as highly credible. Unfortunately with respect to smoking the evidence on the use of fear appeals is not clear or consistent (e.g. Leventhal, 1964, 1966, 1967). For the present it is an approach that is best avoided.

The third element in the model is the subject himself. There is a growing literature on factors that differentiate individuals in their sus-ceptibility to persuasion (e.g. Janis *et al.*, 1959). Here I am more concerned with techniques that focus on the activity of the subject rather than the part played by the communicator or the communica-tion. Given the difficulties associated with direct attempts by teachers to change attitudes to smoking, can we find a technique which minimizes the part played by the teacher? The most important technique in this respect is probably that of role-playing or the debate, which could easily be used in schools. It is known (Culbertson, 1957) that where a subject advocates before others a position to which he is in fact opposed, he is much more likely to change his attitude in that direction than those who simply observe, namely the audience, though even the observers change more than do control subjects who are not present. Attitude change is also substantially greater when the sub-ject has to improvise the speech opposed to his own position than when he simply reads a prepared script (Janis & King, 1954; King & Janis, 1956). It has also been shown (Scott, 1957, 1959) that the attitude change can be increased when the subject wins the debate on the basis of audience votes. Unfortunately it is unlikely that the teacher can rely on the fact that his heavy smokers advocating a non-smoking position will prove to be the better debaters in the eyes of the other pupils and alternative possibilities of manipulating the audience raise severe moral problems. This technique has great potentialities for school use, especially when it is remembered that in the experi-ments quoted the attitudes involved were central, multiplex and connected—military service, the Negro, night hours for women students and the de-emphasis of football.

It is on the subject that many theories of attitude change have focused. Katz's (1960) functional approach to the study of attitude change indicates that in order to know how to change attitudes the communicator must be aware of the motivational bases and the functional significance of attitudes. It recognizes that attitudes meet individual psychological needs and thus to some degree it is con-cerned with individual differences, especially when we bear in mind that the same attitude in different individuals may be related to different needs and serve different functions. Katz distinguishes four types of attitude function.

The instrumental function

The attitudes assist in the maximization of rewards and the achieve-ment of individual goals.

E.g. the pupil develops favourable attitudes to smoking and smokes in order to facilitate his acceptance in a group of peers where he per-ceives that smoking is normative.

The ego-defensive function

These attitudes protect the self from unacceptable inner impulses or from the acknowledgment of outside threatening forces.

E.g. the pupil develops positive attitudes to smoking and smokes in order to overcome feelings of insecurity or inferiority.

The value-expressive function

These attitudes bring satisfaction to the individual since they are an expression of his core values.

E.g. the pupil wishes to be an adult and begins to acquire values and behaviour patterns that he associates with adult status. Smoking is perceived as consistent with and as a means of expressing this adult status.

The knowledge function

Attitudes are acquired as part of the individual's need to make sense out of his experience of the world.

No smoking example is relevant here.

One of the main virtues of Katz's theory is that it proposes particular conditions appropriate to the changing of attitudes serving these different functions. To change attitudes with an instrumental function one must (a) change the need which the attitudes help to realize, or (b) help the subject to recognize that the attitudes and associated behaviours no longer satisfy the need, or (c) persuade the subject that there are better or preferable means of meeting the need. To punish the smoking pupil or to lecture him on the relationship between lung cancer and smoking is unlikely to change his attitude unless such techniques can come to terms with the instrumental functions of the attitude. Greater success would be likely if the teacher could show that (a) group acceptance is not important, or (b) smoking does not increase his chance of group acceptance, or (c) other attitudes and behaviours will promote group acceptance more effectively. Courses (b) and (c) offer most scope, since it may be that smoking is not an important group norm and smoking is unlikely to be a major status-enhancing behaviour. Should smoking be a central group norm then the teacher will have to deal with the problem as a group phenomenon, not merely an individual one.

To punish, or to give information on the health hazards of smoking, or to appeal to the instrumental functions of non-smoking (e.g. the saving in money) are approaches that are unlikely to influence the pupil whose favourable attitudes to smoking serve ego-defensive functions. Punishment and the arousal of fear are likely to increase the defensiveness and thus produce a boomerang effect, and the giving of information is equally dangerous since the information is likely to be distorted. A good example of such distortion is evident in studies of racial prejudice (Allport, 1954):

Mr X: The trouble with Jews is that they only take care of
their own group.

Mr Y: But the record of the Community Chest campaign
shows that they give more generously, in proportion to their numbers, to the general charities of the
community, than do non-Jews.

Mr X: That shows they are always trying to buy favour and
intrude into Christian affairs. They think of nothing but
money; that is why there are so many Jewish Bankers.

Mr Y: But a recent study shows that the percentage of
Jews in the banking business is negligible, far smaller
than the percentage of non-Jews.

Mr X: That's just it; they don't go in for respectable business; they are only in the movie business or run
night clubs.

Where attitudes have ego-defensive roots they are much more susceptible to change if the threat, actual or perceived, can be removed or if the subject can be helped to develop insight into his own motives. In our situation the teacher would need to create a therapeutic relationship in which the pupil can come to terms with the feelings of insecurity that underpin his smoking.

In the case of attitudes with value-expressive functions attitude change is most easily effected if the subject can be brought to feel dissatisfied with the relevant core values or aspect of his self-image and thus remove the attitude's motivational base. Alternatively the subject may be led to recognize that the attitudes are in fact inappro-priate to his values. Thus the pupil who smokes because he sees this as an essential aspect of being grown up is most open to attitude change if he can be persuaded that smoking is not really expressive of mature, adult values and that he has fallen for a mere symbol which is not consistent with his aspirations to maturity.

Attitudes serving the knowledge function are influenced by new and additional information that improves or increases the individual's ability to structure his experience. Giving information is not the best means of converting habitual smokers who are relatively content to be so, but giving information may be appropriate to non-smokers, since by increasing their knowledge about the financial and health disadvantages of smoking it may help to foster unfavourable attitudes and thus inoculate them with greater resistance against those ubiquitous forces pressuring them to do so.

In short, Katz's theory reinforces and expands our common-sense knowledge that attitudes to smoking cannot easily be changed and strengthens our caution about the application of particular attitude change techniques to groups of subjects. A technique will have limited success unless it takes into account the functions served by the same attitude in different individual subjects.

A second important theory which emphasizes the internal psychological processes of the individual is Leon Festinger's (1957) theory of cognitive dissonance. This theory has its roots in the more general 'balance theories' stemming from Heider (1958). Although the theory has now become rather complex in the light of various criticisms and difficulties, the basic ideas are very simple. The theory is concerned with the consistency or consonance of cognitive elements. I smoke but I also know that smoking is bad for my health. These two cognitions are held to be *dissonant* because the obverse of one element would follow from the other. The theory posits that when dissonance arises, it is psychologically uncomfortable and it motivates the person to seek to reduce the dissonance, and restore consonance, and to avoid situations that are likely to increase the dissonance. The larger the dissonance, the greater will be the pressure on the person to reduce it.

Dissonance can be resolved in various ways, though sadly the theory does not tell us under what conditions particular solutions will be sought. The most simple way is to change one of the cognitions. I can reduce my dissonance by giving up smoking and the resulting two cognitions, 'I do not smoke' and 'Smoking is bad for health' are now consonant. Or I might disbelieve the evidence that smoking causes lung cancer, which will also restore consonance and allow me to go on smoking in psychological peace. Alternatively I can add further cognitions in support of one of the two dissonant cognitions. For example, I could support my desire to go on smoking by arguing with myself that smoking reduces my appetite and prevents me from getting fat and that smoking helps me to concentrate on my work. Alternatively I could support the other side by recognizing that if I stopped smoking I would save a considerable amount of money.

We can explain some of the results of earlier experiments in terms of dissonance theory. Take, for instance, the experiment on poetry by Aronson *et al* (1963). The girls were told that someone else disagreed with their evaluation of a poem. The two cognitions were

1 I think this poem is very poor.
2 X thinks very highly of it.

When the girls were told that X was T. S. Eliot, they were in a dissonant situation. Eliot, a revered poet and critic, considered it to be a good poem and it was difficult for the girls to dismiss Eliot's opinion. The easiest way for the girls to resolve the dissonance was by changing the first cognition, their own opinion, so that it was closer to Eliot's opinion, which is what they did. The girls who were told that W was another student were not in a particularly dissonant situation since there is no reason why any girl should value the other student's opinion more than her own. Since there is little dissonance there is little pressure to change her opinion, which is what happened.

A second illustration is the experiment by Aronson & Mills (1959). College women volunteered to participate in a discussion on the psychology of sex. The girls were asked to take a preliminary 'embarrassment test'. Half (the mild initiation group) read aloud five sex-related words. The other half (the severe initiation group) read aloud twelve obscene words and two vivid descriptions of sexual activities from modern novels. They were all then asked to listen to a tape-recording of a discussion which had been intentionally designed to be extremely dull. The subjects were then asked to rate the discussion in terms of its interest. Those who took the severe initiation rated it as much more interesting than did the mild initiation subjects. These results are easily explained by dissonance theory. The cognitions of the mild initiation subjects were:

1 I had to take a silly little test before I could join the group.
2 The discussion was very dull.

Since these cognitions are not dissonant there is no pressure on the girls to change their actual opinions of the discussion. For the severe initiation subjects the cognitions were:

1 I underwent a most unpleasant and embarrassing experience in order to join the group.
2 The ensuing discussion was very dull.

These two cognitions are dissonant and the easiest way to resolve it is to find the discussion of greater interest in order to justify having gone through the initiation.

If you, the reader, now feel that you understand dissonance theory, then explain the results of the experiments on role-playing given earlier in terms of the theory.

The implication of the theory for a teaching situation is the general proposition that pupils will not change their opinions unless the teacher can successfully create dissonance in them. The ability of teachers to change their pupils' attitudes rests on their ability to provoke dissonance in the pupils. One of the best ways to arouse dissonance is to persuade the subjects voluntarily to comply with a request and commit themselves to behaviour that is dissonant with their attitudes. In such a case the dissonance might then be restored by changing the attitudes so that they are consonant with the behaviour. In other words, teachers must persuade the pupils to commit themselves to behaviour which is out of line with their attitudes but in line with what the teacher wants and then hope that the pupils will resolve their dissonance by changing their attitudes to match the behaviour. We can take as an example an experiment that is related to the age-old problem of school dinners where children refuse to eat some vegetables on the grounds that they dislike them. Teachers usually try to solve this either by arguing that the vegetable is nutritious, which fails to tempt the child to eat it, or by derogating the child for his food fads, which is no more successful, or by compelling the child to eat it under threat of punishment, which may successfully get the child to eat it but fails to alter his dislike of it. Brehm (1960) discovered which vegetables the pupils disliked. Then allegedly as part of some consumer research he persuaded the pupils, with a small reward, to eat some of the disliked vegetable. Once the children had complied behaviourally, they were in a dissonant situation since the cognitions now are:

1   I dislike this vegetable.
2   I have eaten some of it.

One way of resolving this dissonance is to change the attitude of the vegetable to greater liking, and this is precisely what some of the children did. We can also see why children who are compelled by the teacher to eat the vegetable do not change their attitudes, for here the two cognitions

1   I dislike the vegetable.
2   I ate it because he made me do so.

are not dissonant. Dissonance can be aroused only when the pupil voluntarily commits himself as a result of persuasion; it is not aroused by coercion. This is probably the most valuable lesson of dissonance theory for teachers.

The method of induced compliance and behavioural commitment is not, unfortunately, the panacea to attitude change that at first sight it seems. The reason is that most situations offer alternative means of reducing dissonance apart from changing attitudes in the desired direction and it is not easy to restrict even the most obvious alternative. In Brehm's experiment the pupils might claim that they ate the vegetable because it was good for them. This justification may be sufficient to reduce the dissonance and thus the motivation for attitude change. When the pupils had eaten a lot of the disliked vegetable, Brehm showed them a 'research report' on the vitamin content of the vegetable. When the report suggested that the vegetable contained many valuable vitamins there was less attitude change towards increased liking than when the report stated that the vegetable contained few vitamins, since in the latter case the pupils could not use the report to justify having eaten the vegetable in the first place.

The persuader himself may unintentionally provide the subject with a means of justifying his compliance and thus reducing the dissonance. When the persuader offers the subject a reward in order to induce the compliant behaviour he takes the risk that the subject may be able to use the reward as a means of overcoming the dissonance created by the compliance—'I did it for the reward'. This has been demonstrated in several experiments (especially the famous one by Festinger & Carlsmith, 1959) which show, in apparent contradiction to reinforcement theory, that if the reward successfully induces compliance then the larger the reward offered, the smaller the amount of attitude change effected. This is because when the reward is very large the subject can justify his behaviour on the basis of the reward; but if the reward is small he will have greater difficulty explaining away his conduct as a function of the reward, and since dissonance cannot be reduced by this method the likelihood of attitude change is increased. In other words if a reward is used to induce compliance it must be just large enough to induce it successfully but not large enough to serve as a rationalization for having complied. In practice this creates many difficulties, especially when several subjects are involved. A reward that is just large enough to induce compliance in one subject may fail to do so in a second subject and may be perceived as extremely large by a third. In the case of smoking, what will induce non-smoking behaviour in a mild smoker may be far too small an incentive for a heavy smoker.

Another significant aspect of the persuader's behaviour is evident in the experiments of Smith (1961) and Zimbardo (1965). These two experiments are similar to Brehm's, though they probably have less relevance to school dinners. The subjects were persuaded to eat fried grasshoppers. In one condition the experimenter was warm, pleasant and friendly. In the second condition he was aloof and somewhat unpleasant. When the subjects had complied and eaten the food, the subjects with the unpleasant communicator increased their liking for the grasshoppers much more than did the subjects in the other condition. The subjects with the pleasant communicator were able to reduce their dissonance by justifying their behaviour on the grounds that they ate the food because the communicator had been so pleasant—'He was so nice I didn't want to offend him.' This course of action was closed to subjects in the unpleasant condition.

These experiments suggest that in persuading the subjects into compliant behaviour the communicator must beware of providing, in the form of rewards or his general manner, means by which the subjects can rationalize their conduct. The more a subject feels his compliance results from his own free choice rather than from external pressures, the greater the chance that the dissonance will have to be reduced by attitude change.

The fourth element in the model of attitude change is the social group of which the subject is a part. Many experiments on attitude change treat the subject as a separate individual and ignore his group affiliations. Yet it is known that group membership factors play an important role in the formation of attitudes, so it is reasonable to infer that the group may either facilitate, or reinforce resistance to, attitude change attempts. In the chapter on group dynamics it was pointed out that groups often filter incoming information so that members are to some degree protected from what is not consistent with group values and norms. More important, members are under pressure to conform to the group's dominant values and norms, being rewarded for the expression of the 'right' attitudes and penalized for the expression of the 'wrong' attitudes.

On the basis of this knowledge we can make some guesses about the ways in which group membership may influence a subject's susceptibility to attitude change in particular circumstances. It would seem reasonable to suggest that where the communicator attempts to change attitudes in a direction that conflicts with group values, a group member will be more resistant to such attempts than when the advocated change does not conflict with group values. It is unlikely that a subject will be easily persuaded to adopt attitudes that will threaten his acceptance in the group. If a pupil is a member of an anti-school group, then the teacher who by definition will hold pro-school attitudes will not be perceived as a highly credible communicator. It is an obvious point, yet many teachers try to cajole difficult pupils into 'good pupil' attitudes and behaviour without acknowledging that for such a pupil to do so he would have to be willing to forego his status or membership in the anti-school group. If, on the other hand, the pupil belongs to a pro-school group, then the teacher will possess greater credibility.

Since to many pupils the teacher may not be a highly credible communicator, it may be possible to use the pupils' peers as communicators with a higher credibility rating. This was attempted by one of my own students, Roy Ainsworth (1968). Sixth-form boys in grammar schools, nearly 60 per cent of whom were smokers, listened to an anti-smoking communication speech on a tape-recorder. Half the subjects were led to believe that the speech summarized the views of a sixth form association in conjunction with the editorial staff of *Sixth Form Opinion*; the other subjects were told that the speaker was conveying the views of a Headteachers' Association in conjunction with Regional Medical Officers of Health. Ainsworth found evidence in support of his hypothesis that a peer-based communication would be more effective in changing attitudes to smoking than would an authority-based communication, and the effect was still evident two weeks after the communication.

It would seem reasonable to conclude that the more a subject values his membership in a group, the more resistant he will be at attempts to change his attitudes in ways that conflict with group values and norms. This is demonstrated in an experiment by Kelley & Volkart (1952). An adult criticized camping and woodcraft activities to Boy Scouts in New England, when obviously such activities are highly normative to Boy Scouts. It was found that the less a boy valued his membership in the Scouts, the more likely he was to change his attitudes in the advocated direction. Those boys who valued their Scout membership very highly are of particular interest since with them there was a boomerang effect, that is, they were more in favour of woodcraft and camping after the communication than before it. In other words, if the communicator attacks the central norms of a group, those members who set great store on their membership will respond to the attack with increased conformity to the group norm. Thus the teacher with an anti-school group of pupils in his class may find that he is reinforcing the group norm every time he attacks it.

An experiment by Kelley & Woodruff (1956) cleverly demonstrates how group membership can be used to facilitate attitude change. The subjects, members of a 'progressive' Teachers' College, listened to a recording of a speech of a professor of education challenging the assumptions of progressive methods. The experimenters argued that if the students could be led to believe that other college students had been influenced by the speech then they too would change their attitudes, since change would no longer be restrained by their perception of the college norm. In the recording the professor's speech was punctuated by applause. Half the subjects were informed that the audience consisted of prestigeful college members and the other half were informed that the audience was made up of townspeople, whose enthusiasm for the speech would be perceived by the students as irrelevant to their perception of the group norms in the college. There was evidence of much greater attitude change away from progressive educational ideas in the first condition.

It is also possible that the influence of group membership in facilitating or hindering attitude change can be affected by whether or not the subject's group membership is made salient. Charters & Newcomb (1958) have shown that when Catholics were placed in a group and made aware that they were all Catholics, they answered a questionnaire on religious matters, under instructions to give their personal opinions, in a way that was much closer to the orthodox Catholic position than did Catholics who completed the same questionnaire but had not been previously reminded of their membership in the Catholic Church. Similar effects may arise in school situations. A teacher may find that a pupil responds quite well to anti-smoking persuasion that takes place in a private teacher–pupil interview, but then discover that the pupil does not live up to his promises to stop smoking when he is with his friends once more. It may be that the pupil made his promises compliantly in order to please teacher and was thus insincere. But it is also possible that in the interview the pupil genuinely accepted the teacher's persuasion and made sincere promises, forgetting temporarily the importance of peer group smoking norms. Back with his friends it is the teacher's influence that is forgotten because it is no longer salient.

A subject's group membership is also important when we consider the use of 'discussion' as a technique of attitude change. Yet discussion does not, as is sometimes thought, of itself make people more open-minded. Often it may have the reverse effect. The way in which a discussion can affect the impact of a communication is demonstrated in an experiment by Mitnick & McGinnies (1958). The subjects, high school pupils, were shown a film on racial tolerance. Half the subjects were simply shown the film and half were also allowed to discuss the film afterwards with fellow pupils. All the subjects were influenced by the film but to different degrees. When the discussion groups consisted of low prejudice subjects, the effect of the discussion was to strengthen the impact of the film and there was a greater change towards tolerance than when low prejudice pupils saw the film without discussion. But when the pupils were highly prejudiced the opposite result occurred. Film plus discussion reduced prejudice less effectively than did the film alone, for in the discussion these subjects were able to reinforce their original highly prejudiced attitudes through group support. Thus the influence of anti-smoking communications might be reduced if committed smokers are allowed to discuss the communication.

The classic experiments on the discussion technique are those of one of the fathers of social psychology, Kurt Lewin (1943), who tackled the very practical problem of trying to change the strongly-held traditional food preferences of American housewives as part of the war effort. The object was to persuade them to buy more beef hearts, sweetbreads and kidneys. Some of the housewives attended lectures in which the vitamin and mineral values of these meats were emphasized as well as economic advantages. Preparation of the meats was discussed in detail and recipes were distributed. Much of the same information was given to the housewives in a second treatment, a group discussion. At the end of the discussion the housewives were asked to indicate, by a show of hands, whether or not they intended to serve one of these meats at home. A follow-up study showed that only 3 per cent of the 'lecture' housewives actually did serve one of the meats whereas 32 per cent of the 'discussion-decision' housewives did so. Similar studies showed the superiority of the discussion over lecture in attempts to change attitudes to the consumption of milk, codliver oil and orange juice. Lewin believed that the discussion helped to unfreeze the old attitudes and open the way to new attitudes which would then be frozen in by the group decision.

On the basis of Lewin's studies it is difficult to disentangle the factors which led to the superiority of the discussion-decision method. The attitude change may have resulted from one or a combination of the following: (i) the discussion, (ii) the group decision, (iii) the public commitment to a decision, or (iv) the perception that most other women in the group were committed to a decision. Bennett (1955) tried to examine the effect of these factors separately and concluded that only two of the factors, namely the act of making a decision and the perceived degree of group consensus, are really important. Unfortunately different studies have yielded conflicting results. Pennington, Haravey & Bass (1958) show that discussion itself can make a difference and it has been shown by Hovland, Campbell & Brock (1957) that school pupils who are publicly committed to a position are much more resistant to counter-propaganda than are pupils whose commitment is private.

In short, we do not as yet know how the four factors operate under different conditions. With reference to our smoking problem the best plan would be to play safe and include all four.

(a) Have a discussion about the effects of smoking and why one should give the habit up. (Avoid putting all the committed smokers into one discussion group lest they simply reinforce their present attitudes. Also try to make sure that not all the group leaders are committed smokers, since they may be closed to influence from low status non-smokers and may indeed influence them towards smoking.)

(b) Let the pupils make a decision at the end of the discussion—hopefully some will decide to give up, especially if in the discussion useful points (e.g. health hazards and potential financial gains) have been raised.

(c) Make sure the commitment not to smoke is as public as possible. (You need not be timid in this. Remember Bernard Braden once offered to announce the names of converts to non-smoking on his television show when he discovered that telling the general public that he was giving up actually cured him. Could not the pupils be encouraged to announce their decision on a notice-board, or in Assembly, or in the school magazine?)

(d) The more smokers who commit themselves to giving up and the higher their status in the informal group, the greater the chance of obtaining some long-term success.

Implicit in the latter part of our discussion of dissonance theory is a suggestion of the significance of the relationship between the communicator and the subject in the process of attitude change. Our original model is too simple, for it suggests that the communicator and the subject are linked merely by the communication. In conceiving the communicator and the subject as quite separate variables, each with distinctive characteristics, (the communicator's 'credibility' and the subject's 'persuasibility') it fails to take into account the general relationship between communicator and subject apart from the situation in which the communication takes place. Yet this general relationship may be a most important factor. For our immediate purposes the main implication may be that certain techniques are more relevant to some general relationships than to others. The theory which takes this general relationship as central is Herbert Kelman's analysis of three processes of social influence (1961). For each of the three processes, termed compliance, identification and internalization, Kelman describes some of the antecedents characterizing the relationship and some of the consequents. (In these brief descriptions I shall continue to use the terms communicator and subject, though Kelman himself does not.)

Compliance. Here the subject accepts influence from the communicator because he wishes to achieve a favourable reaction from him. The communicator has control over the rewards and punishments which he can mete out to the subject and the subject accepts the influence in order to increase the flow of rewards (e.g. approval) and to decrease the flow of punishments (e.g. disapproval) to him.

> He does not adopt the induced behaviour. . . because he
> believes in its content, but because it is instrumental in the
> production of a satisfying social effect. What the individual
> learns, essentially, is to say or do the expected thing in
> special situations, regardless of what his private beliefs may
> be.

The communicator's power rests on his control of the rewards and punishments by means of which he can limit the choice of behaviour of the subject. The induced behaviour is adopted by the subject as part of the situation from which he cannot escape as a means of getting along within it, so he conforms only when the communicator exercises surveillance over him. Moreover if the subject finds a better method of influencing the flow of rewards and punishments to him he will change his behaviour accordingly. Once out of the situation, the induced behaviour need no longer be continued.

Identification. The subject accepts influence as part of a satisfying self-defining relationship with the communicator. The subject is involved in a role relationship which forms part of his self-image. Because the self-definition is anchored in the role relationship, influence is accepted as a means of establishing or maintaining the desired role relationship. The subject may try to be like the communicator or he may conform to a reciprocal role whose expectations are clearly defined for him by the other. Yet the subject does not accept the influence because it is intrinsically satisfying, though unlike the compliant subject he does believe in the attitudes, opinions and actions he adopts. The communicator's power rests on the fact that he is needed by the subject in order to derive a satisfying self-definition with reference to him. The induced behaviour will persist whenever the role relationship is salient or relevant. The subject's behaviour will change whenever he discovers a better means of maintaining the role relationship.

> The individual is not primarily concerned with pleasing the other, with giving him what he wants (as in compliance), but he is concerned with meeting the other's expectations for his own role performance. Thus, opinions adopted through identification do remain tied to the external source and dependent on social support. They are not integrated with the individual's value system, but rather tend to be isolated from the rest of his values. . . .

Internalization. In this case influence is accepted by the subject because it is congruent with his value system, because it is demanded by his own values or because it is likely to assist in the maximization of his values. The communicator's power rests on his credibility, trustworthiness or expertise. Since it becomes part of, and integrated with, the subject's existing value system, the adopted behaviour does not stem from situational constraints or a role relationship and becomes independent of the communicator.

> The behaviour will tend to occur whenever these values are activated by the issues under consideration in a given situation, quite regardless of surveillance or salience of the influencing agent.

The induced behaviour will be abandoned only when it conflicts with basic values or prevents the realization of such values.

Following this summary of Kelman's three processes, we must now develop an application of the scheme to teacher–pupil relationships.

Compliance. Here the relationship is based on the teacher's control over classroom rewards and punishments and the pupil's desire to influence the flow of these rewards and punishments relative to himself. We have, in fact, one form of a 'pleasing teacher' orientation. The pupil operates on the basis of a pleasure principle, seeking to maximize his profits in a situation which is unattractive but from which he cannot escape. The teacher operates on a fear principle; if the pupils conform they are rewarded but often they conform from fear of the punishments that are threatened for non-conformity. Because the pupils conform only under surveillance, the teacher feels the pupils are never to be trusted.

A careful study of this type of relationship reveals that it is a vicious circle. The teacher's reliance on rewards and punishments makes it difficult for the pupils to move into any other sort of relationship, for their behaviour must be constantly orientated towards these rewards and punishments. On the other hand the teacher feels unable to alter his own relationship with the pupils since he does not trust them and experience has taught him that if he abandons his reliance on the reward–punishment system the pupils will no longer conform to his expectations of the pupil role and he will be 'let down'. Typically—but by no means exclusively—this situation arises where the pupils are working class with minimal commitment to middle class values. They care little for school; they have no desire to be 'good' pupils; they feel no strong attraction to the teacher; their values are not congruent with those of the teacher or the school.

Identification. This is more typical of a teacher–pupil relationship which arises when the pupils are middle class, or working class but with middle-class values and aspirations. The pupil desires, or comes to desire, to establish and maintain the 'good pupil' role. Once the pupil has accepted the desirability of this classroom role relationship, he also takes over many of the teacher's assumptions, perceptions and evaluations. They develop common ground. The teacher can operate on a love principle, chastising recalcitrant pupils not with threat of punishment but by inducing a sense of anxiety, guilt or fear of loss of love. The pupils become dependent on and sensitive to the teacher's role expectations as well as to the rewards he offers. When the pupils fail to conform because the role relationship is not salient, the teacher responds with surprise or disappointment at their unexpected and inappropriate behaviour and corrects by means of an appeal to the role relationship. This relationship, like that of compliance, contains many features of the 'pleasing teacher' orientation. In compliance the pupil produces pleasing behaviour in order to avoid punishment. In identification the pupil exhibits pleasing behaviour in order to maintain or enhance his good pupil image and his role relationship with the teacher. For this reason it is not at all easy to distinguish the two types of relationship in practice. Many classrooms contain pupils with both types of relationship and the same pupil may swing from one type of relationship with the teacher to the other on different occasions and under particular circumstances.

Internalization. From the teacher's point of view this is the ideal teacher–pupil relationship, for the pupil can be trusted. Neither threats nor appeals to the relationship are needed to induce conformity or the acceptance of influence. If an appeal is made it will be an appeal to *rationality*. Teacher and pupils share essentially the same values and the teacher's task is to help the pupils to make their values more congruent or to show how a given course of action will assist in the realization of such values. When such rational appeals are made in relationships characterized by compliance or identification they tend to fail because the necessary similarity of values does not exist. Typically this relationship arises with the 'good' sixth form pupil or undergraduate.

There are obvious implications for the process of attitude change in schools that can be derived from Kelman's analysis. A teacher can rely on appeals to reason or his credibility, trustworthiness and expertise only if the pupils are at the internalization stage. It is pointless to make appeals to the relationship if the pupils are in a state of compliance. Many new teachers do make such appeals to reason or to the relationship without realizing that they are likely to fail if the pupils have a compliant orientation. The ultimate problem is how a teacher–pupil relationship can be changed from compliance or identification to internalization. Kelman offers no solution to this problem. We do not even know whether a change from compliance to internalization needs to pass through a stage of identification.

My own solution to the problem has already been propounded in a previous chapter, where I suggest that the 'pleasing teacher' orientation (compliance and identification) can be eliminated on the teacher's initiative if he moves away from an approval-based relationship to one of acceptance. It is through acceptance that a relationship characterized by internalization can be established. This is one of the reasons why I developed the Rogerian approach at such length.

Our survey of attitude research and theory has not produced any simple or easy answers. Admittedly our principal example, attitudes to smoking, was a fairly complex one, involving physical addiction. This was intentional since I suspect that the majority of attitudes that teachers would like to change are no less complex. And at the risk of depressing the reader I must confess that as I write this conclusion I am still smoking happily. Our conclusion reaffirms our original view, that attitudes are very difficult to change, though our journey has led us to discover some of the reasons why this is so. When one takes the wider view it is very fortunate that attitudes are so resistant to change for were it otherwise we should all be remarkably plastic and chameleon-like. We turn to attitude theory and research for assistance when we want to change attitudes which for various reasons we regard as undesirable, forgetting that we also want the attitudes which we happen to regard as desirable to be highly resistant to change. Yet we have learned a great deal of negative value—what *not* to do if we want to change attitudes. We have learned, I hope, that teacher exhortations and sermons, so dear to the profession, are unlikely to change pupils' attitudes and may have an unintended boomerang effect. We have learned how not to waste much time and effort.

We have not, however, dealt at all with the moral questions about attitude. These moral problems are not to be dismissed lightly, even though they are beyond my competence, and perhaps the reader's, as well as beyond the scope of this book. Moral questions arise not when we talk of changing pupils' attitudes to school, or to a subject, or even to smoking, but when we discuss the topics of 'brainwashing' or 'indoctrination'. Psychologists have never shown much enthusiasm about defining these terms, though they have provided fascinating descriptions of indoctrination attempts (e.g. Lifton, 1961). The definition and analysis of the terms has been a task for the philosophers. The difficulty is that indoctrination and brainwashing are insult terms to be applied to the activities of those of whom we disapprove. It is the Communists and other out-groups who indoctrinate; we are educating, or bringing children up properly, or giving them the right values, or leading them to the truth. Yet the same moral problems are involved just as, in general, are the same techniques.

In the popular view, indoctrination is to be identified by its *methods* —interrogation, confession, self-criticism, isolation, degradation, attack on personal identity and so on. Some philosophers (e.g. Wilson, 1964) have argued that it is the *content* which marks indoctrination since it is concerned with beliefs that have no rational foundation in evidence. Others (e.g. Hare, 1964) have argued that it is the *aim* which distinguishes indoctrination. On this view it arises when the aim is that the subject should come to hold a belief unshakeably whatever the forces or evidence to the contrary. This debate among the philosophers has not been satisfactorily resolved but the three aspects of aim, method and content do provide a useful framework in which the moral issues can be discussed.

Aim. My own view is that if any person attempts to influence another person in such a way that he will come to hold a belief or attitude unshakeably, then he is behaving immorally. The teacher who tried to convince the pupil that smoking is invariably wrong, both for himself and others, would thus be open to censure, for he is acting against this assumption. The reasons behind this assumption are two-fold. First I would argue that to try to persuade someone into holding an attitude unshakeably is an unjustified interference with an individual's human rights, since it consists of an imposition of one person's attitudes on another. Second, when a person does come to hold an attitude unshakeably, then he becomes closed to evidence against that attitude and prejudiced against persons holding contrary attitudes. Thus a pupil who is persuaded that smoking is undesirable

397

and wrong may persist in this attitude even when cigarettes become cheap and no longer injurious to health. He retains the attitude even when the rationale is no longer tenable. He would also be intolerant of those people who did smoke. My own view, whilst recognizing the fact of and the need for attitude change, would exclude the induction of unshakeable beliefs since the ideal is not of a society where people all hold the 'right' attitudes with absolute conviction but a society whose members possess coherent sets of values and attitudes but who are also sufficiently flexible and open to experience to change those attitudes in the light of changing evidence, experience and circumstances and who are tolerant of others possessing quite different attitudes.

There are many difficulties in this position. As teachers we are constantly trying to change attitudes in line with the central values of our culture, our subculture, and our individual predilections. Certainly we want the pupils to share attitudes which are held in high esteem by most people in our culture. (We regularly cheat here. As Harold Entwistle has pointed out, most of us approve of religious education, but we would be horrified if pupils began to shape their conduct on the injunctions of the Sermon on the Mount.) We want our pupils to develop, for example, attitudes of honesty. We can justify this on the grounds that honesty is for the good of society and the good of the individual. But in taking this position we must recognize that by the same argument the communists can justify training their children to hold unshakeable beliefs in the communist way. Almost any kind of attitude manipulation could be justified on the grounds that it is for the subject's own good, and nowhere is this more true than in the case of children. In my view we should not aim at training a pupil to hold any belief or attitude unshakeably, however desirable or fundamental it appears to be. Rather we should aim at helping the pupil to develop his own set of values, attitudes and beliefs, but with a full awareness of the whole range of possible orientations, of the arguments that can be sustained for and against each, and of the personal and social consequences of any particular orientation. This is not to say, of course, that the teacher must not hold his own distinctive set of attitudes, but that rather than seeking to impose them on the pupils he should present his own attitudes along with other possible sets of attitudes. This is difficult in the case of very young children, who will not perhaps have the capacity to choose. Inevitably we bring them up to share our own values and attitudes. But there is a difference between intending that they should go on holding these attitudes into maturity and intending that with greater maturity they will be free to shed former attitudes in favour of new ones or to retain the old attitudes because so far as is possible they have been free to do so. A further difficulty in training for flexibility is that we may find ourselves with a person who is highly susceptible to social influence or who is likely to change his attitudes merely because it seems expedient to do so. In trying to avoid the induction of unshakeable attitudes we might create a person with no firmly held attitudes at all. The 'liberal' stance abounds with its own problems.

Method. My own view would be that we should rule as immoral those methods or techniques of attitude change which clearly lack respect for the individual's integrity. On these grounds we could rule out degradation. It would be much more difficult to make judgments on the sort of techniques that we have discussed in this chapter, for although we could regard them as inherently possible they could in fact all be used immorally. It is at this point that we have to ask the teacher for his aim in using a particular technique. We know that public commitment to a position is likely to be accompanied by a congruent attitude change. Whether or not we are justified in using this technique depends on the teacher's motives and the teacher–pupil relationship. It does seem to me morally dubious if teacher cajoles a pupil into committing himself publicly to a renunciation of smoking simply because the teacher recognizes this as an effective technique to stop the pupil smoking, which for whatever reasons the teacher feels would be a desirable state of affairs. It is morally dubious because the teacher is employing the technique in order to manipulate the pupil to the teacher's own ends. It is much less dubious if the pupil tells the teacher that he wants to give up smoking and is seeking the teacher's help and as a result the teacher tells the pupil that public commitment may assist in that endeavour. In this case the teacher is assisting the pupil to use a technique to realize his own ends. Of course, many cases could not be decided as clearly as these. Suppose a pupil has negative attitudes to pupils of other races, believes his attitudes are justified and does not want to change them, and believes in his right to express these attitudes in various ways, including physical attacks. What does the teacher do? He will regard it as his duty to prevent the open expression of such attitudes, especially physical violence, and will try to inhibit such expression by means of punishment, recognizing that punishment is unlikely to change the attitudes or to prevent the expression of these attitudes when the pupil is out of surveillance by the teacher. He may also try to change the attitudes, even if it is only by the doubtful method of giving lectures. Is the teacher justified in trying to change these attitudes? Is any technique justifiable in pursuit of this goal? Have we not ruled as immoral the aim to change the pupil's attitudes so that he comes to believe in racial tolerance and harmony unshakeably, and the use of any technique to achieve this aim? Our teacher is not precluded from action so long as (*a*) he is clear that he does not want to change the attitudes unshakeably—which might require considerable self-analysis—and (b) he does not intend to use any techniques which are morally unacceptable *per se*, i.e. which deny the integrity of the subject. Yet teachers do not always apply these strictures before attempting to change their pupils' attitudes. Even where they are taken into account by no means all the moral problems have been solved.

Content. It is distressing for a teacher to come face to face with beliefs and attitudes in children that seem to the teacher to be irrational or immoral or socially divisive or harmful to the individual's happiness. This fact alone does not give the teacher the right to change those attitudes. Some teachers do act as if there were no moral problems and may go to almost any length to change those attitudes as long as outsiders to the classroom do not complain. Sometimes the teacher may have the active support of superiors or parents, but even this is no adequate justification. The main moral problem stems from the fact that the existence of different attitudes in pupils does not mean they are unworthy, or improper, or less valid. Particularly in the classrooms where the teacher has so much power and the pupils are so young it is very easy to equate 'different' with 'wrong'. Even where the teacher adopts such a view, he will not be indifferent to pupils' attitudes, partly because at times these different attitudes will find expression in disruptive or anti-social classroom behaviour, partly because their expression (e.g. lack of interest in a subject) will make it difficult for the teacher to execute his role, and partly because the teacher will recognize that the possession of certain attitudes will bring the pupil into conflict with society. If certain attitudes result in disruptive behaviour, the teacher has the right to take action that will inhibit the disruption, but not the right necessarily to change the attitudes themselves. Similarly, that a pupil's attitudes make him difficult to teach or are likely to get him into trouble elsewhere does not give the teacher an adequate justification for attempted change. Yet the teacher's concern for the pupil's present and future welfare will prevent him from adopting a policy of 'live and let live' to the pupil. Once again the teacher's motives become very important. The teacher does not need to care for the pupil in order to define his attitudes as 'wrong' and to take action to try to bring them in line with the teacher's 'right' attitudes. Indeed if the teacher really does care then such a perspective would be as foreign to him as would the desire to make the pupils into other versions of himself. The teacher who cares will set about attitude change not in terms of making the pupil conform but in terms of creating conditions in which the pupil will be helped to develop greater insight into his own current attitudes, to view them with greater objectivity and personal freedom, and to change them if necessary in the light of such reflection. The uncaring teacher does not trust pupils; things must be done to them by the teacher if they are to grow up with the right, that is the teacher's, attitudes. The caring teacher does trust the pupils and believes that if he can create the right conditions the pupils will be able to explore and develop their attitudes, passing through and perhaps ending up with attitudes that are different from his own; but always he is confident that the pupils are struggling to grow and that, providing the right conditions are created, then the growth will not in the long term be inimical to individual and social happiness. The uncaring approach takes an intolerant short-term perspective—the right attitudes must be put in now. The caring approach takes a tolerant long-term perspective—the pupil must be encouraged and helped to develop his own attitudes throughout his whole life, of which school is just the first part.

These considerations also apply when we are more concerned with the development of new attitudes rather than the changing of existing ones. The uncaring teacher soon becomes highly selective in what he regards as appropriate or relevant to the induction of the 'right' attitudes, and he need not be too concerned with the rationality or the evidence for these attitudes. The 'right' attitudes are in his view self-evidently so and even to question them is heresy. The caring teacher, concerned with the pupil's growth and freedom to choose is anxious to bring to the pupils' attention all available views and helps the pupils to judge their tenability in the light of rationality, sifting and weighing the evidence, debating advantages and disadvantages. When so described the differences seem very marked. In the day-to-day realities of the classroom the issues are much more blurred and confused. I know too well from my own experience as a teacher how easy it is to view and act towards pupils from the uncaring perspective and how hard it is to keep the caring approach as an ideal. But then teaching is a profession for idealists and idealists need both understanding of human relationships and powers of empathy if they want to realize their ideals.

Recommended reading

J. A. C. BROWN, *Techniques of Persuasion*, Penguin, 1963.

A. R. COHEN, *Attitude Change and Social Influence*, Basic Books, 1964.

M. JAHODA & N. WARREN (eds), *Attitudes*, Penguin, 1966.

R. J. LIFTON, *Thought Reform and the Psychology of Totalism*, Penguin, 1961.

12　Staff relationships

In spite of extensive research into teacher–pupil relations and pupil–pupil relations, there has been almost no systematic research into teacher–teacher relationships. Since teachers seem to be even more sensitive and resistant to research on teacher–teacher relations than to investigations of teacher–pupil relations, it is not surprising that few researchers have had the courage to embark on this difficult field. Yet the social relationships of teachers form an important part of being a teacher; it is the teacher's colleagues who in many respects control and influence his induction into the profession. The teacher's conception of himself, his values and attitudes to many aspects of education may, as I have indicated in earlier chapters, be influenced by his relationships with his colleagues and his superiors and thus influence the teacher's behaviour in the classroom and his relationship with the pupils. Life in the staffroom and its impact upon the teacher constitutes one of the most significant gaps in our knowledge of social processes within the school. This is particularly true with respect to a teacher's first post, since it is at this point that a teacher's conception of himself as a teacher is most firmly developed.

Within the school the teachers are part of a formal organization; each teacher has his place in the pyramid of authority. At the top of the hierarchy stands the head. Beneath him comes the deputy head, followed by the heads of departments. It is the head and the deputy head who command the highest salaries. The heads of department are given responsibility allowances that vary in size according to the formal status of the subject and to the size of the department. Science and English are large departments with high status whereas the music or domestic science departments are smaller in size and status. This differential is reflected in the financial rewards given to the respective heads of department. Within the larger departments, other members of staff are given special responsibility allowances of various size, and in recent years special allowances have sometimes been paid to some members of staff for pastoral rather than academic responsibilities—the heads of houses. At the base of the hierarchy come the junior teachers with no special allowances—and a few of the less competent teachers whose work has never in the eyes of the head merited an allowance. Formal status can also be influenced by seniority, either in terms of age or length of service to the school. Two teachers who have the same formal status in terms of allowances may differ in formal status because one teacher is older and has been at the school longer than the second teacher. Generally speaking, the formal status hierarchy coincides with the power structure, but there are exceptions. Formally speaking the head's secretary and the school caretaker have little power, but in practice their actual power is so great that they are persons whom the staff—and the head—must learn to cultivate with great care.

In the primary school where there is a small number of teachers on the staff there may be few differences in formal status. In a school with a head and three assistant teachers there is little scope for the formation of elaborate friendships and cliques. In the larger secondary school there develops an intricate informal organization of friendship groups and cliques. Usually, this informal organization interpenetrates with the formal organizational hierarchy. These informal groupings often centre round the teachers' subject specialisms; teachers within one department can be expected to become friends not only because they interact frequently but also because they teach the same subject and have similar academic interests. Many groups are sex-based, sometimes being encouraged by different staff-rooms for male and female teachers. Other groups may develop because of a similarity in age or seniority, where the members are also similar in formal status. Not uncommonly the armchairs and the warmest corner of the staffroom are by tradition reserved for such senior members of staff. Some groups develop around common interests or activities or attitudes—the group which plays bridge or chess after lunch, the group whose members eat sandwiches together in preference to the school dinner, the group of teachers who all supervise sporting activities and so on. I know of one school where one staff-room was entirely given over at lunch time to those teachers who enjoy crosswords; in turn the clues to the crossword in *The Times*, the *Guardian* and the *Daily Telegraph* were publicly read out to an enthusiastic audience. My presence was ignored until I managed to solve a fairly difficult clue. Groups with an attitudinal focus are common—school politics (the pro- and anti-head groups), national politics (the Tories and the Socialists), religion (Protestant teachers in a Roman Catholic school), education (radicals and conservatives). Some friendships are influenced by proximity of residence and the giving of lifts to school. In some cases the teachers and their spouses form social groups outside school in the evenings and at the weekend.

Typically, then, whilst there are often a few confirmed non-joiners in every staff-room, most teachers belong to several overlapping groups. Social relationships are very fluid as the teachers pass in and out of these different groups at different times of the day. Although in a few cases the teachers do become conscious of themselves as members of a particular group, many of these groupings are of a less clearly defined order. Yet they do exist and most teachers both like and associate with some members of staff more than with others. In some schools overt antagonisms grow up between groups, sometimes to the point when conflicts and rivalries can make life in the staff-room distinctly unpleasant. More typically occasional arguments flare up and subside; the basic uneventfulness of staff-room life is not disturbed to a large degree by the fact that two members of staff are not on speaking terms. Teachers are highly sensitive to life in the staff-room and the quality of teacher–teacher relationships; they have no difficulty in distinguishing between a 'good staff' and a 'bad staff'.

The formal status of teachers in the formal hierarchy is really a form of prestige status as defined in Chapter 9. In the informal relationships there develops a form of informal status which may or may not be congruent with the formal status. Some heads of department are almost universally unpopular and carry little influence with colleagues, whereas a teacher of relatively low formal status may prove to be of very high sociometric and social power status. The discovery of such incongruencies forms an important task for a new teacher—and especially for a new headteacher.

Within all these groups there develops a set of values and norms, and a status structure. Since these are highly variable, it is as difficult to make generalizations about the informal groups of teachers as it is about the informal groups of pupils. However, there are often some general staff-room values and norms to which all are pressured to conform and to which the majority do conform. Like all group norms they are not instantly obvious. They are perceived and learned and then taken for granted so that experienced teachers find it extremely difficult to specify them. Some of these have been hinted at in earlier parts of the books. It would now be useful to draw them together.

The first concerns the autonomy of the teacher, a value which is very basic within the profession. The value refers principally to classroom autonomy, as was noted earlier; intrusions into the privacy of the classroom are resented. This means that matters which are relevant to the teacher's classroom role are very delicate. Teaching methods, discipline, the teacher's 'philosophy of education' are, according to the autonomy value, a matter of personal choice. No headteacher or teacher can impose his view in these respects on others, except in the very broadest terms, without infringing this value. This is part of the reason why teachers frequently gossip about one another's practices and beliefs but show little enthusiasm for an open serious discussion of them. Such a discussion would make the differences between teachers in beliefs and practices quite explicit, would involve the making of judgments about the beliefs and practices of others, thus provoking some inter-teacher hostility, and would create pressures toward an undesired consensus and uniformity. Most teachers prefer to go their own way, within obvious limits, and in order to defend this right they have to be willing to allow all other teachers to go their own way too. With the development of team teaching and allied innovations the traditional conception of teacher autonomy is breaking down and one of the most useful bi-products will be the encouragement of greater serious discussion about educational problems between teachers in the same school. At present too many teachers find they can discuss and debate educational issues seriously only in the relative anonymity of conferences.

The second dominant value is that of loyalty to the staff group, to which I referred in the discussion of Goffman's concept of the team in an earlier chapter. Teachers soon learn not to talk about one another in the presence of pupils and not to inform the head or other superiors about a teacher's activities.

Within most staff-rooms there is a mediocrity norm, which like the mediocrity norm among the pupils seems to prohibit too great an enthusiasm and too great an effort. Teachers should not arrive too early at school; they should not spend all their lunch time marking books or preparing lessons; they should not supervise too many extracurricular activities. Such behaviour prevents a teacher from being 'one of the boys'. When teachers do contravene this norm they find themselves teased by their colleagues and are laughingly accused of seeking promotion from the head. Sometimes teachers seem very anxious to deny the existence of this norm, but it exists in most staff-rooms in some form. Its presence is betrayed by the fact that most teachers prefer the teacher who is constantly late and who cunningly manages to escape various duties (so long as they do not then have to be undertaken by another teacher) to the teacher who is so keen and efficient that he makes all the other teachers feel a degree of incompetence or guilt.

An associated norm is the norm of cynicism. Teachers are not expected by their colleagues to be enthralled by the job of teaching or rapturous about the pupils. It is much more normative to grumble in the staff-room—about the head, about the pupils, about the facilities. The irrepressible enthusiasm and idealism of some student and beginning teachers is taken by their more experienced colleagues as a sign of professional immaturity which time will soften. Many staff-rooms contain a particularly cynical teacher who, if he is witty, is rarely an unpopular figure. Combined with this cynicism goes a degree of anti-intellectualism. Teachers are not expected by their colleagues to be talking about educational issues unless it is of a narrow curricular focus or concerned with the latest policies of the head, the Local Authority or the Minister of Education. Nor are teachers expected to read books on education, which is regarded as a student activity. One is also rarely expected to talk about 'high culture'—plays, books, films, opera and classical music. Typical staff-room conversation is much more likely to be about the previous night's television or the local football team.

Perhaps the most favourite topic for staff-room discussion is the pupils. They provide an endless source of amusement and outrage. Such informal conversations have two important effects. First, they pressure the teachers into reaching a consensus about the 'goodness' or 'badness' of particular pupils or classes. Once a pupil or class have become categorized, then teachers will happily offer further tales exemplifying this agreed categorization. Second, teachers who have not met a particular class or pupil before cannot help but be influenced by these established reputations. One of the main functions of teachers' staff-room gossip is to create preconceptions and expectations in the new teacher's mind prior to an actual teacher interaction. Typically the new teacher is warned by well-meaning colleagues about difficult classes and pupils before he actually meets them and it is extremely difficult not to be influenced by such warnings. Pupils and classes have their reputations made quite as much within the staff-room as within the classroom itself.

A few teachers play very little part in the informal life of the staff-room. A few of them have a 'legitimate' reason for their withdrawal—coaching a sports team, producing a school play and so on. But there is no doubt that the majority of teachers who decline to enter this informal life are perceived by their colleagues as 'odd'. It is true, of course, that in many cases such teachers are among the least competent on the staff. Because the staff see so little of them, they are easily labelled and made the target of gossip. Elizabeth Richardson (1967) has wisely commented on this. She points out that these labels cannot possibly do justice to the great complexities of the human personality, nor can such labelling contribute to that collective understanding and co-operation that is needed among a school staff. Why then do teachers use these labels and act as if they wanted to keep people fixed in these labels? Miss Richardson suggests that we use our colleagues to externalize the conflicts that we cannot handle in ourselves.

406

it is not easy for us to face in ourselves the good and the bad schoolmaster, the strong and the weak classroom personality, the perceptive and the blind student of human relations. We therefore look for mirror images, not of our total selves, but of the irreconcilable parts, the bits we claim and the bits we reject. . . In this way, the members of a staff group try to deal with their own internal conflicts by splitting off the good and the bad parts of themselves. They can all identify with the good teacher into whom they have projected their own best qualities, and use him or her as a sort of ego-ideal or ideal self. At the same time, they get rid of their own ineffectiveness by projecting it into someone who seems less competent than they feel themselves to be, using him as a scapegoat to carry their burdens.

Headteacher–teacher relationships

Usually the headteacher is not part of the informal organization of the teaching staff. He is unlikely to mix socially with the teachers in the staff-room at break or lunchtime. In an informal survey I once conducted, in only one instance did the head regularly participate in informal staff-room activities—the head belonged to the bridge-playing group. A head is more likely to ask members of staff to come to see him in his room, even when the nature of the discussion is not particularly private. Many heads also converse with teachers on the corridor or pay visits to teachers in the classroom. As a result the head tends to be isolated from the teachers' informal social groups and activities and the teachers do not usually regard him as 'one of us'. He may reinforce this perception of him by his own behaviour, for example by knocking on the staff-room door before entering on his rare visits. Most heads would probably not take exception to the statement of Tosh (1964).

A Headmaster is a lonely figure, and no man likes to be lonely. No matter what the temptation, however, he must never be too friendly with any member of his Staff, and he must never say anything about a Teacher to another Teacher. . . A Headmaster may unbend with his Deputy, but even here a certain distance must be kept.

Edmonds (1968) makes a similar point.

You will miss the gregarious comforts of the staff-room and sometimes feel the separateness, even the loneliness any new Head may expect to have. If appointed from the same staff, it may not be so noticeable, though the old familiarity will not be advisable.

The function of this aloofness in the head is presumably to preserve his authority, and many of the techniques he adopts to achieve
this end are similar to those employed by teachers in maintaining their
social distance from the pupils. He will remain fairly formal in his
relationships with his teachers—though this does not mean that he
will be unfriendly—by using such devices as referring to teachers by
their surnames rather than their first names. In the privacy of his
study he may use first names but he does not usually expect this usage
to be reciprocal. Again, as was clear in teacher–pupil relationships,
he will be aware that he likes some individuals or groups more than
others, especially when such persons are seen by him as more loyal or
more effective or as supporting policies which he is anxious to
institutionalize. But in public he will try to give the impression of
fairness and impartiality; he does not wish to be accused of favourit-
ism if it can be avoided.

The need of the headteacher to preserve his authority springs from
the autocratic nature of the relationship between head and teachers.
By autocratic I am not referring to the head's style or manner but to
the distribution of power in the relationship. Autocracy arises in a
relationship where one person exercises far greater power over the
other. The head's autocracy is often formulated in the school's
articles of government. The articles of one school, for instance,
state:

> The Head Master shall control the internal organisation,
> management and discipline of the School, shall exercise
> supervision over the arrangements for school meals, so far
> as is necessary for the purposes of school discipline and
> organisation, and over the teaching and non-teaching staff,
> other than the Clerk to the Governors, and shall have power
> to suspend pupils from attendance for any cause which he
> considers adequate.

Naturally certain checks on this enormous power are built into the
articles. When any pupil is suspended the head must inform the
parents of their right of appeal to the Governors and a report must be
made to the Governors. It is in fact the Governors who form the
principal formal power check. The articles make the head's subordin-
ation to the Governors quite explicit.

> All proposals and reports affecting the conduct and curricu-
> lum of the School shall be submitted by the Head Master to
> the Governors. . . . Suitable arrangements shall be made for
> enabling the teaching staff to submit their views or proposals
> to the Governors through the Head Master.

The first of these limitations on the head's power is much less specific than it is explicit. Since the Governors are often local councillors or laymen with little intimate knowledge of educational matters, meeting no more than two or three times in a year, the head's 'proposals and reports' must of necessity be brief and superficial and the Governors' approval will be mostly a technical formality. Governors' meetings can assume a slightly farcical quality since the pretence of dealing with important business is executed most seriously, and in this respect the Governors' experience on committees of Local Government serves them in good stead. In general they assume that the head they appointed will be fully competent to run the school smoothly and efficiently. They have little choice but to trust him. The second limitation on the head's power is, to say the least, naïve in its conception. It is extremely unlikely that teachers will convey their dissenting views or complaints to the Governors when this means doing so through the head. In practice teachers are much more likely to regard their Union as a better defender of their rights, but unless the cause of the complaint is extremely serious, teachers have in effect no means of objecting to the head's use or abuse of his power. They content themselves with staff-room grumbling or subtle personal feuds.

Given that the headteacher—teacher relationship is autocratic in structure, it is a different matter whether or not the head exercises his power in an authoritarian manner. He may do so or he may not. Most teachers with whom I have discussed the matter consider that heads tend to lean towards authoritarianism. Certainly it would be agreed by all that almost all heads act in an authoritarian manner on occasions. Partridge (1966) describing one school paints a picture which cannot be considered very unrepresentative.

Most major decisions and even some minor ones are taken by the Headmaster. The place and function of the formal staff meetings illustrate this. Full formal meetings are summoned irregularly at the discretion of the Head or Deputy; thus whereas some schools hold staff meetings either every week or fortnightly, this School may hold them only two or three times a term; in the first seven weeks of one term there was no kind of staff meeting at all. Any democratic co-ordination of the work of the School is, of course, impossible. Staff meetings during the past year tended to be called for one or a combination of three purposes. They were not called for the staff to discuss educational issues or problems, or make important decisions relevant to the running of the School. During this year, such staff meetings as were held were either to ratify decisions already taken, or as public recruiting drives, when the Headmaster or his Deputy wished the teachers to undertake some extramural voluntary work, or some course of action within the School time-table but beyond the scope of the academic curriculum, or to deliver a reprimand to individual members or to the staff as a whole for some error or infringement of duty.

streaming was discussed at a long staff meeting, and it was decided by a majority vote to experiment during the following year by abolishing it internally. Detailed plans were drawn up and a display organized for one of the 'open days' to explain the proposed new system to parents. At the beginning of the new school year the Headmaster and his Deputy went back on all that had been previously agreed by the staff, streaming tests were given to the new intake and all other classes though adjusted as usual for a new year were left unchanged. As far as could be ascertained no public explanation was given, or asked, the Headmaster's inference being as always that it was his school to run in the way he chose, whatever the preferences or wishes of his staff.

Many heads, though not as extreme as this perhaps, view themselves as the policy-makers of the school and the staff as executives whose job it is to put this policy into operation. Staff meetings, called on his initiative, can soon be conceived as part of this process, or, even worse, as a mere exercise in public relations, since the head need not take into account or be bound by views expressed or decisions agreed at such a meeting. The autocratic structure lends itself to an authoritarian style and few heads manage to avoid the danger.

Many heads are aware that the authoritarian image is currently in disfavour within the teaching profession, and increasingly also in the eyes of parents and other members of the public. In my experience the most authoritarian heads have become very defensive, propounding at great length the arguments in favour of such an approach and challenging critics to offer viable alternatives which, when offered, are rejected on the grounds that they are so absurd that they do not merit serious discussion. A shrewder course has been taken by the majority of heads who have changed their image according to the times. Following the pattern of greater informality in teacher–pupil relationships, they have become more informal and more friendly in their relationships with teachers. Since authoritarianism is traditionally associated with a formal, aloof, cold and dictatorial manner, the change to an informal and friendly style can be taken as evidence of movement away from authoritarianism. The contemporary head is more likely to foster an image of himself as more of an equal, calling his teachers by their first names and buying them a pint of beer at the local pub, than to cultivate an image of himself based on public school headmasters of Victorian vintage or even on the paternalism of latter-day grammar school heads. But a change of image is not necessarily associated with a shift in power and in my view there is little evidence of such a power shift. Autocracy can survive such changes of image relatively unscathed; the despotism remains, whether or not it is exercised benevolently.

Associated with this change in image is the espousing by heads of the notion of 'consultation'. This is a very tricky concept, which can mean what one wants it to mean. Consultation may involve a full discussion at a staff meeting, with policy being determined by consensus or majority vote. It can also mean making allowances for the opinions of the deputy head or a few senior teachers who are known to be in favour of the policy anyway. In an extreme case it can mean obtaining staff views and then ignoring them. When a head tells us that he is keen on consultation, he tells us nothing. The same may be said of 'delegation', the other fashionable concept. If the head delegates only minor or routine tasks and decisions, there will be no meaningful transference of power.

Of major interest to the social psychologist is the effect of the autocracy on the relations between the head and the staff. With respect to the teachers the principal effect of the autocracy resides in their dependence on the head. They are dependent on the head in three main areas. First, they are dependent on him for internal promotion, since the distribution of the additional monetary allowances is at the head's discretion. These allowances can make a substantial difference to a teacher's income and are thus regarded as important. Second, teachers have to rely on the head for a reference where a change of school or an external promotion is sought. Unless the head provides at least a satisfactory reference, a candidate for a new post is unlikely to find himself on the short-list. Since such references are highly confidential, a teacher may find it difficult to discover how good a reference he has been given. Third, the teacher depends on the head for various favours. Some of these favours can be of great significance. Recently, for instance, many Local Education Authorities have decided to leave the distribution of money to different departments within the school to the head's discretion. There are many other favours, great and small, at the head's disposal and by withholding them the head can make life difficult and unpleasant for a teacher. In these three ways it will pay the teacher to keep in the head's 'good books'.

The result is a replication at a higher level of what we have observed as a principal phenomenon of teacher–pupil relationships, namely 'pleasing teacher'. The three laws now become laws of pleasing headteacher.

First law: find out what pleases and displeases the head.

Second law: bring to the head's attention those things which please him and conceal from him those things which displease him.

Third law: remember that it is a competitive situation. The teacher must try to please the headteacher and avoid displeasing him more than do other teachers.

The teacher is under pressure to publicize to the head those aspects of his behaviour which will elicit the head's approval. If the head is keen on 'good discipline' then acquire the reputation of being an excellent disciplinarian and do not talk to him about child-centred approaches; if the head is keen on the extra-curricular life of the school, found a few societies and clubs and do not be seen heading for the car park at one minute past four; if the head is keen on traditional methods and is proud of the school's examination record, teach accordingly and temper or mask your enthusiasm for creative approaches to teaching and learning.

The third law is very important, especially when we remember that internal promotion is highly competitive. The teacher must, if possible, try to avoid giving his colleagues the impression that he is a sycophant or a 'head's man', since this may endanger his relationship with his colleagues and their willingness to co-operate with him. If a teacher is seen visiting the head's study too frequently, he will soon hear about it, though he can dissipate suspicion if he indicates that he was summoned. Similarly, a wise teacher does not boast that the head consults him constantly. The ideal is to be known to be in favour with the head and perhaps to 'have his ear' but at the same time to be known as to some degree a staff-room critic of the head. It is not an easy game to play.

Perhaps the most difficult aspect to the art is that of pleasing the head without giving *him* the impression that one is sycophantic. Various techniques are possible here. It helps, for example, if a head of department has several quite incompetent teachers under him, since he can then, allegedly in the interests of the pupils, bring the failings of his associates to the head's attention and make his own star the brighter. He can also by implication claim all the pupils' success as his own personal achievement.

It is in the direct, face-to-face relationship with the head that the greatest complexities arise, since the teacher must ingratiate himself without giving the head the impression that ingratiation is taking place. Successful ingratiation must not be recognized by the target person or it will fail in its objective.

Ingratiation is very complex as recent studies by social psychologists have shown (Jones, 1964). Here I shall be concerned with the

412

four basic tactics. The first, and most obvious, is that of *flattery*, paying compliments or expressing high positive evaluations to the target person. This is a difficult technique since most of us are wary about the sincerity of other people's compliments but it has great potentiality since if I can convey to you that I like and approve of you, the chances are that you will reciprocate the liking (see Chapter 8). The flatterer has to be careful not to be too extravagant with his compliments since excess will give grounds for suspicion that the compliments are not selflessly motivated. The best approach for a teacher is to flatter the head in those areas where the head is least secure or certain in his views of his strengths and weaknesses, e.g. some new policy introduced by the head about which he has confessed certain reservations and where the correctness of his decision is uncertain. It is pointless to flatter the head on his obvious strengths, and positively dangerous to flatter him on those weaknesses that he himself fully recognizes. This was fully known to Lord Chesterfield when he wrote to his son:

> Cardinal Richelieu, who was undoubtedly the ablest statesman of his time, or perhaps of any other, had the idle vanity of being thought the best poet too; he envied the great Corneille his reputation and ordered a criticism to be written upon the 'Cid'. Those, therefore, who flattered skilfully, said little to him of his abilities in state affairs, or at least but *en passant*, and as it might naturally occur. But the incense which they gave him, the smoke of which they knew would turn his head in their favour, was as a *bel esprit* and as a poet. Why? Because he was sure of one excellency, and distrustful as to the other. You will easily discover every man's prevailing vanity, by observing his favourite topic of conversation; for every man talks most of what he has most a mind to be thought to excel in. Touch him but there, and you touch him to the quick.

It is also dangerous to flatter the head when he is fishing for compliments, since he will be very suspicious of sycophancy on such occasions. One of the most successful techniques is for the teacher to make flattering remarks to someone—e.g. the deputy head—who is certain to report the flattery to the head. No flattery seems more genuine than indirect flattery.

The second tactic is *conformity in opinion or behaviour*. This follows the principle that we like those who seem to share our values, attitudes, opinions or interests (see Chapter 8). Of the two, behavioural conformity is the more effective since it cannot be dismissed as mere words; thus imitation becomes the sincerest form of flattery. Unfortunately when the conformity is very high, it tends to arouse suspicion. The teacher who always agrees with the head, whether genuinely or not, soon acquires the reputation of being a 'yes man' in everyone's eyes. If the teacher genuinely does agree with the head often, he can avoid arousing suspicion by sprinkling his agreement with a few disagreements, especially on irrelevant or unimportant matters. A further device is to disagree with the head initially and then switch to agreement later. The head can compliment himself on his perspicacity and the teacher on his openness to persuasion when facts prove the point. As in flattery, the indirect reporting of one's conformity, or conformity on matters where the head is uncertain, are likely to be especially effective.

The third tactic is *self-presentation* or what is popularly called 'putting one's best foot forward'. The teacher has to present himself to the head as approximating closely to the head's ideal of the 'good' teacher. He must highlight his own strengths and virtues and mask his weaknesses and vices. I have already illustrated this in the discussion of the three laws. The danger in the tactic is that the teacher may appear to be boastful. A slightly exaggerated modesty or self-deprecation can thus be most effective.

The fourth tactic is *rendering favours*. The teacher who can always be relied upon to carry out efficiently any essential task, from entertaining the Governors to pacifying the caretaker, is likely to find himself regarded as the head's right-hand man. The norm of reciprocity can be expected to pay its usual dividends.

The deputy head is in a specially vulnerable position. The head expects the deputy to be loyal to him, but the staff also wish to claim his primary loyalty. The head expects him to keep him informed of staff attitudes and intentions; but the staff too wish to be informed about the head and if he does not do so he will be carefully excluded from staff secrets and gossip for fear that he will inform the head. The deputy then is under pressure to be a double-informer and to develop techniques of double-ingratiation. It is difficult to do so without being found out.

Now I do not want to suggest that all teachers are ruthlessly self-seeking ingratiators, expert in the art of deceiving gullible heads. I am suggesting that this is one important but rarely considered element characterizing power-dependence relationships such as that of headteacher–teacher; that heads vary enormously in their insight into such ingratiation as do teachers in their ingratiation skills; and that teachers' ingratiation varies from rare and unconscious tendencies to ingratiation to highly skilled, intentional and systematic ingratiation. One head once told me that none of his teachers ever tried to ingratiate themselves with him. In my view the head was remarkably unperceptive and/or his teachers were very skilled in the art.

414

Having considered some effects of the autocracy on the teachers' behaviour, we must now consider the effects on the head. The most obvious is that it is difficult for the head to know when the staff are ingratiating and when they are giving their true opinions. It means that the head may lack feedback on his own behaviour, or that such feedback will be confused, since he will tend to find his own opinions reflected back at him. The conspiracy of silence or ingratiating assent can prevent the head from knowing the true views of his staff. Because he is looking for assent he will be tempted to take the evidence at its face value. It would probably be nearer the mark if he worked on the assumption that mild assent signified indifference or mild dissent and that the expression of mild dissent cloaked strongly dissenting views. Most schools, fortunately, contain two types of teacher who are not inclined to ingratiation. The first is the teacher who is so secure in his knowledge that the head has a high opinion of him that he can afford to tell the truth. Such a teacher can offer important feedback to the head. The second is a teacher who, being a senior member of staff with a large allowance but no aspirations to promotion, can afford to tell the truth because of his lack of dependence. Such teachers do not always function as avenues of feedback to the head since their honesty on past occasions may have led them to conflict with the head's policies. The head may regard them as 'thorns in his flesh' and if he does so he severs their feedback potentiality, since he ascribes little legitimacy to their views. The head may also gain feedback from a deputy who is privy to staff views and gossip, but obviously unless the staff trust the deputy he will obtain little information to report to the head.

An associated effect of the autocracy is the head's tendency to view events from his own perspective rather than the teachers'. I have already pointed out that student teachers can be remarkably insensitive to the pupils' perspective, in spite of the fact that they themselves so recently held this role, and the same is true of heads with respect to teachers. Suppose the head calls a staff meeting and with all sincerity invites the honest comments of his staff on a policy he intends to introduce. He is, he says, anxious to have an open and free discussion. From his perspective there is no reason why the staff should not enter fully into the discussion. If the teachers are silent or relatively unresponsive, the head may infer that the staff are indifferent to his proposal. The truth may be that the teachers are unhappy about the idea but afraid of speaking against it lest they be thought to be leading a dissenting or rebel group. If one member of staff has the courage to express a dissenting view, the head may reply with a counter-argument. From his point of view this may be what he means by a discussion. Yet the teachers' perception may be very different. The dissenting teacher may feel that he was 'slapped down' by the counter-argument and be inhibited from pressing his case any further. Other teachers may share this perception, noting cynically how effectively the head put him in his place. They too will be inhibited from offering support. The head, however, still working on the assumption of an open discussion, may infer that only one teacher dissents and that his own counter-argument has been accepted as unanswerable. A condensed version of this situation is given in this account, describing overt behaviour only, by Margaret Phillips (1965):

> Staff meetings are always held in the Headmistress's room. There is usually a certain amount of time to wait before everyone is assembled and this waiting is always silent. The staff, usually very talkative, very positive, very individualistic, turns into a set of dumbstruck people who strongly resemble a group of students. The Headmistress's suggestion and remarks are received without comment or with a lukewarm murmur of assent. *After* the meeting, in the safety of the staff room, opinions are freely expressed by all and sundry.

By taking the different perspectives of the different participants in events such as this we are able to offer a meaningful analysis of the events and obtain useful insights into their origins and dynamics. Unfortunately the participants themselves perceive the events only through their own perspective.

Let us take another illustration. Suppose a head decides to drift along the corridors of his school so that he knows what is going on. He may hope that in this way he can discover when some teachers, particularly new ones, are in difficulties with a class and offer his assistance. The teachers, however, may perceive such behaviour as an attempt to spy on them or to assess their classroom competence. Or suppose the head decides to call teachers into his office for a regular interview. By this method he may hope to express his interest in the teachers' welfare and give the teachers an opportunity for an informal chat. The teachers, however, may feel that they are being called to account and that they are being subjected to a regular interrogation.

A head often forgets that teachers are supersensitive to his behaviour, and it is because of this that they can infer or attribute motives and intentions which have no real existence in the head's mind. This may in turn influence the teachers' behaviour towards the head which he, because of his ignorance of its real roots, must explain on other grounds. Even when misunderstandings and misinterpretations of this sort do not result in open conflicts, they can be significant barriers to understanding, respect and co-operation. The teachers' supersensitivity is evident when the teachers use certain cues in the head's behaviour as a basis for 'reading in' certain meanings which have little foundation and are certainly unintended by the head. That the head has been seen speaking angrily to a pupil at the beginning of the day may start a rumour, which will circulate with great rapidity, that the head is in a bad mood. Rumours about the head's moods acquire significance when we remember that teachers often feel that they have to choose the 'right moment' before they can approach him with a problem or a proposal. Yet such rumours can be based on the most slender of evidence. Unless the head can match the teachers' supersensitivity to him with an equivalent supersensitivity to them, the relationship can become distorted in unnecessary or even harmful ways.

Another effect of autocracy on the head is for him to respond to the teachers in a similar way to the response of teachers to pupils. He may institute or insist upon policies on the ground that it is for the staff's own good, that he knows their interests better than they do, that it is his job to run the school as he thinks best even though the staff offer opposition, that the staff are too immature or inexpert to exercise the necessary foresight or responsibility. Usually such arguments are rationalizations which cover up the head's inability or unwillingness to achieve his ends by more onerous and time-consuming but more justifiable means. I also fear that some heads, having shed such relatively crude justifications for their authoritarianism, are looking to industrial experience for 'management techniques' which are more subtle but at the same time essentially manipulative in their approach to human relations.

My own view is that heads should be abolished in their present form. This is hardly a radical proposal when we remember that Edward Short, an ex-headmaster Minister of Education, publicly argued in October 1968 for the head as a chairman of staff. I see no reason why the head should not be elected by the staff on a short-term basis, say one to three years, with the possibility of further terms of office. Everyone acknowledges that schools need some person to act as a chairman of staff meetings, to supervise school administration, to maintain links with the Local Education Authority, parents and other bodies, and to make day-to-day decisions. But I do not see why he should not be accountable to the teachers themselves. Until he is I do not believe that our schools can be considered to be democratic institutions or that the staff will be willing to take that corporate responsibility which is the mark of a vital human organization.

Of course, many teachers prefer the head as he is and resist changes of this sort. The head is a convenient scapegoat when things go wrong, an obvious target for blame. Many teachers willingly yield many responsibilities to him—after all, he is paid for it—and at the same time take pleasure in accusing him of being a dictator. He is also a useful filter between themselves and parents or other outsiders.

He can also be a useful ultimate deterrent to the pupils. The heads themselves are not the most ardent propagandists for their own abolition. They argue that without a head school policy could be determined only at the cost of endless committee meetings; that there would be no continuity of leadership; that it takes years to learn to become a good head; that staff turnover would rise under any other system; that staff like being told what to do and would be highly insecure under an elected head; that the present type of head best serves the need for change and innovation in schools. Some of these arguments contain important truths; others are unwarranted surmises or assertions to justify the status quo. Under the system of an elected head many old problems would remain and new ones would arise. But it does not seem likely, as has sometimes been asserted, that there would be a shortage of teachers willing to be elected, not even if we removed the present salary differential for heads. Nor do I think that staff would treat such an election at all lightly and so be easily swayed into choosing the 'wrong' man. Even if they did, disaster would not inevitably follow, since the election would not be irrevocable and staff would retain the main threads of control during the head's period of office. The present system assumes that a head will be good at his job. Inevitably mistakes are made under the present appointment system since there are no clear criteria by which we can select the right man from among the many applicants with appropriate experience and the highest qualifications and references, just as there are no clear criteria by which we can judge the 'goodness' of a head once he has been in office for some time. Yet there are many schools, even if they form a small minority in the country as a whole, which are forced to carry or suffer a failed headteacher until the day he retires, dies or is 'kicked upstairs'. And before this he can do untold harm to the school, the teachers and the pupils.

Many heads concede that they will not be effective heads for up to thirty years, as occurs in the present system. Some would genuinely like to give way to younger men, but feel unable to do so since resignation would lead not only to a decline in status, but also a decline in salary and, most important, to a reduced pension. Some heads, too, would like to return in the later years of their career to the classroom, for not all heads are disingenuous in their protestations that they would like more time to teach. We need to offer such heads an acceptable way out.

There are some heads, and it is very important to recognize their existence, who in practice act as if they had been elected rather than appointed. They are familiar with the dangers inherent in their position and avoid them with self-honesty and skill and considerable personal cost. If one could judge the system by them alone, there would be few grounds for criticism or complaint. But they are still relatively rare. I fear that those heads who most vociferously defend the present system are the very persons who under an elected system would never find themselves in a headship. In seeking to preserve the status quo they make demands which are, to everyone but themselves, patently absurd. Take, for instance, the conclusions of a conference of heads, as it was reported in *The Times Educational Supplement* a short time ago.

Head teachers must have absolute power in the running of their schools. Heads, it was said, should always be careful to canvass teachers' opinions before making decisions, and should be willing to listen to criticism. The ideal school was clearly one presided over by an all-powerful but tactful, reasonable, just and courteous head and staffed by enthusiastic teachers willing to voice their views but also willing to listen submissively to wiser counsels.

Change is coming and its main source is new entrants to the profession. Working in a University Department of Education I have been struck by the growing number, though still a very small minority, of student teachers in initial training who reject many of the assumptions behind our present system. Many of their values and ideas, both educational and social, conflict with those of traditional teachers and heads; they are less willing to accept the head's authority; and they are reluctant to ingratiate. If they do move into the profession in increased numbers, and provided they stay and retain their ideals, they will disturb the present system in fundamental ways. Their presence in schools will mean, among other things, that heads will no longer be able to rely on the 'united front' to outsiders which has been customary. Already we have had cases where teachers, agreeing with the pupils but disagreeing with the head, refuse to keep their dissent within the confines of the staff-room but openly declare their dissent to the pupils. I suspect that this is only the beginning of more large-scale conflicts which will serve as an impetus to change.

It does not follow that the existence of fundamental differences in values and opinions within the profession will necessarily produce harmful conflict. But if the conflict is to be beneficial, heads as presently constituted will need the capacity to understand how the situation looks from a more radical teacher perspective, for understanding is the necessary first step to dialogue. It will do no one any good if the head is so locked inside his own assumptions that he reacts with horror and incomprehension to suggestions based on different assumptions. Each side is then reduced to claiming that the other side is 'arrogant' and each will be right. The traditionalists will continue to invoke the 'minority of militant agitators and anarchists' and the radicals will continue to inveigh against 'petty Hitlers'. Neither can grant the other integrity if neither can move beyond one limited perspective or ideology. Consider the misunderstanding that arises about the term *authoritarian*, a splendidly emotional word if ever there was one. Many heads deny that they are authoritarian, implicitly contrasting their own behaviour with that of the heads they served under years before. The young insist on the legitimacy of their charge since, having a much shorter experience, the tag authoritarian is derived from a different base-line. Many heads could reduce the frequent misunderstandings and entrenched positions if they could recognize the assumptions on which their perspective is based. 'Loyalty to the school' may be being defined as loyalty to the head; 'professional' may mean accepting the code of conduct the head himself followed as a teacher and continues to expect. From the young, radical teacher's perspective, 'loyalty to the school' may be defined in terms of a particular educational and social ideology which assumes greater rights for teachers and pupils; 'professional' may mean refusing to compromise with a sincerely held but quite different set of educational values. Had we more skill in taking the perspective of the other, so recognizing the meaning and validity of the values, opinions and conduct to the person who holds that perspective, then although we do not possess a single solution to our problems, at least we have relationships within which solutions have a greater chance of being found.

Addendum to Chapter 6

The last section of Chapter 6, on voluntary schooling, was based on a paper I wrote in 1969, but which was not accepted for publication by several popular educational periodicals. Naturally I tend to the view that this rejection was based not on the worthlessness of the ideas I was putting forward, but rather on the fact that the climate of educational opinion, as measured by the views of these editors, was not ready for radical ideas, even though my paper had been received with great interest at conferences. My own thinking about voluntary schooling, as Chapter 6 makes clear, stemmed from my reading of the work of Rogers and Holt and from my experience in a downtown secondary modern school. Since this book was completed, the 'de-schooling' writers from the United States have begun to make some impact in Britain. Had I been familiar with their writings earlier, the last part of Chapter 6 would have taken a different form. I now take this opportunity at proof-stage of adding a brief comment on their work.

The 'de-schoolers' are Ivan Illich (*Deschooling Society*, Caldar and Boyars, 1971), who is often regarded as the high priest of the movement; Paul Goodman (*Compulsory Miseducation*, Penguin, 1971), whose book was first published in the United States in 1962 and who can thus justifiably claim to be the movement's founder; and Everett Reimer (*School is Dead*, Penguin, 1971), who has been an associate of Illich. These writers can be seen as an offshot of what have been called the New Romantics, including such men as Edgar Friedenberg, Jonathan Kozol, Herbert Kohl and John Holt. The 'de-schoolers' merit the title Romantics since, in spite of major differences in their views, they share some common attitudes and assumptions. They are vitriolic in their criticism of contemporary society which they condemn for its materialism, conformity and mass culture. They are passionate in their pleas for greater justice and freedom. They are for individual self-realization and against human beings as anonymous cogs in a complex social and technological machine. The educational system itself is seen as essentially harmful. Only lip-service, they claim, has been paid to the notions of equality of opportunity and child-centredness. Schools persist as bastions of privilege and authoritarianism. Schools make slaves of their pupils, converting the middle classes who opt in to passive inmates who must learn to play the system in the service of their instrumental goals of social climbing and qualification hunting, but rejecting the working classes who drop out and labelling them as social and individual deviants and failures. They all recognize that the costs of education in Western society are rising at a rate which cannot be sustained. They share the view that these costs in any case represent a gigantic waste of money. More education will mean worse education.

The 'de-schoolers' are fully aware that the extension of compulsory education derives in part from the political necessity of keeping young people off the streets because there are no jobs for them. It is a difficulty for de-schooling, and for a system of voluntary schooling, that the world of work is increasingly less available as a viable alternative for those who do not want to go to school. Work experience for the young is possible only if industry and commerce are willing to sacrifice some of their efficiency, i.e. profits, in the interests of young people. Rising trends in unemployment, especially among the young and the over-fifties, may not, however, mark the beginning of a period of vaster unemployment on entering the cybernetic age, as some critics have predicted. Men, especially those who espouse the Protestant ethic, will perhaps not be so readily deprived of their work. It may be that there will arise a new breed of Luddites, who could be more successful than their forbears, especially if they have the collective strength and support of the trade unions and a vote-conscious government behind them. I suspect, too, that we tend to underestimate man's capacity for creating new forms of work or for devising new variations on traditional occupational roles. Those who can shed the Protestant ethic and espouse the new emergent values, and who would thus yield themselves to a permanent state of leisure in the 'alternative society' may for the next few decades remain a small minority.

The 'de-schoolers', like many others, are disturbed by the gross inequality in the distribution of educational resources in our society and the monopolization of education and learning and teaching by educational institutions. For example, they see schools as a regressive form of taxation, since the poor contribute to their maintenance whilst it is the middle classes who benefit most from them. This is because the middle class child goes to school longer and because costs increase with the level of schooling. Reimer suggests that we might channel education funds directly into personal educational accounts which would then be spent when and how the individual desired. Whilst some educational facilities might be offered free, most would be self-financing, receiving their income from the educational accounts of satisfied clients. Reimer recognizes that this is no panacea and that profiteering might have to be controlled. It seems likely to me that the middle classes would soon set up private establishments which would receive financial support from parents in addition to educational account finances, thus supplying superior educational resources and personnel to preserve existing class differentials in education. I am tempted to agree with Reimer that it might nevertheless be an advance on our present system.

What the 'de-schoolers' would do with our present educational structure and institutions is not always clear. Illich advocates the total de-institutionalization of education and in a recent publication has described those who make suggestions like Reimer's as 'the most dangerous category of educational reformer'. The main concern of the 'de-schoolers' seems to be to set up *alternatives* to schools, where education would be de-formalized and placed in the hands of people who would not be licensed as teachers. For example, one might promote centres where access to information, records, machines and other educational resources would be freely available to those who wanted to consult them, somewhat along the lines of a free public library system. They all seem keen to promote networks of peers who would teach one another and learn together co-operatively. Whilst the 'de-schoolers' all proclaim extremely attractive ideas for alternatives to school, they tell us very little about how such innovations would be introduced and sustained, and even less about the means by which institutionalized education would be brought to the point of collapse. They seem to be curiously unwilling to discuss the social and political realities which entrench educational institutions. If schools are challenged, they will, I think, show that they are far from being dead opponents, and an Illich-type rhetoric would be a most flimsy weapon in the ensuing battle.

All the de-schoolers are advocating fundamental changes in Western society's culture and structure as well as in the educational system, and I think they are right when they suggest that there are urgent educational changes needed which cannot wait until other social and economic changes have paved the way. I think they are wrong in implying that massive educational reforms can be effected without at least a few major concomitant social and economic changes. Given the present power structure, we can expect few radical innovations from Parliament, from the Department of Education and Science, from local Chief Education Officers and their Education Committees. Certainly if the reforms are to be made by teachers and are likely to affect them, then changes will be neither sudden nor dramatic. Piecemeal reform and slow, sparse innovation is the more likely development. Educational and social evils do go hand in hand, and we need a growing awareness of this combined with a willingness to attempt radical and speedy solutions. The 'de-schoolers'' indictment of the status quo and the alternatives they put forward make a basic contribution to this process of renewal. They are prophets of doom whose function is to make aware and arouse interest, and we would be foolish to expect clearcut and detailed schemes for social and educational change. Like prophets of doom, they preach their pessimistic message because they have a romantic and idealistic conviction that their visions are not beyond realization.

If teachers, learners, administrators, politicians and the general public—or at least a substantial minority of them—are to prove to be responsive to the vision, then we shall need to consider more detailed schemes by which we can move from our present position towards a realization of the vision. In my view—which I am confident would not be shared by the 'de-schoolers' in their desire for instant and thorough revolution—the early stages would consist of changes *within* the present institutional framework of education. This accounts for my attachment to the notion of voluntary schooling, which is not an abolition of educational institutions but a fundamental internal reformation. Naturally I must acknowledge that my conception of a voluntary system is not to be realized merely by a *legal* change. Parliament could legalize voluntary schooling with only superficial changes in the nature of schooling. Schools could continue virtually as at present, responding to the voluntariness of attending with no structural or role changes but rather with an attitude of 'take it or leave it' to pupils. Reformation must come from below before we can expect reform from above, which would in any case have only a marginal impact in the absence of the former. We have, I trust, learned our lesson over comprehensive schools. So we need an awareness and conviction among teachers and others, who would then form the necessary pressure groups to bring about the appropriate changes. Voluntary schooling I see as the first major step, for it is at least possible to see such a reform as viable within the next few decades. Total de-schooling does not appear to be a viable alternative within the foreseeable future, nor am I convinced that this is necessarily the most desirable long-term outcome.

In the meantime it is highly possible that the impact of the 'de-schoolers' on practising teachers will be marginal. Teachers are highly sceptical about extreme positions. The radical and revolutionary gospel of the 'de-schoolers' may well not be heard, simply because they may be labelled as extremists by a majority that is unwilling to listen. That has always been the fate of prophets of doom. Their disciples and supporters will most easily be found in

Colleges and Departments of Education, whose legitimacy in the eyes of teachers and others is already very low. No doubt a few disciples will take action, but as long as these experiments in alternatives are conducted *outside* the state educational institutions—as in the case of the recent Free School at Liverpool—they offer no greater threat than the pioneering work of A. S. Neill at Summerhill.

In the last fifteen years social scientists have exposed many of the inequalities and injustices in our society, and not least in the educational system, where they can claim some of the credit for the abolition of the eleven-plus, the development of comprehensive schools and so on. But we must recognize that these reforms are like a pin-prick in the main problems, even though the public has been persuaded by the politicians that these are the main problems. The 'de-schoolers' inhibit our self-congratulation and complacency. In the meantime, I think there are two educational developments which give grounds for hope. The first is the experimental work going on in a few primary schools. At this stage a pupil would, in my own conception of voluntary schooling, be following a self-selected and self-directed course, exploring as few or as many aspects of life as he wishes. There would be no set curriculum; no class teaching of subjects; no attempts to 'integrate'; no age-grading; no streaming. The teacher would be a resource provider, offering help, advice, and encouragement to further exploration. A few primary schools come very close to this pattern now. The second is the changes that are occurring among the young generation of teachers, to which I referred at the end of Chapter 12. In their case, the New Romantics are preaching to the converted, or at least to the sympathetic. In my view they will soon emerge as a potent force—there are already signs of such a burgeoning among the supporters of *Rank and File* within the National Union of Teachers. The teaching profession is nursing its own version of 'the alternative society' within its bosom, and before long it will sting.

Bibliographical index

Figures in square brackets refer to page numbers in the present book

ABRAVANEL, E. 'A psychological analysis of the concept of role', unpublished master's thesis, Swarthmore College, 1962, quoted in Brown, R., *Social Psychology*, Free Press, 1965. [51]

AINSWORTH, R. 'An experimental study of attitudes and attitude change in sixth-formers with respect to smoking', unpublished dissertation for the Diploma in the Advanced Study in Education, University of Manchester, 1968. [389]

ALLEN, E. A. 'Attitudes to school and teachers in a secondary modern school', *British Journal of Educational Psychology*, vol. 31, 1961, pp. 106–9. [163]

ALLPORT, G. W. 'Attitudes', in Murchison, C. (ed.), *A Handbook of Social Psychology*, Clark University, 1935. [376]

ALLPORT, G. W. *The Nature of Prejudice*, Addison-Wesley, 1954. [384]

AMIDON, E. J. & GIAMATTEO, M. 'The verbal behaviour of superior elementary teachers', 1966, in Amidon and Hough (1967). [136]

AMIDON, E. J. & HOUGH, J. B. *Interaction Analysis: Theory Research and Application*, Addison-Wesley, 1967. [227, 335]

AMIDON, E. J. & HUNTER, E. 'Verbal interaction in the classroom: the Verbal Interaction Category System', in Amidon and Hough (1967). [132]

AMIS, KINGSLEY *I Want it Now*, Jonathan Cape, 1968. [120–2, 129]

ANDERSON, H. H. & BREWER, H. M. 'Studies of teachers' classroom personalities', *Applied Psychology Monographs*, Stanford University Press, 1945, 1946. [141f]

ANDERSON, N. H. 'Primacy effects in personality impression formation using a generalized order effect paradigm', *Journal of Personality and Social Psychology*, vol. 2, 1965, pp. 1–9. [47]

ARGYLE, M. *Social Interaction*, Methuen, 1966. [129]

ARGYLE, M. *The Psychology of Interpersonal Behaviour*, Penguin, 1967. [30, 116]

ARONSON, E. & MILLS, J. 'The effect of severity of initiation on liking for a group,' *Journal of Abnormal and Social Psychology*, vol. 59, 1959, pp. 177–81. [386]

427

ARONSON, E., TURNER, J. A. & CARLSMITH, J. M. 'Communicator credibility and communication discrepancy as determinants of opinion change', *Journal of Abnormal and Social Psychology*, vol. 67, 1963, pp. 31–6. [380, 385]

ASCH, S. E. 'Forming impressions of personality', *Journal of Abnormal and Social Psychology*, vol. 41, 1946, pp. 258–90. [35ff, 46f, 49]

ASCH, S. E. 'Effects of group pressure upon the modification and distortion of judgments' (1951), in Cartwright, D. & Zander, A. *Group Dynamics*, Tavistock, 1960 (2nd ed.). [301f]

ASCH, S. E. *Social Psychology*, Prentice-Hall, 1952. [101f, 105]

AUSUBEL, D. P., SCHIFF, H. M. & ZELENY, M. P. 'Validity of teachers' rating of adolescents' adjustment and aspirations', *Journal of Education Psychology*, vol. 45, 1954, pp. 394–405. [157]

BACKMAN, C. W. & SECORD, P. F. 'The effect of perceived liking on interpersonal attraction', *Human Relations*, vol. 12, 1959, pp. 379–84. [280]

BALES, R. F. 'Some uniformities of behaviour in small social systems', in Swanson, G. E., Newcombe, T. M. & Hartley, E. L. (eds) *Readings in Social Psychology* (rev. ed.), Holt, Rinehart & Winston, 1952. [313, 322]

BALES, R. F. 'The equilibrium problem in small groups', in Parsons, T., Bales, R. F. & Shils, E. A. (eds) *Working Papers in the Theory of Action*, Free Press, 1953. [313, 322]

BARBIANA, SCHOOL OF *Letter to a Teacher*, Penguin, 1970. [176, 190f, 227]

BARKER, R. G. & WRIGHT, H. F. *One Boy's Day*, Harper, 1951. [178f]

BARKER LUNN, J. C. *Streaming in the Primary School*, N.F.E.R., 1970. [160, 276, 319]

BARTLETT, F. C. *Remembering*, Cambridge, 1932. [32]

BATESON, G., JACKSON, D. D., HALEY, J. & WEAKLAND, J. 'Toward a theory of schizophrenia', *Behavioural Science*, vol. 1, 1956, pp. 251–64. [65]

BECKER, H. S., 'Social class variations in pupil–teacher relationships', *Journal of Educational Sociology*, vol. 25, 1952, pp. 451–65 [232]

BELOFF, H. 'Two forms of social conformity: acquiescence and conventionality', *Journal of Abnormal and Social Psychology*, vol. 56, 1958, pp. 99–104. [299]

BENNETT, E. B. 'Discussion, decision, commitment and consensus in "group decision" ', *Human Relations*, vol. 8, 1955, pp. 251–73. [392]

BERG, L. *Risinghill: Death of a Comprehensive School*, Penguin, 1968. [76f]

BERGER, P. L. *Invitation to Sociology*, 1963 (Penguin, 1966). [96]

BERNE, E. *Games People Play*, Deutsch, 1966 (Penguin, 1967). [117]

BIDDLE, B. J. & THOMAS, E. J. *Role Theory: concepts and research*, Wiley, 1966. [71, 92]

BION, B. R. *Experience, in Groups*, Tavistock, 1961. [336]

BLAU, P. M. *The Dynamics of Bureaucracy*, University of Chicago Press, 1955. [123ff]

BLAU, P. M. *Exchange and Power in Social Life*, Wiley, 1964. [123]

BLISHEN, E. *Roaring Boys*, Thames & Hudson, 1955. [256f]

BLISHEN, E. *This Right Soft Lot*, Thames & Hudson, 1969. [260f]

BLUMER, H. 'Sociological analysis and the "variable" ', *American Sociological Review*, vol. 21, 1956, pp. 683–90. [21f]

BLUMER, H. 'Society as symbolic interaction', in Rose, A. M. (ed.) *Human Behaviour and Social Processes*, Routledge & Kegan Paul, 1962. [11]

BLUMER, H. 'Sociological implications of the thought of George Herbert Mead', *American Journal of Sociology*, vol. 71, 1966, pp. 535–48. [11, 174]

BLYTH, W. A. C. *English Primary Education*, Routledge & Kegan Paul, 1965. [142]

BOAS, G. *A Teacher's Story*, Macmillan, 1963. [235]

BONNEY, M. E. 'Relationships between social success, family size, socio-economic home background, and intelligence among school children in grades III to V', *Sociometry*, vol. 7, 1944, pp. 26–39. [306]

BONNEY, M. E. 'A sociometric study of the relationship of some factors to mutual friendship on the elementary, secondary and college levels', *Sociometry*, vol. 9, 1946, pp. 21–57. [273]

BONNEY, M. E. 'Sociometric study of agreement between teacher judgements and student choices, in regard to the number of friends possessed by high school students', *Sociometry*, vol. 10, 1947, pp. 133–46. [328]

BRADFORD, L. P., GIBB, J. R. & BENNE, K. D. *T-group Theory and Laboratory Method*, Wiley, 1964. [335]

BREHM, J. W. 'Attitudinal consequences of commitment to unpleasant behaviour', *Journal of Abnormal and Social Psychology*, vol. 60, 1960, pp. 379–83. [387]

BRIM, O. G. 'The parent–child relation as a social system', *Child Development*, vol. 28, 1957, pp. 343–64. [30]

BRITTAIN, C. V. 'Parent vs. peer-group cross-pressures', *American Sociological Review*, vol. 28, 1963. [343]

BROOKOVER, W. B., THOMAS, S. & PATTERSON, A. 'Self-concept of ability and scholastic achievement', *Sociology of Education*, vol. 37, 1964, pp. 271–8. [60]

BROPHY, J. E. & GOOD, T. L. 'Teachers' communication of differential expectations for children's classroom performance', *Journal of Educational Psychology*, vol. 61, 1970, pp. 365–74. [59f]

BROWN, J. A. C. *Techniques of Persuasion*, Penguin, 1963. [401]

BROWN, R. *Social Psychology*, Free Press, 1965. [54f]

BROXTON, J. A. 'A test of interpersonal attraction predictions derived from balance theory', *Journal of Abnormal and Social Psychology*, vol. 63, 1963, pp. 394–7. [280]

BRUNER, J. S. *Toward a Theory of Instruction*, Norton, 1966. [178]

BRUNER, J. S., GOODNOW, J. J. & AUSTIN, G. A. *A Study of Thinking*, Wiley, 1956. [50]

BRUNER, J. S., SHAPIRO, D. & TAGIURI, R. 'The meaning of traits in isolation and in combination', in Tagiuri, R. & Petrullo, L. (eds) *Person Perception and Interpersonal Behavior*, Stanford University Press, 1958. [37]

BRUNER, J. S. & TAGIURI, R. 'The perception of people', in Lindzey, G. (ed.) *Handbook of Social Psychology* (vol. II), Addison-Wesley, 1954. [37]

BUSH, R. N. 'A study of student–teacher relationships', *Journal of Educational Research*, vol. 35, 1942, pp. 645–56. [163]

BUSH, R. N. *The Teacher-Pupil Relationship*, Prentice-Hall, 1954. [149, 157]

BYRNE, D. 'The influence of propinquity and opportunities for interaction on classroom relationships', *Human Relations*, vol. 14, 1961a, pp. 63–9. [269f]

BYRNE, D. 'Interpersonal attraction as a function of need affiliation and attitude similarity', *Human Relations*, vol. 14, 1961b, pp. 283–9. [271]

BYRNE, D. 'Interpersonal attraction and attitude similarity', *Journal of Abnormal and Social Psychology*, vol. 62, 1961c, pp. 713–15. [280]

BYRNE, D. & BLAYLOCK, B. 'Similarity and assumed similarity of attitudes between husbands and wives', *Journal of Abnormal and Social Psychology*, vol. 67, 1963, pp. 636–40. [280]

BYRNE, D. & BUEHLER, J. A. 'A note on the influence of propinquity on acquaintanceships', *Journal of Abnormal and Social Psychology*, vol. 51, 1955, pp. 147–8. [269]

CARTER, L., HAYTHORN, W. & HOWELL, M. 'A further investigation of the criteria of leadership', *Journal of Abnormal and Social Psychology*, vol. 45, 1950, pp. 350–8. [318]

CARTER, L. & NIXON, M. 'Ability, perceptual, personality and interest factors associated with different criteria of leadership', *Journal of Psychology*, vol. 27, 1949, pp. 377–88. [318]

CARTWRIGHT, D. & ZANDER, A. *Group Dynamics*, Tavistock, 1963 and subsequent editions. [336]

CHARTERS, W. W. & NEWCOMB, T. M. 'Some attitudinal effects of experimentally increased salience of a membership group', in Maccoby, E. E., Newcomb, T. M. & Hartley, E. L. (eds) *Readings in Social Psychology*, Holt, Rinehart & Winston, 1958. [390]

CHOWDHRY, K. & NEWCOMB, T. M. 'The relative abilities of leaders and non-leaders to estimate opinions of their own groups', *Journal of Abnormal and Social Psychology*, vol. 47, 1952, pp. 51–7. [322]

CLAIBORN, W. L. 'Expectancy effect in the classroom: a failure to replicate', *Journal of Educational Psychology*, vol. 60, 1969, pp. 377–83. [58]

COGAN, M. L. 'Research on the behaviour of teachers: a new phase', *Journal of Teacher Education*, 1963, vol. 14, pp. 138–43. [133]

COHEN, A. K. *Delinquent Boys*, Free Press, 1955. [6f]

COHEN, A. R. *Attitude Change and Social Influence*, Basic Books, 1964. [401]

COLEMAN, J. S. *The Adolescent Society*, Free Press, 1961. [339f, 353ff, 369, 372]

COOLEY, C. H. *Human Nature and the Social Order*, Scribner, 1902. [9]

COOLEY, C. H. *Social Organization*, Scribner, 1909. [290]

CRUTCHFIELD, R. S. 'Conformity and character', *American Psychologist*, vol. 10, 1955, pp. 191–8. [302f]

CULBERTSON, F. M. 'Modification of an emotionally held attitude through role-playing', *Journal of Abnormal and Social Psychology*, vol. 54, 1957, pp. 230–3. [382]

DAHLKE, H. O. 'Determinants of sociometric relations among children in elementary schools', *Sociometry*, vol. 16, 1953, pp. 327–38. [306]

DAVIS, K. 'The sociology of parent–youth conflict', *American Sociological Review*, vol. 5, 1940, pp. 523–35. [349]

DEUTSCH, M. & SOLOMON, L. 'Reactions to evaluations by others as influenced by self-evaluation', *Sociometry*, vol. 22, 1959, pp. 93–112. [280]

DOUGLAS, J. W. B. *The Home and the School*, Macgibbon & Kee, 1964. [160]

DOUVAN, E. & ADELSON, J. *The Adolescent Experience*, Wiley, 1966. [344, 372]

EDMONDS, E. L. *The First Headship*, Blackwell, 1968. [407]

ELLIS, H. F. *The World of A. J. Wentworth, B.A.*, Penguin, 1964. [257f]

EMERSON, R. M. 'Power-dependence relations', *American Sociological Review*, vol. 27, 1962, pp. 31–40. [307, 312]

ENTWISTLE, H. *Child-centred Education*, Methuen, 1970a. [214]

ENTWISTLE, H. *Education, Work and Leisure*, Routledge & Kegan Paul, 1970b. [227]

EPPEL, E. M. & EPPEL, M. *Adolescents and Morality*, Routledge & Kegan Paul, 1966. [343]

EPPERSON, D. C. 'A re-assessment of indices of parental influence in "The Adolescent Society"', *American Sociological Review*, vol. 29, 1964, pp. 93–6. [339]

EVANS, K. M. *Sociometry and Education*, Routledge & Kegan Paul, 1962. [306]

FARLEY, R. *Secondary Modern Discipline*, Black, 1960. [266]

FESTINGER, L. *A Theory of Cognitive Dissonance*, Harper & Row, 1957. [385ff]

FESTINGER, L. & CARLSMITH, J. M. 'Cognitive consequences of forced compliance', *Journal of Abnormal and Social Psychology*, vol. 58, 1959, pp. 203–10. [388]

FESTINGER, L., SCHACHTER, S. & BACK, K. *Social Pressures in Informal Groups*, Harper & Row, 1950. [268]

FINLAYSON, D. S. & COHEN, L. 'The teacher's role: a comparative study of the conceptions of college of education students and head teachers', *British Journal of Educational Psychology*, vol. 37, 1967, pp. 22–31. [90f]

FLANDERS, N. A. 'Diagnosing and utilizing social structures in classroom learning', in Henry, N. B. (1960). [132]

FLANDERS, N. A. 'Some relationships between teacher influence, pupil attitudes, and achievement', in, Biddle, B. J. & Ellena, W. J. (eds) *Contemporary Research on Teacher Effectiveness*, Holt, Rinehart & Winson, 1964; reprinted in Amidon & Hough (1967). [136, 261f]

FLANDERS, N. A. 'Teacher influence in the classroom', in Amidon & Hough (1967). [136f]

FORD, J. *Social Class and the Comprehensive School*, Routledge & Kegan Paul, 1969. [273, 278]

FORD, J., YOUNG, D. & BOX, S. 'Functional autonomy, role distance and social class', *British Journal of Sociology*, vol. 18, 1967, pp. 370–81. [274]

FORSTER, E. M. 'Notes on the English Character' (1920), in *Abinger Harvest*, Penguin, 1967. [138]

FORSTER, E. M. *Aspects of the Novel*, 1927 (Penguin, 1962). [39]

FRENCH, J. R. P. & RAVEN, B. 'The bases of social power', in Cartwright, D. (ed.) *Studies in Social Power*, 1959, reprinted in Cartwright, D. & Zander, A. *Group Dynamics*, Tavistock, 1960. [308f, 313]

GALLAGHER, J. J. 'Social status of children related to intelligence, propinquity and social perception', *Elementary School Journal*, vol. 58, 1958, pp. 225–31. [274]

GARFINKEL, H. *Studies in Ethnomethodology*, Prentice-Hall, 1967. [99f]

GETZELS, J. & GUBA, E. G. 'Social behaviour and the administrative process', *School Review*, vol. 65, 1957, pp. 423–41. [98]

GETZELS, J. & THELEN, H. A. 'The classroom as a unique social system', in Henry, N. B. (1960). [87]

GIBB, C. A. 'The principles and traits of leadership', *Journal of Abnormal and Social Psychology*, vol. 42, 1947, pp. 267–84. See also Gibb, C. A., *Leadership*, Penguin Modern Psychology Series, 1969. [316]

GNAGEY, W. J. 'Effects on classmates of a deviant student's power and response to a teacher-exerted control technique', *Journal of Educational Psychology*, vol. 51, 1960, pp. 1–8. [330]

GOFFMAN, E. *The Presentation of Self in Everyday Life*, Doubleday Anchor, 1959. [110ff, 118, 129, 405]

GOFFMAN, E. *Encounters*, Bobbs-Merrill, 1961. [98f, 164]

GOFFMAN, E. *Stigma*, Prentice-Hall, 1963a. [9]

GOFFMAN, E. *Behavior in Public Places*, Free Press, 1963b, [115f, 186]

GOLD, M. 'Power in the classroom', *Sociometry*, vol. 21, 1958, pp. 50–60. [313]

GORDON, C. W. *Social System of the High School*, Free Press, 1957. [363ff, 372]

GOULDNER, A. W. 'The norm of reciprocity: a preliminary statement', *American Sociological Review*, vol. 25, 1960, pp. 161–78. [124]

GRIFFIN, J. H. *Black Like Me*, Collins, 1962 (Panther, 1964). [9]

GRONLUND, N. E. 'The relative ability of home-room teachers and special-subject teachers to judge the social acceptability of pre-adolescent pupils', *Journal of Educational Research*, vol. 48, 1955, pp. 381–91. [328]

GROSS, N. MASON, W. S. & MCEACHERN, A. W. *Explorations in Role Analysis*, Wiley, 1958. [92]

HALLWORTH, H. J. 'Sociometric relationships among grammar school boys and girls between the ages of eleven and sixteen years', *Sociometry*, vol. 16, 1953, pp. 39–70. [315]

HALLWORTH, H. J. 'Teachers' personality ratings of high school pupils', *Journal of Educational Psychology*, vol. 52, 1961, pp. 297–302. [158]

HALLWORTH, H. J. 'A teacher's perception of his pupils', *Educational Review*, vol. 14, 1962, pp. 124–33. [158]

HAMBLIN, R. L. 'Leadership and crises', *Sociometry*, vol. 21, 1958, pp. 322–35. [318]

HARDING, D. W. *Social Psychology and Individual Values*, Hutchinson, 1953. [141]

HARE, R. M. 'Adolescents into adults', in Hollins, T. B. (ed.) *Aims in Education*, Manchester University Press, 1964. [397]

HARGREAVES, D. H. *Social Relations in a Secondary School*, Routledge & Kegan Paul, 1967. [66, 158, 192, 272f, 322ff, 336, 340, 364]

HARGREAVES, D. H. 'Interpersonal relations and the social system of the school', in *Development in Learning*, vol. III: *Contexts of Education*, ed. J. F. Morris & E. A. Lunzer, Staples Press, 1969. [165]

HARGREAVES, D. H. 'The delinquent subculture and the school', in Carson, W. G. & Wiles, P. (eds) *Crime and Delinquency in Britain*, Robertson, 1971. [192]

HARRISON, N. 'Attitudes and Friendship Patterns in a Comprehensive School', unpublished M.Ed. thesis, University of Manchester, 1970. [275]

HARVEY, O. J. & CONSALVI, C. 'Status and conformity to pressure in informal groups', *Journal of Abnormal and Social Psychology*, vol. 60, 1960, pp. 182–7. [321]

HEBER, R. F. 'The relationship of intelligence and physical maturity to social status of children', *Journal of Educational Psychology*, vol. 47, 1956, pp. 158–62. [306]

HEIDER, F. 'Social perception and phenomenal causality', *Psychological Review*, vol. 51, 1944, pp. 358–74. [42]

HEIDER, F. 'Attitudinal and cognitive organization', *Journal of Psychology*, vol. 21, 1946, 107–12. [281]

HEIDER, F. *The Psychology of Interpersonal Relations*, Wiley, 1958. [38, 281, 288, 385]

HEMPHILL, J. K. & WESTIE, C. M. 'The measurement of group dimensions', *Journal of Psychology*, vol. 29, 1950, pp. 325–42. [293]

HENRY, J. 'Docility, or giving teacher what she wants', *Journal of Social Issues*, vol. 11, 1955, pp. 33–41. [178]

HENRY, J. *Culture Against Man*, Tavistock, 1966. [180, 184f, 227]

HENRY, N. B. *Dynamics of Instructional Groups*, 59th Year Book of the National Society for the Study of Education, Chicago University Press, 1960.

HIGBEE, K. L. 'Fifteen years of fear arousal: research on threat appeal', *Psychological Bulletin*, vol. 72, 1969, pp. 426–44. [381]

HINES, B. *A Kestrel for a Knave*, Michael Joseph (Penguin, 1969). [193f]

HOGGART, R. *The Uses of Literacy*, Penguin, 1957. [19]

HOLLANDER, E. P. 'Conformity, status and idiosyncrasy credit', *Psychological Review*, vol. 65, 1958, pp. 117–27. [321]

HOLLANDER, E. P. 'Competence and conformity and acceptance of influence', *Journal of Abnormal and Social Psychology*, vol. 61, 1960, pp. 365–9. [321]

HOLLANDER, E. P. *Leaders, Groups and Influence*, Oxford, 1964. [321]

HOLLANDER, E. P. & WILLIS, R. H. 'Some current issues in the psychology of conformity and nonconformity', *Psychological Bulletin*, vol. 68, 1967, pp. 62–76. [299]

HOLLINS, T. H. B. 'Teachers' attitudes to children's behaviour', unpublished M.Ed. thesis, University of Manchester, 1955. [156]

HOLLY, D. N. 'Profiting from a comprehensive school', *British Journal of Sociology*, vol. 6, 1965, pp. 150–7. [364]

HOLT, J. *How Children Fail*, Pitman, 1964 (Penguin, 1969). [174, 182, 186ff, 197, 227]

HOMANS, G. C. *The Human Group*, Routledge & Kegan Paul, 1951. [268ff]

HOMANS, G. C. *Social Behaviour: its Elementary Forms*, Routledge & Kegan Paul, 1961. [123ff, 321]

HORNEY, K. *Neurosis and Human Growth*, Routledge & Kegan Paul, 1951. [27]

HOVLAND, C. I. 'Reconciling conflicting results derived from experimental and survey studies of attitude change', *American Psychologist*, vol. 14, 1959, pp. 8–17. [375]

HOVLAND, C. I., CAMPBELL, E. H. & BROCK, T. 'The effects of "commitment" on opinion change following communication', in Hovland, C. I. (ed.) *The Order of Presentation in Persuasion*, Yale, 1957. [392]

HOVLAND, C. I. & PRITZKER, H. A. 'Extent of opinion change as a function of amount of change advocated', *Journal of Abnormal and Social Psychology*, vol. 54, 1957, pp. 257–61. [380]

HOVLAND, C. I., HARVEY, O. J. & SHERIF, M. 'Assimilation and contrast effects in reactions to communication and attitude change', *Journal of Abnormal and Social Psychology*, vol. 55, 1957, pp. 244–52. [380]

HOVLAND, C. I. & WEISS, W. 'The influence of source credibility on communication effectiveness', *Public Opinion Quarterly*, vol. 15, 1951, pp. 635–50. [379]

HOY, W. K. 'Pupil control ideology and organizational socialization: a further examination of the influence of experience on the beginning teacher', *School Review*, vol. 77, 1969, pp. 257–65. [90]

HOYLE, E. *The Role of the Teacher*, Routledge & Kegan Paul, 1969. [149]

HUNTER, E. *The Blackboard Jungle*, Constable, 1955. [87f, 167f, 232, 236ff, 332ff]

HYMAN, H. H. 'The psychology of status', *Archives of Psychology*, No. 269, 1942. [340]

ISHERWOOD, C. *Down There on a Visit*, Methuen, 1962. [16f]

JACKSON, B. *Streaming: an Education System in Miniature*, Routledge & Kegan Paul, 1964. [160]

JACKSON, B. & MARSDEN, D. *Education and the Working Class*, Routledge & Kegan Paul, 1962. [18, 362]

JACKSON, J. M. 'Structural characteristics of norms', in Henry, N. B. (1960). [297]

JACKSON, P. W. *Life in Classrooms*, Holt, Rinehart & Winston, 1968. [162, 178, 184f, 227]

JACKSON, P. W. & LAHADERNE, H. M. 'Scholastic success and attitudes toward school in a population of sixth graders', *Journal of Educational Psychology*, vol. 58, 1967, pp. 15–18. [157]

JACOBS, E. B. 'Attitude change in teacher education', *Journal of Teacher Education*, vol. 19, 1968, pp. 410–15. [90]

JAHODA, M. 'Conformity and independence: a psychological analysis', *Human Relations*, vol. 12, 1959, pp. 99–119. [299]

JAHODA, M. & WARREN, N. (eds) *Attitudes*, Penguin, 1966. [401]

JAMES, W. *The Principles of Psychology*, Holt, 1890. [8, 18, 20]

JANIS, I. L. & FESHBACK, S. 'Effects of fear-arousing communications', *Journal of Abnormal and Social Psychology*, vol. 48, 1953, pp. 78–92. [381]

JANIS, I. L. & KING, B. T. 'The influence of role-playing on opinion change', *Journal of Abnormal and Social Psychology*, vol. 49, 1954, pp. 211–18. [382]

JANIS, I. L. et al. *Personality and Persuasibility*, Yale, 1959. [382]

JECKER, J. D., MACCOBY, N. & BREITROSE, H. S. 'Improving accuracy in interpreting non-verbal cues of comprehension', *Psychology in the Schools*, vol. 2, 1965, pp. 239–44. [335]

JECKER, J. D. & LANDY, D. 'Liking a person as a function of doing him a favor', *Human Relations*, vol. 22, 1969, pp. 371–8. [280]

JENNINGS, H. H. *Leadership and Isolation*, Longmans, 1950. [315]

JONES, E. E. *Ingratiation*, Appleton-Century-Crofts, 1964. [412]

JONES, E. E., DAVIS, K. E. & GERGEN, K. J. 'Role-playing variations and their informational value for person perception', *Journal of Abnormal and Social Psychology*, vol. 63, 1961, pp. 302-10. [98]

JONES, E. E. & DAVIS, K. E. 'From acts to dispositions: the attribution process in person perception', in Berkowitz, L. (ed.) *Advances in Experimental Social Psychology*, vol. II, Academic Press, 1965. [44]

JONES, E. E. & DE CHARMS, R. 'Changes in social perception as a function of the personal relevance of behaviour', *Sociometry*, vol. 20, 1957, pp. 75-85. [52]

JONES, E. E. & THIBAUT, J. W. 'Interaction goals as bases of inference in interpersonal perception', in Tagiuri, R. & Petrullo, L. (eds) *Person Perception and Interpersonal Behavior*, Stanford University Press, 1958. [102]

KATZ, D. 'The functional approach to the study of attitude change', *Public Opinion Quarterly*, vol. 24, 1960, pp. 163-204. [382ff]

KELLEY, H. H. 'The warm-cold variable in first impressions of people', *Journal of Personality*, vol. 18, 1950, pp. 431-9. [48-50]

KELLEY, H. H. 'Two functions of reference groups', in Swanson, G. E., Newcomb, T. M. & Hartley, E. L. *Readings in Social Psychology*, Holt, Rinehart & Winston, 1952. [340]

KELLEY, H. H. & VOLKART, E. H. 'The resistance to change of group-anchored attitudes', *American Sociological Review*, vol. 17, 1952, pp. 453-65. [390]

KELLEY, H. H. & WOODRUFF, C. L. 'Members' reactions to apparent group approval of a counter-norm communication', *Journal of Abnormal and Social Psychology*, vol. 52, 1956, pp. 67-74. [390]

KELLEY, J. 'A study of leadership in two contrasting groups', *Sociological Review*, vol. 11, 1963, pp. 323-36. [306]

KELMAN, H. C. 'Processes of opinion change', *Public Opinion Quarterly*, vol. 25, 1961, pp. 57-78. [393ff]

KELMAN, H. C. & HOVLAND, C. I. ' "Reinstatement" of the communicator in delayed measurement of opinion change', *Journal of Abnormal and Social Psychology*, vol. 48, 1953, pp. 327-35. [379]

KELVIN, P. *The Bases of Social Behaviour*, Holt, Rinehart & Winston, 1970. [3, 30, 316f, 376]

KING, B. T. & JANIS, I. L. 'Comparison of the effectiveness of improvised verses, non-improvised role-playing in producing opinion changes', *Human Relations*, vol. 9, 1956, pp. 177-86. [382]

KLEIN, J. *The Study of Groups*, Routledge & Kegan Paul, 1956. [336]

KOMAROVSKY, M. 'Cultural contradictions and sex roles', *American Journal of Sociology*, vol. 52, 1946, pp. 184-9. [355]

KOUNIN, J. S. & GUMP, P. V. 'The ripple effect in discipline', *Elementary School Journal*, vol. 59, 1958, pp. 158-62. [134, 329]

KOUNIN, J. S., FRIESEN, W. V. & NORTON, A. E. 'Managing emotionally disturbed children in regular classrooms', *Journal of Educational Psychology*, vol. 57, 1966, pp. 1-13. [256f]

KRECH, D., CRUTCHFIELD, R. S. & BALLACHEY, E. L. *Individual in Society*, McGraw-Hill, 1962. [301, 378]

KUHN, M. H. 'The reference group reconsidered', *Sociological Quarterly*, vol. 5, 1964, pp. 6-21. [12]

435

KUHN, M. H. & MCPARTLAND, T. S. 'An empirical investigation of self-attitudes', *American Sociological Review*, vol. 19, 1954, pp. 68–76. [19]

LA BRUYÈRE, JEAN DE *Characters*, Penguin, 1970. [268, 285]

LAING, R. D. *The Politics of Experience and the Bird of Paradise*, Penguin, 1967. [108–10, 129]

LAING, R. D., PHILLIPSON, H. & LEE, A. R. *Interpersonal Perception*, Tavistock, 1966. [40, 108–10]

LAMBERT, P. 'The "successful" child: some implications of teacher-stereotyping', *Journal of Educational Psychology*, vol. 56, 1963, pp. 551–3. [157]

LEMERT, E. M. 'Paranoia and the dynamics of exclusion', *Sociometry*, vol. 25, 1962, pp. 1–20. [57]

Letter to a Teacher, see Barbiana, School of

LEVENTHAL, H. & NILES, P. 'A field experiment on fear arousal with data on the validity of questionnaire measures', *Journal of Personality*, vol. 32, 1964, pp. 459–79. [381]

LEVENTHAL, H. & WATTS, J. C. 'Sources of resistance to fear-arousing communications on smoking and lung cancer', *Journal of Personality*, vol. 34, 1966, pp. 155–75. [381]

LEVENTHAL, H., WATTS, J. C. & PAGAON, F. 'Effects of fear and instructions on how to cope with danger', *Journal of Personality and Social Psychology*, vol. 6, 1967, pp. 313–21. [381]

LEWIN, K. *Resolving Social Conflicts*, Harper, 1948. [141]

LEWIN, K. 'Group decision and social change', (1943) in Maccoby, E., Newcomb, T. M. & Hartley, E. L. (eds) *Readings in Social Psychology*, Holt, Rinehart & Winston, 1958. [391ff]

LEWIN, K., LIPPITT, R. & WHITE, R. K. 'Patterns of aggressive behaviour in experimentally created social climates', *Journal of Social Psychology*, vol. 10, 1939, pp. 271–99. [136, 140–2, 240]

LIEBERMAN, S. 'The effects of changes in roles on the attitudes of role occupants', *Human Relations*, vol. 9, 1956, pp. 385–402. [79]

LIFTON, R. J. *Thought Reform and the Psychology of Totalism*, 1961 (Pelican, 1967). [397, 401]

LIPPITT, R., POLANSKY, N. & ROSEN, S. 'The dynamics of power', *Human Relations*, vol. 5, 1952, pp. 37–64. [307, 309, 313f]

LUCHINS, A. S. 'Experimental attempt to minimize the impact of first impressions', in Hovland, C. I. (ed.) *The Order of Presentation in Persuasion*, Yale, 1957. [47–9]

LUNBERG, G. A. & BEAZLEY, V. ' "Consciousness of kind" in a college population', *Sociometry*, vol. 11, 1948, pp. 59–74. [274]

MCALHONE, B. *WHERE on Drugs*, Advisory Centre of Education, Cambridge, 1970. [368]

MACCOBY, E. E. 'Role-taking in childhood and its consequences for social learning', *Child Development*, vol. 30, 1959, pp. 239–52. [12, 30]

MACCOBY, E. E. 'The taking of adult roles in middle childhood', *Journal of Abnormal and Social Psychology*, vol. 63, 1961, pp. 493–503. [12]

MCCORD, J., MCCORD, W. & THURBER, E. 'Some effects of paternal absence on male children', *Journal of Abnormal and Social Psychology*, vol. 64 1962, pp. 361–9. [24]

MCCULLERS, C. *Member of the Wedding*, 1946 (Penguin, 1962). [337]
MCDONALD, L. *Social Class and Delinquency*, Faber & Faber, 1969. [274]
MCINTYRE, D., MORRISON, A. & SUTHERLAND, J. 'Social and educational variables relating to teachers' assessments of primary school pupils', *British Journal of Educational Psychology*, vol. 36, 1966, pp. 272–9. [160]
MACKENZIE, R. F. *The Sins of the Children*, Collins, 1967. [168]
MACMURRAY, J. *Persons in Relation*, Faber & Faber, 1961. [10f]
MAISONNEUVE, J. 'Selective choices and propinquity', *Sociometry*, vol. 15, 1952, pp. 135–40. [269]
MARCH, J. G. & SIMON, H. A. *Organizations*, Wiley, 1958. [293]
MARCH, W. *October Island*, 1952, (Penguin, 1960). [22]
MATZA, D. *Becoming Deviant*, Prentice-Hall, 1969. [174]
MAYS, J. B. *The Young Pretenders*, Michael Joseph, 1965. [337]
MEAD, G. H. *Mind, Self and Society*, University of Chicago Press, 1934. [9–12, 14f, 22, 30, 99, 203, 341]
MEAD, M. *Coming of Age in Samoa*, 1928 (Penguin, 1943). [337]
MEAD, M. *Growing up in New Guinea*, 1930 (Penguin, 1942). [337]
MEREI, F. 'Group leadership and institutionalization', *Human Relations*, vol. 2, 1949, pp. 23–39. [320]
MERTON, R. K. *Social Theory and Social Structure*, Free Press, 1949 (rev. ed., 1957). [56, 62, 340, 376]
MERTON, R. K. 'The role-set: problems in sociological theory', *British Journal of Sociology*, vol. 8, 1957, pp. 106–20. [73]
MICHAEL, W. B., HERROLD, E. E. & CRYAN, E. W. 'Survey of student–teacher relationships', *Journal of Educational Research*, vol. 44, 1951, pp. 657–74. [163]
MILES, M. B. 'The T-group and the classroom', in Bradford, Gibb & Benne (1964). [178]
MILGRAM, S. 'Some conditions of obedience and disobedience to authority', *Human Relations*, vol. 18, 1965, pp. 57–76. [303]
MITCHENER, J. A. *Hawaii*, Secker & Warburg, 1966. [22]
MITNICK, L. L. & MCGINNIES, E. 'Influencing ethnocentrism in small discussion groups through a film communication', *Journal of Abnormal and Social Psychology*, vol. 56, 1958, pp. 82–90. [391]
MORRIS, J. F. 'The development of adolescent value-judgments', *British Journal of Educational Psychology*, vol. 28, 1958, pp. 1–14. [343]
MORTON-WILLIAMS, R. & FINCH, S. *Schools Council Enquiry I: Young School Leavers*, H.M.S.O., 1968. [81–3, 219]
MUSGROVE, F. 'Parents' expectations of the junior school', *Sociological Review*, vol. 9, 1961, pp. 167–80. [76]
MUSGROVE, F. *Youth and the Social Order*, Routledge & Kegan Paul, 1964. [343ff, 362, 372]
MUSGROVE, F. & TAYLOR, P. H. 'Teachers' and parents' conception of the teacher's role', *British Journal of Educational Psychology*, vol. 35, 1965, pp. 171–8. [75]
MUSGROVE, F. & TAYLOR, P. H. *Society and the Teacher's Role*, Routledge & Kegan Paul, 1969. [75, 83, 226]
MUSS, R. E. 'Differential effects of studying versus teaching on teachers' attitudes', *Journal of Educational Research*, vol. 63, 1969, pp. 185–9. [90]

MUSSEN, P. H. & DISTLER, L. 'Masculinity, identification and father–son relationships', *Journal of Abnormal and Social Psychology*, vol. 59, 1959, pp. 350–6. [24]

NEUGARTEN, B. L. 'Social class and friendship among school children', *American Journal of Sociology*, vol. 51, 1956, pp. 305–13. [273]

NEW, M. I. 'Social integration in a non-selective comprehensive school, two grammar schools and two technical high schools in an industrial city', unpublished M.Ed. thesis, University of Manchester, 1967. [273]

NEWCOMB, T. M. *Social Psychology*, Dryden Press, 1950. [340]

NEWCOMB, T. M. 'Attitude development as a function of reference groups: the Bennington study', in Maccoby, E. E., Newcomb, T. M. & Hartley, E. L. (eds) *Readings in Social Psychology*, Holt, Rinehart & Winston, 1952, 1958. [363]

NEWCOMB, T. M. 'Varieties of interpersonal attraction', in Cartwright, D. & Zander, A. (eds) *Group Dynamics*, Tavistock (rev. ed.), 1960. [286f]

NEWCOMB, T. M. 'The acquaintance process', Holt, Rinehart & Winston, 1961. [280–4, 288]

NILES, F. S. 'The influences of parents and peers on adolescent girls', unpublished M.Ed. thesis, University of Manchester, 1968. [343]

NORTHWAY, M. L. 'Outsiders', *Sociometry*, vol. 7, 1944, pp. 10–25. [306]

OLMSTED, M. *The Small Group*, Random House, 1959. [336]

OPPENHEIM, A. N. 'Social status and clique formation among grammar school boys', *British Journal of Sociology*, vol. 6, 1955, pp. 228–45. [273]

OTTAWAY, A. K. C. *Learning Through Group Experience*, Routledge & Kegan Paul, 1966. [335]

PARSONS, T. 'The school class as a social system: some of its functions in American society', *Harvard Educational Review*, vol. 29, 1959, pp. 297–318; reprinted in Halsey, A. H., Floud, J. & Anderson, C. A. *Education, Economy and Society*, Free Press, 1961, and in Grinder, R. E. (ed.) *Studies in Adolescence*, Macmillan, 1963. [16, 155]

PARTRIDGE, J. *Middle School*, Gollancz, 1966; published also by Penguin Books under the title *Life in a Secondary Modern School*. [409f]

PAYNE, D. E. & MUSSEN, P. H. 'Parent–child relations and father identification among adolescent boys', *Journal of Abnormal and Social Psychology*, vol. 52, 1956, pp. 359–62. [24]

PENNINGTON, D. F., HARAVEY, F. & BASS, B. M. 'Some effects of decision and discussion on coalescence, change and effectiveness', *Journal of Applied Psychology*, vol. 42, 1958, pp. 404–8. [392]

PEPITONE, A. 'Attributions of causality, social attitudes and cognitive matching processes', in Tagiuri, R. & Petrullo, L. (eds) *Person Perception and Interpersonal Behavior*, Stanford University Press, 1958. [43]

PETERS, R. S. *The Concept of Education*, Routledge & Kegan Paul, 1967. [206]

PHILLIPS, M. *Small Social Groups in England*, Methuen, 1965. [416]

PIAGET, J. *The Moral Judgment of the Child*, Routledge & Kegan Paul (1932), 1968. [14]

PICKLES, H. C. 'Pupils' reaction to a teacher's verbal comments', unpublished dissertation for the Diploma in the Advanced Study in Education, University of Manchester, 1970. [158]

POLANYI, M. *The Tacit Dimension*, Routledge & Kegan Paul, 1966. [259]

POTASHKIN, R. 'A sociometric study of children's friendship', *Sociometry*, vol. 9, 1946, pp. 48–70. [307]

PRECKER, J. A. 'Similarity of valuings as a factor in selection of peers and near-authority figures', *Journal of Abnormal and Social Psychology*, vol. 47, 1952, pp. 406–14. [271]

PROCTOR, C. H. & LOOMIS, C. P. 'Analysis of sociometric data', in Jahoda, M., Deutsch, M. & Cook, S. W. (eds) *Research Methods in Social Relations*, Dryden Press, 1951. [306]

RABINOWITZ, W. & ROSENBAUM, I. 'Teaching experience and teachers' attitudes', *Elementary School Journal*, vol. 60. 1960, pp. 313–19. [90]

REDL, F. *When We Deal with Children*, Free Press, 1966. [247, 266]

REDL, F. & WATTENBERG, W. W. *Mental Hygiene in Teaching*, Harcourt Brace, 1951. [143, 149, 244]

REDL, F. & WINEMAN, D. *Controls from Within*, Free Press, 1952. [244]

REDL, F. & WINEMAN, D. *The Aggressive Child*, Free Press, 1957. [247]

RICHARDSON, E. *The Environment of Learning*, Nelson, 1967. [406f]

RICHARDSON, H. M. 'Commonality of values as a factor in friendships of college and adult women', *Journal of Social Psychology*, vol. 11, 1940, pp. 303–12. [271]

RILEY, M. W., RILEY, J. W. & MOORE, M. E. 'Adolescent values and the Riesman typology', in Lipset, S. M. & Lowenthal, L. *Culture and Social Character*, Free Press, 1961. [345]

ROGERS, C. R. *Client-centred Therapy*, Houghton Mifflin, 1951. [202ff]

ROGERS, C. R. *On Becoming a Person*, Constable, 1961. [202ff]

ROGERS, C. R. *Freedom to Learn*, Merrill, 1969. [202ff, 227, 265]

ROSENBERG, M. *Society and the Adolescent Self-image*, Princeton, 1965. [361]

ROSENTHAL, R. & JACOBSON, L. *Pygmalion in the Classroom*, Holt, Rinehart & Winston, 1968. [57–68]

RUBINGTON, E. & WEINBERG, M. S. (eds) *Deviance: The Interactionist Perspective*, Macmillan, 1968. [230]

RUDD, W. G. A. & WISEMAN, S. 'Sources of dissatisfaction among a group of teachers', *British Journal of Educational Psychology*, vol. 32, 1962, pp. 275–91. [172]

RUDDOCK, R. *Roles and Relationships*, Routledge & Kegan Paul, 1969. [92]

SCHACHTER, S. 'Deviation, rejection, and communication', *Journal of Abnormal and Social Psychology*, vol. 46, 1951, pp. 190–207. [298]

SCHOFIELD, M. *The Sexual Behaviour of Young People*, Longmans, 1965; revised for Penguin Books, 1968. [346]

SCHUTZ, A. *Der sinnhafte Aufbau der sozialen Welt*, Springer, 1932. Translated by Walsh, G. & Lehnert, F. under the title *The Phenomenology of the Social World*, Northwestern University Press, 1967. [33, 111, 249f]

SCHUTZ, A. 'Concept and theory formation in the social sciences', *Journal of Philosophy*, vol. 51, 1954, pp. 257–73. [176f]

SCHUTZ, A. *Reflections on the Problem of Relevance*, Yale University Press, 1970. [34]

SCHUTZ, A. *On Phenomenology and Social Relations*, selected writings ed. by H. R. Wagner, University of Chicago Press, 1970. [105f]

SCOTT, M. B. & LYMAN, S. M. 'Accounts', *American Sociological Review*, vol. 33, 1968, pp. 46–62. [196]

SCOTT, W. A. 'Attitude change through reward of verbal behaviour', *Journal of Abnormal and Social Psychology*, vol. 55, 1957, pp. 72–5. [382]

SCOTT, W. A. 'Cognitive consistency, response reinforcement, and attitude change', *Sociometry*, vol. 22, 1959, pp. 219–29. [382]

SCOTT, W. A. 'Attitude change by response reinforcement; replication and extension', *Sociometry*, vol. 22, 1959, pp. 328–35. [382]

SECORD, P. F. 'Facial features and inference processes in interpersonal perception', in Tagiuri, R. & Petrullo, L. (eds) *Person Perception and Interpersonal Behavior*, Stanford, 1958. [37]

SHERIF, M. 'The concept of reference group in human relations', in Sherif, M. & Wilson, M. O. (eds) *Group Relations at the Crossroads*, Harper, 1953. [340]

SHERIF, M. *Group Conflict and Co-operation*, Routledge & Kegan Paul, 1966. [279]

SHERIF, M., WHITE, B. J. & HARVEY, O. J. 'Status in experimentally produced groups', *American Journal of Sociology*, vol. 60, 1955, pp. 370–9. [318]

SHERIF, M. & SHERIF, C. W. *Reference Groups*, Harper, 1964. [296, 313f, 318, 321f, 336, 355f, 372]

SHIBUTANI, T. 'Reference groups as perspectives', *American Journal of Sociology*, vol. 60, 1955, pp. 562–70. [340]

SHIBUTANI, T. *Society and Personality*, Prentice-Hall, 1961. [15,30]

SILBERMAN, M. L. 'Behavioural expression of teachers' attitudes toward elementary school students', *Journal of Educational Psychology*, vol. 60, 1969, pp. 402–7. [59]

SMITH, E. E. 'The power of dissonance techniques to change attitudes', *Public Opinion Quarterly*, vol. 25, 1961, pp. 626–39. [388]

SOLES, S. 'Teacher role expectations and the internal organization of secondary schools', *Journal of Educational Research*, vol. 57, 1964, pp. 227–35. [149]

SORENSON, A. G., HUSEK, T. R. & YU, C. 'Divergent concepts of teacher role: an approach to the measurement of teacher effectiveness', *Journal of Educational Psychology*, vol. 54, 1963, pp. 287–94. [142]

SPARK, MURIEL *The Prime of Miss Jean Brodie*, Macmillan, 1961. [86]

SPROTT, W. J. H. *Human Groups*, Penguin, 1958. [30, 290, 336]

STANISLAVSKI, C. *An Actor Prepares*, 1926 (Penguin, 1967). [115]

START, K. B. 'Substitution of games performance for academic achievement as a means of achieving status among secondary school children', *British Journal of Sociology*, vol. 17, 1966, pp. 300–5. [364]

STENHOUSE, L. (ed.) *Discipline in Schools: A Symposium*, Pergamon, 1967. [266]

STOGDILL, R. M. 'Personal factors associated with leadership: a survey of the literature', *Journal of Psychology*, vol. 25, 1948, pp. 35–71; reprinted in Gibb, C. A. (ed.) *Leadership*, Penguin Modern Psychology Series, 1969. [316]

STRICKLAND, L. H. 'Surveillance and trust', *Journal of Personality*, vol. 26, 1958, pp. 200–15. [43f]

SUGARMAN, B. 'Involvement in youth culture, academic achievement and conformity in school', *British Journal of Sociology*, vol. 18, 1967, pp. 151–64. [339, 345, 369]

SULLIVAN, H. S. *Conceptions of Modern Psychiatry*, White Psychiatric Foundation, 1940. [12, 21]

TAGIURI, R. 'Social preference and its perception', in Tagiuri, R. & Petrullo, L. (eds) *Person Perception and Interpersonal Behavior*, Stanford, 1958. [68, 258]

TAYLOR, P. H. 'Children's evaluations of the characteristics of the good teacher', *British Journal of Educational Psychology*, vol. 32, 1962, pp. 258–66. [163]

TERSON, P. 'The tragedy of the worn-out games teacher', *Guardian*, 20 July 1967. [79f]

THELEN, H. A. *Dynamics of Groups at Work*, Chicago, 1954. [149f, 152f]

THEODORSON, G. A. 'The relationship between leadership and popularity roles in small groups', *American Sociological Review*, vol. 22, 1957, pp. 58–67. [314]

THIBAUT, J. W. & KELLEY, H. H. *The Social Psychology of Groups*, Wiley, 1959. [123]

THIBAUT, J. W. & RIECKEN, H. W. 'Authoritarianism, status and the communication of aggression', *Human Relations*, vol. 8, 1955, pp. 95–120. [42f]

THOMAS, W. I. *The Child in America*, Knopf, 1928. [56, 104]

THOMAS, W. I. *The Unadjusted Girl*, Little, Brown, 1931. [104]

THORPE, J. G. 'An investigation into correlates of sociometric status within school classes', *Sociometry*, vol. 18, 1955, pp. 49–55. [306]

TIEDEMAN, S. C. 'A study of pupil–teacher relationships', *Journal of Educational Research*, vol. 35, 1942, pp. 657–64. [163]

TINBERGEN, N. *The Study of Instinct*, Clarendon Press, 1951. [10]

TOOGOOD, J. E. 'The selection of children for responsibility in the junior school,' unpublished dissertation for the Diploma in Educational Guidance, University of Manchester, 1967. [158]

TOSH, N. D. F. *Being a Headmaster*, MacLellan, 1964. [407]

TRAPP, E. P. 'Leadership and popularity as a function of behavioural predictions', *Journal of Abnormal and Social Psychology*, vol. 51, 1955, pp. 452–7. [322]

TRILLING, L. *The Liberal Imagination*, Secker & Warburg, 1951. [3]

TROW, W. C. 'Role function of the teacher in the instructional group', in Henry, N. B. (1960). [143]

TURNER, R. H. 'Role-taking: process versus conformity', in Rose, A. M. (ed.) *Human Behaviour and Social Processes*, Routledge & Kegan Paul, 1962. [97]

UPDIKE, J. *The Centaur*, Deutsch, 1963. [209]

VENESS, T. & BRIERLEY, D. W. 'Forming impressions of personality: two experiments', *British Journal of Social and Clinical Psychology*, vol. 2, 1963, pp. 11–19. [36]

VERBA, S. 'Leadership: affective and instrumental' (1961), in Backman, C. W. & Secord, P. F. (eds) *Problems in Social Psychology*, McGraw-Hill, 1966. [314]

VIDAL, GORE *In a Yellow Wood*, Dutton, 1947. [20]

VIDAL, GORE *Washington, D.C.*, Heinemann, 1967. [44]

WALLBERG, H. J. 'Personality-role conflict and self-conception in urban practice teachers', *School Review*, vol. 76, 1968, pp. 41–9. [90]

WALLER, W. *The Sociology of Teaching*, Wiley, 1932. [144, 147, 149, 227, 266]

WALLER, W. 'The rating and dating complex', *American Sociological Review*, vol. 2, 1937, pp. 727–34. [347]

WALLIN, P. 'Cultural contradictions and sex roles: a repeat study', *American Sociological Review*, vol. 15, 1950, pp. 288–93. [355]

WALSTER, E. & FESTINGER, L. 'The effectiveness of "overheard" persuasive communications', *Journal of Abnormal and Social Psychology*, vol. 65, 1962, pp. 395–402. [379]

WAUGH, EVELYN *Decline and Fall*, 1928 (Penguin, 1937). [230ff]

WEINSTEIN, E. A. & DEUTSCHBERGER, P. 'Some dimensions of altercasting', *Sociometry*, vol. 26, 1963, pp. 454–66. [119]

WHYTE, W. F. *Street Corner Society*, Chicago, 1943. [299, 318f, 321f]

WICKMAN, E. K. *Children's Behaviour and Teachers' Attitudes*, Commonwealth Fund, 1928. [156]

WILLANS, G. & SEARLE, R. *Down with Skool!* Parrish, 1958. [183]

WILLIAMS, H. A. *The True Wilderness*, Constable, 1965. [27]

WILLIAMS, J. R. & KNECHT, W. W. 'Teachers' ratings of high school students on "likeability" and their relation to measures of ability and achievement', *Journal of Educational Research*, vol. 56, 1962, pp. 152–9. [156f]

WILMOTT, P. *Adolescent Boys in East London*, Routledge & Kegan Paul, 1966. [343, 347]

WILSON, J. 'Education and indoctrination', in Hollins, T. H. B. (ed.) *Aims in Education*, Manchester University Press, 1964. [397]

WINDER, P. 'Reference group values, academic motivation and the attainment of secondary school boys', unpublished M.Ed. thesis, University of Manchester, 1970. [344f]

WINSLOW, C. N. 'A study of the extent of agreement between friends' opinions and their ability to estimate the opinions of each other', *Journal of Social Psychology*, vol. 8, 1937, pp. 433–42. [271]

WISHNER, J. 'Re-analysis of "impression of personality"', *Psychological Review*, vol. 67, 1960, pp. 96–112. [37]

WITHALL, J. 'The development of a technique for the measurement of social-emotional climates in classrooms', *Journal of Experimental Education*, vol. 17, 1949, pp. 347–61. [131]

WONG, L. EVAN, & BAGLEY, C. 'Conformity, attitude change, and group membership in adolescent girls', *Moral Education*, vol. 2, 1970, pp. 115–21. [328f]

WRIGHT, B. D. & TUSKA, S. A. 'From dream to life in the psychology of becoming a teacher', *School Review*, vol. 76, 1968, pp. 253–93. [90]

WRIGHT, D. S. 'A comparative study of the adolescent's conceptions of his parents and teachers', *Educational Review*, vol. 14, 1962, pp. 226–32. [60]

YABLONSKY, L. 'An operational theory of roles', *Sociometry*, vol. 16, 1953, pp. 349–56. [97]

YOUNG, M. & WILLMOTT, P. *Family and Kinship in East London*, Routledge & Kegan Paul, 1957. [18]

ZIMBARDO, P., WEISENBERG, M., FIRESTONE, I. & LEVY, B. 'Communicator effectiveness in producing public conformity and private attitude change', *Journal of Personality*, vol. 33, 1965, pp. 233–55. [388]

Subject index

Acceptance 207
Accounts 196f
Altercasting 119
Attitude: definition of 367f; functions of 383ff; and group membership 389ff
Attribution processes 37ff
Authority of the teacher 139, 242
Autocracy 366, 408

Balance theory 281ff
Boomerang effect 380, 384, 390, 397

Car culture 355f
Categorization 50ff, 64–8, 159, 289f, 377, 406
Civil inattention 116
Classroom climate 136ff
Clown 175, 325f
Cognitive dissonance theory 385
Communicator credibility 379ff
Compliance 192f, 197, 242, 300, 387f, 393ff
Conformity 299–304, 320, 393ff, 413f
Consensual validation 21, 285
Constructs, common-sense and social scientific 176f
Control techniques 244ff

Dating 347f, 363
Definition of the situation 56, 103–22, 137, 139, 154–98, 232ff, 229; teacher's definition of the situation 154–61, 173; pupil's definition of the situation 161–4, 178–98
Delinquent orientation 192f, 369
Deputy headteacher 77, 402, 414
Disciplinary illusion 234f
Disobedience, types of 185
Dispositional properties 38ff
Distributive justice 126f
Double bind 65
Drinking 341, 367
Drugs 367f

Emotional appeals 381
Empathy 198, 203ff, 214, 261ff, 324–6, 335f, 401, 420
End-anchoring 322
Exchange theory 106, 122–9, 155, 173, 178f, 243, 285f, 312f, 321
Expressive leaders 314
Eye contact 116, 298

First impressions 46ff

Governors 408f
Group, definition of 290–4
Group culture 294–304
Group dynamics and the teacher 327–36
Group norms and sanctions 295ff
Group structure 304–27
Groups, primary and secondary 290

Halo effect 55, 346

Headteacher 73f, 76f, 82, 86f, 90, 95, 112, 146f, 166, 225, 235, 363, 366f, 377, 379, 389, 402–20
Hidden curriculum 184
Humorous decontamination 247

Idealization 92, 112–14
Identification 197, 393ff
Idiosyncrasy credit theory 321
Implicit personality theory 37
Impression formation 35ff
Indifferent orientation 193f
Informal pupil roles 325–7
Ingratiation 412ff
Instrumental leader 314
Intentions 40ff
Interaction 31, 34, 49ff, 56, 93–227; and role 93–109; definition of 101–3; links and chains 134
Interaction analysis 130–7, 335
Internalization 300, 393ff
Interpretive schemes 34, 197, 295

Labelling see Categorization
Latitude of acceptable behaviour 296ff
Laws: of pleasing teacher 179ff; of pleasing headteacher 412ff
Leadership 313ff

Marks 198ff
Motives 40ff, 194, 244f, 250ff, 416

Negotiation 106ff, 118f, 126–9, 137, 147, 164–70
Non-punitive exile 248
Norm of mediocrity 327, 358f, 405

Overlapping 257

Perception 29, 31–68, 107–10, 249–54, 415ff; and first impressions 46–50, 270; and role 51ff; and systems of relevancies 34
Pleasing teacher 178ff, 394f
Pop culture 350, 369
Prefects 310–12
Prestige 310ff, 404
Property 312

Psychegroups and sociogroups 315
Psychological myth 66
Punishment illusion 246f
Pupil power 166f, 366

Reference groups 340ff, 360
Resource 312ff, 318, 365
Ripple effect 134, 329
Rites de passage 337f
Role 69–92; and interaction 93–109; and perception 51ff; and self see Self; of the P.E. teacher 79f; of the student teacher 89–92, 112, 228–67; set 73, 146; strain 20, 74–92; style 99, 149–54
Romantic Love Complex 348f

Scapegoat 326, 407, 417
School counsellor 78, 242
School dinners 378ff
Self: and body 7–9; and ideal self 25f, 407; and looking glass self 9; and role 12–21, 28, 70, 79f; as a social structure 9–21; self-esteem 27f, 147, 180, 186, 188f, 360ff, 407; social sex role 21–5; social self 18, 20; stability of 28–30
Self-directed learning 210ff
Self-evaluation 208ff
Self-fulfilling prophecy 27, 56–68; anti-self-fulfilling prophecy 62; autistic self-fulfilling prophecy 64
Self-presentation 27, 110–15, 147, 180, 182ff, 235ff, 405, 414
Sex 346ff, 368, 386
Signal interference 247
Significant others 12–18, 26, 28, 61ff, 85, 146, 340f
Situational proprieties 115f
Smoking 323, 367, 369, 377ff
Social power 307ff
Social sex role 21–5
Social skills 152–4, 195f, 246–66, 335
Sociological myth 66, 331

Sociometry 305ff
Status hierarchy 304ff
Stereotyping 52–5
Strain towards symmetry 282ff
Streaming 66f, 160f, 272ff, 319f, 322–5, 340, 410
Student teacher *see* Role
Superordinate goal 279
Supersensitivity 416
Symbolic interactionism 15, 21f, 60–2, 99, 174f, 178, 272,

Tact 29, 316

Taking the role of the other 11ff
Teacher effectiveness 151–4, 228–67
Teacher–teacher relations 73f, 91, 113, 146ff, 172, 239f, 262, 402–20
Teaching machines 202
Team teaching 88, 405

Voluntary schooling 221–7

With-it-ness 256f
Working consensus 118, 164–77

Printed in the United States
by Baker & Taylor Publisher Services